Methods in Molecular Biology™

Series Editor
John M. Walker
School of Life Sciences
University of Hertfordshire
Hatfield, Hertfordshire, AL10 9AB, UK

For further volumes:
http://www.springer.com/series/7651

Cell-Free Protein Synthesis

Methods and Protocols

Edited by

Kirill Alexandrov and Wayne A. Johnston

Institute for Molecular Bioscience, The University of Queensland, St. Lucia, QLD, Australia

Humana Press

Editors
Kirill Alexandrov
Institute for Molecular Bioscience
The University of Queensland
St. Lucia, QLD, Australia

Wayne A. Johnston
Institute for Molecular Bioscience
The University of Queensland
St. Lucia, QLD, Australia

ISSN 1064-3745 ISSN 1940-6029 (electronic)
ISBN 978-1-4939-6254-9 ISBN 978-1-62703-782-2 (eBook)
DOI 10.1007/978-1-62703-782-2
Springer New York Heidelberg Dordrecht London

Softcover reprint of the hardcover 1st edition 2014

Printed on acid-free paper

Humana Press is a brand of Springer
Springer is part of Springer Science+Business Media (www.springer.com)

Preface

Advances in Life Sciences and Biotechnology have historically relied on the ability to replicate the building blocks of life in vitro, in order to elucidate their mode of action. Much biotechnological progress in the last 40 years has been focused on developing more efficient analysis and synthesis technologies for both DNA and proteins. However, while orders of magnitude reduction in costs for DNA sequencing and synthesis was achieved during the last decade, the throughput and cost of technologies for protein production and engineering have changed comparatively little.

Cell-free protein expression is a rapid and high-throughput methodology for conversion of DNA-encoded genetic information into protein-mediated biochemical activities. It holds the promise to narrow the technological gap between DNA and protein technologies and provide a platform for broad application of synthetic biology principles in the Life Sciences.

Cell-free technologies have developed in two opposite but complementary directions: scale-up and miniaturization. Scale-up aims to produce preparative amounts of high-value recombinant proteins rapidly and without involvement of a recombinant host. Miniaturization aims to extract the most information out of the smallest amount of the largest possible number of proteins or protein variants at the lowest possible cost. Combination of both directions is expected to provide us with a powerful platform for protein analysis, engineering, and manufacturing.

This book is aimed to bring together the key opinion leaders of cell-free technology development and provide case studies and detailed protocols for application of cell-free methodology. The book aims to cover the main directions in the development of cell-free technologies including several recently developed cell-free systems. The book also presents a number of applications of cell-free systems that range from discovery of biofuel enzymes to in vitro assembly of viruses.

Target groups: Biochemists, bioengineers, biotechnologists, cell biologists, and chemical and synthetic biologists.

St. Lucia, QLD, Australia

Kirill Alexandrov
Wayne A. Johnston

Contents

Contributors

KIRILL ALEXANDROV • *Institute for Molecular Bioscience, The University of Queensland, St. Lucia, QLD, Australia*
EIJI ANDO • *Analytical & Measuring Instruments Division, Clinical & Biotechnology Business Unit, Shimadzu Corporation, Kyoto, Japan*
LAI F. BERGEMAN • *Department of Biochemistry, University of Wisconsin, Madison, WI, USA*
FRANK BERNHARD • *Centre for Biomolecular Magnetic Resonance, Institute for Biophysical Chemistry, Goethe-University of Frankfurt/Main, Frankfurt/Main, Germany*
AARTHI CHANDRASEKARAN • *Joint BioEnergy Institute (JBEI), Emeryville, CA, USA; Sandia National Laboratories, Livermore, CA, USA*
ELIZABETH A. COOK • *Protein Technology Group, Babraham Research Campus, Babraham Bioscience Technologies Ltd, Cambridge, UK*
VOLKER DÖTSCH • *Centre for Biomolecular Magnetic Resonance, Institute for Biophysical Chemistry, Goethe-University of Frankfurt/Main, Frankfurt/Main, Germany*
NATHANIEL L. ELSEN • *Department of Biochemistry, University of Wisconsin, Madison, WI, USA*
TORU EZURE • *Analytical & Measuring Instruments Division, Clinical & Biotechnology Business Unit, Shimadzu Corporation, Kyoto, Japan*
SLAWOMIR FILIPEK • *Faculty of Chemistry, University of Warsaw, Warsaw, Poland*
BRIAN G. FOX • *Department of Biochemistry, University of Wisconsin, Madison, WI, USA*
YASUO FUKAMI • *Graduate School of Science, Kobe University, Nada, Japan; Research Center for Environmental Genomics, Kobe University, Nada, Japan*
DAVID W. GALBRAITH • *School of Plant Sciences and BIO5 Institute, University of Arizona, Tucson, AZ, USA*
UMESH GHOSHDASTIDER • *Centre for Biomolecular Magnetic Resonance, Institute for Biophysical Chemistry, Goethe-University of Frankfurt/Main, Frankfurt/Main, Germany; Faculty of Chemistry, University of Warsaw, Warsaw, Poland*
SABRINA GUILLEMER • *Protéus, Parc Georges Besse, Nîmes, France*
STEFAN HABERSTOCK • *Centre for Biomolecular Magnetic Resonance, Institute for Biophysical Chemistry, Goethe-University of Frankfurt/Main, Frankfurt/Main, Germany*
MINGYUE HE • *The Inositide Laboratory, Babraham Institute, Babraham Research Campus, Cambridge, UK*
Z. HUGH FAN • *Department of Mechanical and Aerospace Engineering, University of Florida, Gainesville, FL, USA; Department of Biomedical Engineering, University of Florida, Gainesville, FL, USA*
HIROAKI IMATAKA • *Department of Materials Science and Chemistry, Graduate School of Engineering, University of Hyogo, Himeji, Japan; Molecular Nanotechnology Research Center, Graduate School of Engineering, University of Hyogo, Himeji, Japan; RIKEN Systems and Structural Biology Center, Tsurumi-ku, Yokohama, Japan*
KIRSTEN JACKSON • *Department of Biomedical Engineering, University of Florida, Gainesville, FL, USA*

WAYNE A. JOHNSTON • *Institute for Molecular Bioscience, The University of Queensland, St. Lucia, QLD, Australia*
LEI KAI • *Centre for Biomolecular Magnetic Resonance, Institute for Biophysical Chemistry, Goethe-University of Frankfurt/Main, Frankfurt/Main, Germany*
TAKASHI KANAMORI • *The Department of Medical Genome Sciences, Graduate School of Frontier Sciences, The University of Tokyo, Kashiwanoha, Kashiwa, Chiba, Japan; GeneFrontier Corporation, Kahiwanoha, Kashiwa, Chiba, Japan*
RUBA KHNOUF • *Biomedical Engineering Department, Jordan University of Science and Technology, Irbid, Jordan*
DONG-MYUNG KIM • *Department of Fine Chemical Engineering and Applied Chemistry, Chungnam National University, Daejeon, South Korea*
TOMOMI KIMURA-SOYEMA • *RIKEN Systems and Structural Biology Center, Tsurumi, Yokohama, Japan*
TOMINARI KOBAYASHI • *Department of Materials Science and Chemistry, Graduate School of Engineering, University of Hyogo, Himeji, Japan*
ANDREW KRALICEK • *The New Zealand Institute for Plant & Food Research Limited, Auckland, New Zealand*
ATSUSHI KUROTANI • • *RIKEN Systems and Structural Biology Center, Yokohama, Japan; Plant Science Center, Yokohama, Japan; Department of Biotechnology and Life Science, Tokyo University of Agriculture and Technology, Tokyo, Japan*
YUTETSU KURUMA • *The Department of Medical Genome Sciences, Graduate School of Frontier Sciences, The University of Tokyo, Kashiwa, Chiba, Japan*
YONG-CHAN KWON • *Department of Fine Chemical Engineering and Applied Chemistry, Chungnam National University, Daejeon, South Korea*
CHOY THENG LOH • *Research School of Chemistry, Australian National University, Canberra, ACT, Australia*
YI MA • *Centre for Biomolecular Magnetic Resonance, Institute for Biophysical Chemistry, Goethe-University of Frankfurt/Main, Frankfurt/Main, Germany; School of Bioscience and Bioengineering, South China University of Technology, Guangzhou, People's Republic of China*
KODAI MACHIDA • *Department of Materials Science and Chemistry, Graduate School of Engineering, University of Hyogo, Himeji, Japan; Molecular Nanotechnology Research Center, Graduate School of Engineering, University of Hyogo, Himeji, Japan*
SHIN-ICHI MAKINO • *Department of Biochemistry, University of Wisconsin, Madison, WI, USA*
JEAN-MICHEL MASSON • *Protéus, Parc Georges Besse, Nîmes, France*
TAMÁS MÉSZÁROS • *Department of Medical Chemistry, Molecular Biology and Pathobiochemistry, Semmelweis University, Budapest, Hungary*
SZILVIA K. NAGY • *Department of Medical Chemistry, Molecular Biology and Pathobiochemistry, Semmelweis University, Budapest, Hungary*
KIYOSHI OZAWA • *School of Chemistry, University of Wollongong, Wollongong, NSW, Australia*
CÉCILE PERSILLON • *Protéus, Parc Georges Besse, Nîmes, France*
DAVIDE PROVERBIO • *Centre for Biomolecular Magnetic Resonance, Institute for Biophysical Chemistry, Goethe-University of Frankfurt/Main, Frankfurt/Main, Germany*
RUHU QI • *Research School of Chemistry, Australian National University, Canberra, ACT, Australia*
GILLES RAVOT • *Protéus, Parc Georges Besse, Nîmes, France*

CHRISTIAN ROOS • *Centre for Biomolecular Magnetic Resonance, Institute for Biophysical Chemistry, Goethe-University of Frankfurt/Main, Frankfurt/Main, Germany*
YOSHIHIRO SHIMIZU • *Laboratory for Cell-Free Protein Synthesis, Quantitative Biology Center, RIKEN, Chuo-ku, Kobe, Hyogo, Japan*
MIKAKO SHIROUZU • *RIKEN Systems and Structural Biology Center, Tsurumi, Yokohama, Japan*
ANUP K. SINGH • *Biotechnology and Bioengineering Department, Joint BioEnergy Institute, Sandia National Laboratories, Livermore, CA, USA*
JAE-KWANG SONG • *Korea Research Institute of Chemical Technology, Daejeon, South Korea*
TAKASHI SUZUKI • *1 Nishinokyo Kuwabaracho, Nakagyo, Kyoto, Japan*
TAICHI E. TAKASUKA • *Department of Biochemistry, University of Wisconsin, Madison, WI, USA*
ALEXANDER A. TOKMAKOV • *RIKEN Systems and Structural Biology Center, Tsurumi, Yokohama, Japan; Graduate School of Science, Kobe University, Nada, Japan; Research Center for Environmental Genomics, Kobe University, Nada, Japan*
TAKUYA UEDA • *The Department of Medical Genome Sciences, Graduate School of Frontier Sciences, The University of Tokyo, Kashiwanoha, Kashiwa, Chiba, Japan*
KIRK A. VANDER MEULEN • *Department of Biochemistry, University of Wisconsin, Madison, WI, USA*
MOTOAKI WAKIYAMA • *RIKEN Systems and Structural Biology Center, Tsurumi, Yokohama, Japan*
JOHNNIE A. WALKER • *Department of Biochemistry, University of Wisconsin, Madison, WI, USA*
XIAO WANG • *Department of Pathology, Shandong University School of Medicine, Jinan, People's Republic of China*
XIAONING WANG • *School of Bioscience and Bioengineering, South China University of Technology, Guangzhou, People's Republic of China*
SHIGEYUKI YOKOYAMA • *RIKEN Systems and Structural Biology Center, Yokohama, Japan; Department of Biophysics and Biochemistry, Graduate School of Science, The University of Tokyo, Bunkyo-ku, Tokyo, Japan*
XRISTO ZÁRATE • *Universidad Autonoma de Nuevo Leon, UANL, Facultad de Ciencias Quimicas, Av. Universidad S/N, Ciudad Universitaria, San Nicolas de los Garza, Nuevo Leon, Mexico*
KONG-NAN ZHAO • *UQ Centre for Clinical Research, The University of Queensland, Herston, Brisbane, QLD, Australia; Institute of Molecular Virology and Immunology, Wenzhou Medical College, Wenzhou, Zhejiang, People's Republic of China*
LIANG ZHAO • *Division of Molecular Genetics and Development, Institute for Molecular Bioscience, The University of Queensland, Brisbane, QLD, Australia*

Chapter 1

Production of Eukaryotic Cell-Free Lysate from *Leishmania tarentolae*

Wayne A. Johnston and Kirill Alexandrov

Abstract

In this chapter, we describe the production and application of a eukaryotic cell-free expression system based on *Leishmania tarentolae*. This single-celled flagellate allows straightforward and inexpensive cultivation in flasks or bioreactors. Unlike many other *Leishmania* species, it is nonpathogenic to humans and does not require special laboratory precautions. An additional reason it is a convenient source organism for cell-free lysate production is that all endogenous protein expression can be suppressed by a single antisense oligonucleotide targeting splice leader sequence on the 5′-end of all protein coding RNAs. We describe simple procedures for cell disruption and lysate processing starting from bioreactor culture. We also describe introduction of genetic information via vectors containing species-independent translation initiation sites (SITS). We consider that such an inexpensive eukaryotic cell-free production system has many advantages when expressing multi-subunit proteins or difficult to express proteins.

Key words *Leishmania tarentolae*, Fluorophores, Nitrogen cavitation, Eukaryote, Bioreactors, SITS, Overlap/extension PCR, mCherry, eGFP

1 Introduction

Decades ago, eukaryotic cell-free protein expression systems played a primary role in the discovery of the genetic code. However, present-day advances in eukaryotic cell-free expression are still relevant due to issues of quality in the commonly used *E. coli* cell-free lysates and problems with cost and complexity in existing eukaryotic cell-free systems. Specifically, the advantages of low cost, scalability, and high productivity by the prokaryotic *E. coli*-based cell-free system are offset by problems in producing multi-domain proteins in active form, as well as supporting their assembly into complexes. A number of eukaryotic cell-free systems have been made commercially available, specifically Wheat Germ Extract (WGE), Rabbit Reticulocyte Lysate (RRL), Insect Cell Lysate (ICE), and HeLa cell lysate (HCL). Although these eukaryotic cell-free systems perform better than *E. coli* cell-free lysate in

Kirill Alexandrov and Wayne A. Johnston (eds.), *Cell-Free Protein Synthesis: Methods and Protocols*, Methods in Molecular Biology, vol. 1118, DOI 10.1007/978-1-62703-782-2_1, © Springer Science+Business Media, LLC 2014

multi-domain protein folding, their production remains complex and expensive. Hence the challenge remains to develop a cheap and scalable eukaryotic cell-free system.

A promising eukaryotic organism that serves as a source for such a cell-free system is the unicellular flagellate *Leishmania tarentolae*, primarily due to two complementary characteristics. Firstly, in promastigote form it is not only suitable for flask-based cultivation in an inexpensive media but can also be grown in bioreactor format for maximal productivity. A second advantage is the presence of identical splice leader sequences on all endogenous mRNAs. These can be targeted by a single antisense oligonucleotide, allowing near-complete suppression of endogenous mRNA translation [1]. Although the best-studied species *Leishmania major* is a major source of human disease, *L. tarentolae* infects only the moorish gecko (*Tarentolae mauritanica*), and hence it can be cultivated without special precautions in laboratory environments. It is already widely used as a transgenic organism for in vivo protein expression [2].

In order to facilitate the development of cell-free systems, Mureev et al. reported the design of universal sequences based on polymeric RNA structures that facilitate translational initiation. These species-independent translation sequences (SITS) are applicable to all eukaryotic cell-free systems but have proved very suitable for introducing genetic information into the *Leishmania* cell-free lysate system [3].

The *Leishmania* cell-free system has proved useful in isolating a nearly complete complement of Rab GTPases [4] and has been used to de-orphanize putative translation initiation factors and phosphatases from the original host organism [5]. Current work is focused on the integration of the *Leishmania* cell-free system with microfluidic-based single molecule spectroscopy, using labeled proteins to map multi-domain protein interaction networks.

This chapter describes the technique for cultivation and disruption of the host organism, lysate preparation, and supplementation for coupled transcription/translation protein expression. We also provide details for template preparation based on plasmid or overlap/extension PCR-based synthesis.

2 Materials

2.1 Manufacture of *Leishmania* Cell-Free Lysate

1. Actively maintained *L. tarentolae* promastigote cultures (*see* **Note 1**).
2. Filter sterilized Terrific broth medium (TB): Bacto-tryptone 12 g/L, yeast extract 24 g/L, glycerol 8 mL/L, glucose 1 g/L, KH_2PO_4 2.31 g/L, K_2HPO_4 2.54 g/L (*see* **Note 2**).

3. Sterile penicillin/streptomycin mix for cell culture (5,000 U/mL penicillin, 5,000 µg/mL streptomycin, or similar). Add to TB medium at 0.2 % v/v just prior to use (*see* **Note 3**).
4. Filter sterilized 0.25 % v/v hemin in 50 % v/v triethanolamine (suggested bovine-derived hemin, e.g., Sigma H5533). Add to TB medium at 0.5 % v/v just prior to use.
5. Disposable sterile tissue culture flasks of 50 mL capacity.
6. Baffled glass sterile culture flasks of 5 L (maximum) capacity, for use with 1 L cultures.
7. Shaking incubator suitable for all flasks above.
8. Laboratory scale bioreactor with temperature, pH, and dissolved oxygen control (optional, if desired all cultivations can be done in 5 L flasks as above).
9. 1.5 mL microcentrifuge tubes.
10. Microcentrifuge.
11. Bench centrifuge capable of spinning desired total *Leishmania* production culture (10 L suggested) at 2,500 × *g*.
12. Centrifuge capable of 10,000 × *g* and 30,000 × *g* spins and centrifuge tubes of sufficient strength to suit.
13. Sucrose elution buffer (SEB): 45 mM HEPES–KOH pH 7.6, 250 mM sucrose,100 mM KOAc, 3 mM $Mg(OAc)_2$.
14. Elution buffer (EB): as SEB but without sucrose.
15. Calibrated pH probe.
16. Vacuum receiver flask of approximately 250 mL capacity.
17. Compressed nitrogen (high purity grade).
18. PD-10 Superdex 25 columns (GE Healthcare).
19. Nitrogen cavitation cell disrupter (suggested Parr Industries model 4635 or 4639).
20. Components of 5× coupled transcription/translation feeding solution (5×FS): 6 mM ATP, 0.68 mM GTP, 22.5 mM $Mg(OAc)_2$, 1.25 mM spermidine, 10 mM DTT, 200 mM creatine phosphate, 100 mM HEPES-KOH pH 7.6, 5 % (v/v) PEG 3000, protease inhibitor at 5× recommended concentration (suggested Complete™ EDTA-free, Roche), 0.68 mM of each amino acid, 2.5 mM rNTP mix (ATP, GTP, UTP, and CTP), 0.05 mM anti-splice leader DNA oligonucleotide; sequence CAATAAAGTACAGAAACTGATACTTATATAGCGTT (*see* **Note 4**), 0.5 mg/mL T7 RNA polymerase, 200 U/mL creatine phosphokinase.
21. Liquid nitrogen.

Table 1
Nonspecific primers required for OE-PCR creation of cell-free expression templates

SITS fragment	5′-GGGTTATTGTCTCATGAGCGGATACATATTTGAATGTATTTAGAAAAATAAA CAAATAGGGGTTCCGCGCACATTTCCCCGAAAAGTGCCACCTGACGTCTAAT ACGACTCACTATAGGGACATCTTAAGTTTATTTTATTTTATTTTATTTTATTTTAT TTTATTTTATTTTATTTTATTTTATTTTATTTAACCATGACAGTAATGTATAAAGT CTGTAAAGACATTAAACACGTAAGTGA-3′
Primer-F2	5′-GGGTTATTGTCTCATGAGCGG-3′

2.2 Creation of Templates for Expression in Leishmania Cell-Free System

For overlap/extension PCR based on genes of interest, it is necessary to design primers based on the ORFs themselves as described in Subheading 3.4. Availability of standard primer synthesis services is required. Additional primers general to all ORFs are required as per Table 1.

2.3 Expression and Visualization of Proteins

1. Apparatus and materials for SDS-PAGE gel analysis of expressed proteins.
2. (Optional) FluoroTect™ $Green_{Lys}$ in vitro Translation Labeling System (Promega) for incorporating BODIPY®-FL fluorescent label into expressed proteins.
3. (Optional) Incubating fluorescence plate reader capable of measuring typical label fluorophores during expression.
4. (Optional) Gel fluorescent scanner capable of measuring typical label fluorophores (eGFP, mCherry, BODIPY®-FL).

3 Methods

3.1 Culture of L. tarentolae for Maintenance and Production of Fermentation Inoculum

Maintenance cultures (50 mL in tissue culture flasks, 74 rpm inclined agitation, 26.5 °C) are kept in exponential growth phase (doubling time 6–7 h) at the approximate concentration range $0.1–1.5 \times 10^8$ cells/mL, via suitable dilutions in fresh TB (plus hemin and antibiotics) every 2–3 days. An alternative is to track culture density as OD600, with the maintenance range approximately 0.25–0.4. Routine biomass measurements may also be taken as OD600 values and converted to cells/mL counts via a standard curve (*see* **Notes 5** and **6**).

For creating a *Leishmania* inoculum suitable for the cell-free lysate production culture, maintenance cultures are expanded successively over two 24 h periods in TB medium supplemented with hemin and antibiotics. Typical dilutions are 20–200 mL (4 × 50 mL tissue culture flasks) and 200 mL to 2 L (2× baffled 5 L flasks with 1 L fill) at 26.5 °C, 74 rpm agitation.

3.2 Culture of Leishmania for Cell-Free Lysate Production

Two methods are described for *Leishmania* lysate production culture. Firstly, batch culture in 5 L culture flasks filled with 1 L medium per flask and secondly, small-scale bioreactor culture (10 L total or similar).

For batch culture in flasks, cultures are expanded into 10 × 1 L flask and grown overnight (26.5 °C, 74 rpm agitation) for 14 h. The target harvest cell density is approximately OD600 = 5.0 (corresponding to approximately 2.0×10^8 cells/mL). A suitable inoculum cell density can be calculated from Monod growth kinetics and a doubling time of 6 h at 26.5 °C (*see* **Note 7**).

For batch culture in bioreactors, inoculum density can be calculated as above but with reduced doubling time of 9 h, to compensate for slower growth at the altered temperature profile used.

Bioreactor parameters:

Aeration: 1vvm with compressed air, oxygen controller set to 10 % of air saturation (*see* **Note 8**).

pH control: control pH to pH7.4 with 1 M HCl and 1 M NaOH (*see* **Note 9**) added via automatic pumping (generally only acid addition will be required).

Temperature control: maintain 26.5 °C for the first hour after inoculation, then reduce temperature set point to 24 °C (*see* **Note 10**).

Harvest time is predicted using the desired final biomass level (suggested OD600 = 5.0) via formula, with bioreactor sampling to verify actual reactor biomass close to predicted harvest time.

3.3 Production of Cell-Free Lysate

1. Harvested cells are pelleted by centrifugation at 2,500 × *g*, with spent medium removed by careful decantation (*see* **Note 11**).
2. Add SEB at 4 °C (*see* **Note 12**) for washing (approximately 20× pellet volume is sufficient; pellet resuspensions can be pooled as appropriate to reduce the number of centrifuge tubes in each washing step). Carefully resuspend the pellet via gentle mixing and pipetting of SEB against the cell pellet (but not directly pipetting the pellet up and down), then spin 2,500 × *g* for 10 min and decant.
3. Repeat previous washing step.
4. Carefully resuspend the twice-washed *Leishmania* pellet in the minimum volume of 4 °C SEB required for resuspension.
5. Remove and spin down a test volume (1.0 mL) of the concentrated *Leishmania* suspension in a 1.5 mL microcentrifuge tube at 2,500 × *g* for 10 min (*see* **Note 13**). Record initial weight of the filled tube. Remove supernatant as carefully and completely as possible. Reweigh tube with pellet only; calculate tube difference (from initial weight) to obtain the weight of supernatant removed.

Assuming supernatant density to be 1.0 g/mL, calculate supernatant volume in mL. Finally, calculate the ratio of pellet volume to total volume in the *Leishmania* cell concentrate as 1.0 mL supernatant volume/1.0. This ratio is typically 0.6 v/v.

6. Dilute the remaining concentrated *Leishmania* suspension to a volumetric ratio of 0.38 v/v (*see* **Note 14**) with additional addition of SEB, based on the volumetric ratio derived in the previous step.
7. The 0.38 v/v cell suspension is disrupted using a precooled nitrogen cavitation device (70 bar nitrogen, 45 min equilibration time prior to disruption at 4 °C). Care must be taken to "fire" the disruption device into a precooled, thick-walled vessel (*see* **Note 15**). A vacuum receiver flask is suitable.
8. Spin disrupted cell suspension at 10,000 ×*g* at 4 °C for 20 min. Carefully remove 2/3 of the supernatant and transfer to a fresh centrifuge tube of suitable mechanical strength for 30,000 ×*g*.
9. Spin at 30,000 ×*g* at 4 °C for 20 min. Carefully remove 2/3 of the supernatant (*see* **Note 16**).
10. Remove sucrose from the 30,000 ×*g* supernatant by gel filtration on PD-10 Superdex 25 columns (GE Healthcare) into fresh elution buffer minus sucrose (EB; 45 mM HEPES–KOH pH 7.6, 100 mM KOAc, 3 mM $Mg(OAc)_2$) at 4 °C.
11. Supplement the lysate with 5×FS (concentrated feeding solution containing the necessary cofactors and enzymes for coupled transcription and translation). Four volumes of feeding solution are mixed with 10 volumes of gel-filtered lysate, with quantities scaled as required (*see* **Note 17**). The resulting 14 volumes represent sufficient supplemented lysate for 20 volumes of total final reaction volume. Magnesium concentration of the 5×FS may need to be varied (*see* **Note 18**). It is also possible to directly freeze the unsupplemented lysate from gel filtration and add feeding solution after thawing, just prior to starting protein expression reactions.
12. The supplemented lysate is aliquoted, snap-frozen in liquid nitrogen, and stored at −80 °C.

3.4 Construction of DNA Templates for Coupled Transcription/Translation in the Leishmania Cell-Free System

In vitro translation using the *Leishmania* cell-free extract requires preparation of template DNA with both the T7 promoter and species-independent translation initiating sequence (SITS). The SITS (Fig. 1) includes a polymeric unstructured region upstream of the start codon and a three-hairpin structure downstream of the start codon. The SITS is analogous to the Internal Ribosome Entry Site (IRES) in other cell-free systems. Although developed for the *Leishmania* cell-free system, it is species independent and can be used for expression in other pro- and eukaryotic cell-free protein

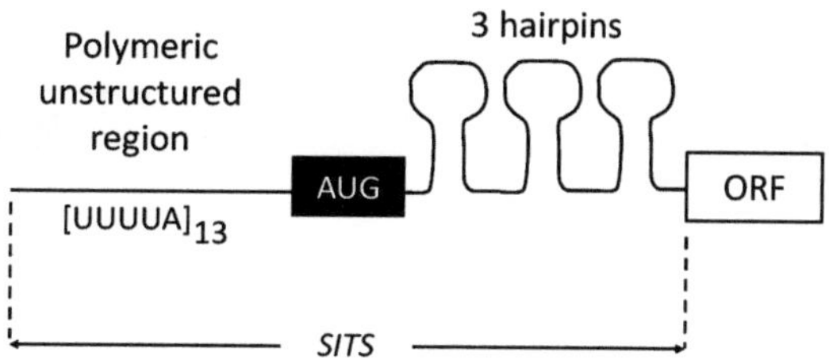

Fig. 1 Structure of SITS in mRNA. Template for this structure is appended to DNA ORFS of interest for coupled transcription translation in the *Leishmania* cell-free system

expression systems [3]. It should be noted that the translation of the added sequence downstream of the start codon results in the addition of 17 amino acids to the N′ terminus of the target protein.

Provided the SITS and T7 promoter are present, the system can be programmed with either linear template prepared via overlap/extension PCR (OE-PCR) [6] or plasmid-based template. The OE-PCR method is rapid and flexible and allows rapid generation of large protein libraries directly from unpurified PCR products. However, a plasmid-based approach is recommended for high yield/high volume expressions and for open reading frames longer than 2,500 bp. A suitable candidate plasmid for *Leishmania* cell-free expression is pLTE (*see* **Note 19**).

The steps for generating OE-PCR-based templates for expression are summarized in Fig. 2 and proceed as follows:

1. Design an ORF gene-specific forward primer with 5′ adapter sequence as shown in Fig. 2b (Primer F1) along with a reverse primer annealing 100–150 bp 3′ of the target ORF stop codon of the donor construct (Primer R1). It is possible to introduce a new stop codon for C-terminal truncations at this point (*see* **Note 20**).
2. PCR amplify the donor template with Primer F1 and R1, thus fusing the ORF to the 5′ adapter sequence for subsequent OE-PCR. Assemble the PCR reaction in 50 μL using high-fidelity DNA polymerase using 1 ng/μL template, primer concentration 250 μM primer, and 35 cycles.
3. For the OE-PCR, combine the unpurified product from the previous step PCR, SITS fragment, Primer F2, and Primer R1 in the quantities listed in Table 2. Sulfate-free PCR reaction buffer is recommended as it allows direct transcription/translation from the crude OE-PCR product. Thermal conditions for the OE-PCR reaction are provided in Table 3.
4. Use gel electrophoresis to determine whether the linear DNA template constructed in OE-PCR is of appropriate size.

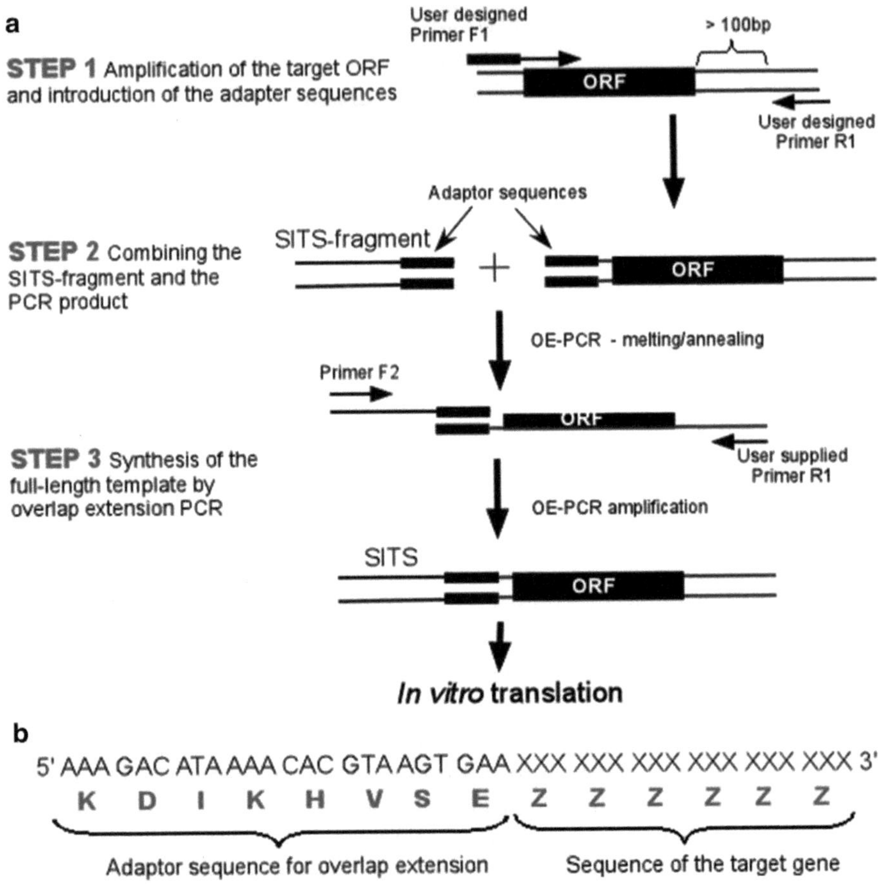

Fig. 2 Schematic representation of overlap/extension PCR construction of templates for *Leishmania* cell-free expression

Table 2
OE-PCR reaction mixture

Reagent	Stock concentration	Final concentration	Volume for 50 µL
Water			To 50 µL
PCR buffer	10×	1×	5 µL
dNTPs	10 mM	0.2 mM	1 µL
Primer F2	10 µM	0.5 µM	2.5 µL
Primer R1	10 µM	0.5 µM	2.5 µL
Completed first PCR	1×	0.05×	2.5 µL
SITS fragment	95 nM	5 nM	2.6 µL
Taq DNA polymerase	5 U/µL	2.5 U/100 µL	0.25 µL

Table 3
OE-PCR reaction conditions

Cycle step	Temp (°C)	Time	# Cycles
Initial denaturation	95	3 min	1
Denaturation	94	30 s	30
Annealing	50	30 s	
Extension	72	1 min per kb	
Final extension	72	5 min	1

Appropriate size will depend on the size of the ORF, i.e., ~ORF + 243 bp (SITS fragment) + 100 bp (depending on Primer R1 design).

3.5 Expression and Visualization of Proteins Using the Leishmania Cell-Free System

In the author's laboratory, expression of target proteins is typically carried out in N- or C-terminal fusion with fluorescent domains. Typically used are enhanced GFP (eGFP), superfolder GFP (sfGFP), or mCherry. Additionally, BODIPY®-FL can be included in cell-free expressions to visualize bands on SDS-PAGE gels and determine whether expressed proteins are of correct size.

1. Thaw an aliquot of supplemented lysate and keep on ice until added to a reaction (*see* **Note 21**).
2. Add template and sterile polished water. For a final reaction volume of 20 μL, the additions are 14 μL supplemented lysate plus 6 μL of DNA templates and water. For plasmid-based templates, an optimal DNA concentration is 20 nM (*see* **Note 22**). If using crude mixture from OE-PCR, use 2–4 μL of template. Include a negative control with no template (*see* **Note 23**).
3. Incubate for 3 h at 27 °C. If expressing using a fluorophore reporter and in an incubating fluorescence reader, the reaction can be terminated when expression of reporter gene (e.g., eGFP) indicates the cessation of protein production (excitation at 488 nm, emission at 507 nm for eGFP).
4. Visualize on an SDS-PAGE gel with suitable size markers. If samples are unboiled, fluorophores in fusion proteins can be directly visualized on the gel by fluorescence scanning; provided samples are not heated prior to gel loading. If using BODIPY®-FL labeling, all expressed bands can be seen regardless of heating (*see* **Note 24**). An example of SDS-PAGE visualization of expressed fusion proteins with and without heating and BODIPY®-FL labeling is presented in Fig. 3.

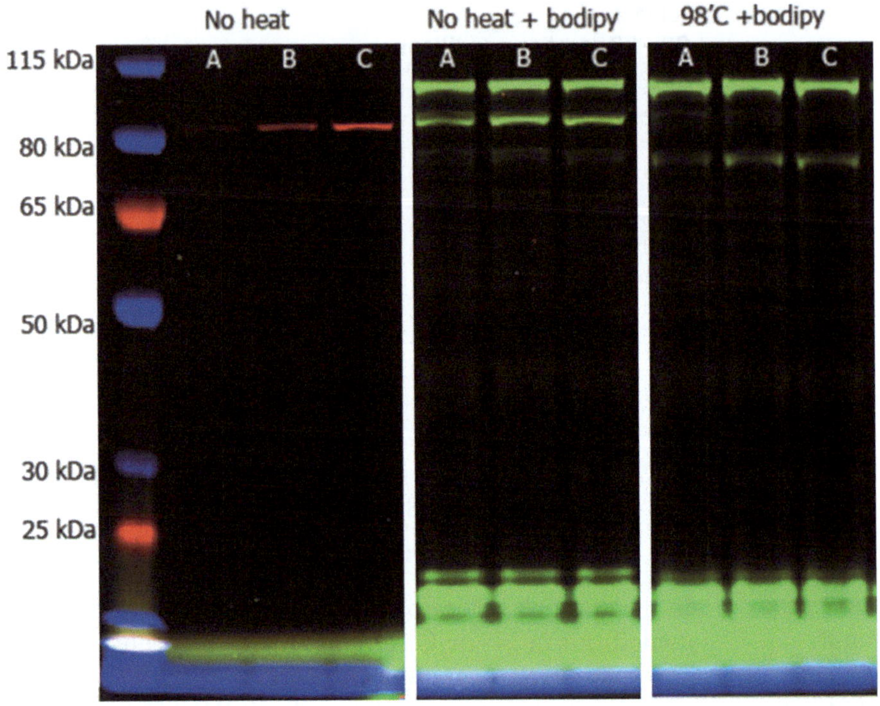

Fig. 3 Expression of Sorting Nexin 1 (Snx1) in the *Leishmania* cell-free system, visualized via SDS-PAGE gel. Snx1 was C-terminally labeled with the mCherry fluorophore. mCherry fluorescence is visualized in the *red channel*, BODIPY®-FL in the *green channel. Lanes A, B,* and *C* represent three *Leishmania* cell-free lysates of varying concentration. *No heat* represents direct loading of reaction plus SDS-PAGE loading buffer onto the gel, in the absence of BODIPY®-FL labeling (i.e., mCherry fluorophore visible only). *No heat+Bodipy* represents both fluorophore and total protein productions. *98 °C+Bodipy* represents total protein production only (heat linearized protein/destroyed fluorophore). The 98 °C linearized protein versus unheated protein migrates at different speeds due to the presence of folded fluorophore in the latter case, both can be seen simultaneously in the *No heat+Bodipy* lanes as some fluorophore linearizes on the gel even in the absence of heating in completed expressions

4 Notes

1. Culture of *L. tarentolae* may be covered by local regulations relating to quarantine. Cultures maintain viability with standard methods of glycerol freezing and −80 °C storage, and such frozen cultures can be transferred between laboratories on dry ice.
2. Medium components, particularly yeast extract and tryptone, should be of the highest grade available. *Leishmania* growth rate, cell size, and final lysate quality can be adversely affected by the lower quality of complex media components.
3. The addition of antibiotics to reduce bacterial contamination of *Leishmania* cultures is not strictly necessary but is recommended as it does not appear to reduce cell-free lysate quality.

Leishmania can be routinely passaged for lysate production over months, and antibiotic addition reduces the chances of contamination during this time.

4. In *Leishmania* spp., all endogenous mRNA messages carry identical leader sequence. Anti-splice leader oligonucleotide blocks translation of endogenous mRNAs and consequently allows expression exclusively of the desired exogenous message in the cell-free lysate [3].
5. Dilution of *Leishmania* OD readings to the instrument linear range (typically <0.3) should be done in fresh TB (including hemin), as dilution of cultures into distilled water or phosphate-buffered saline (PBS) alters cell morphology which markedly change relationship between cells/mL and OD600.
6. Accurate quantification of *Leishmania* biomass at cell disruption is critical for high-quality cell-free lysate production, much more so for routine cell dilution for maintenance cultures. Hemocytometer-based cell counts on diluted culture give a high variation between individual counts and are not recommended for routine cell density tracking. It is suggested to create an OD600 versus cell count standard curve using triplicate (or more) counts at several cell concentrations and use the standard curve to calculate cell count from OD600.
7. Duration of the production culture should not be extended to significantly longer than 14 h, as quality of the lysate was demonstrated to deteriorate using 24 h production cultures.
8. *Leishmania* cultivation for cell-free lysate is grown to lower biomass densities than bacterial cultures, and hence oxygenation demand is lower. Although the sheer force transmitted by conventional Rushton turbines (typical bacterial bioreactor impellor type) can damage *Leishmania* cells and result in biomass loss, rpm levels can be kept low in the oxygen control loop thus avoiding such cell damage. It is recommended to set minimum stirring speed for the bioreactor oxygen control loop to zero (mixing by air lift from sparging only) and set maximum stirring rate to not greater than 150 rpm.
9. Use of ammonium hydroxide for pH control (typically used for bacterial bioreactor cultures in order to replenish dissolved nitrogen used during growth) is not recommended as it results in lower final lysate quality.
10. Reducing the temperature set point after 1 h cultivation can be considered optional but has been demonstrated to increase the activity of resulting lysate.
11. The typical cell pellet is quite loose and requires very careful decantation to avoid cell loss with the discarded supernatant.

12. All steps once *Leishmania* culture is removed from incubation should be done at 4 °C, either in a suitable cold room or alternately with all stock solutions and cell suspensions in water-ice-containing trays on the lab bench. All lysate preparation steps should be followed without delays.
13. Clip the end of a 1 mL pipette tip to allow precise aspiration of viscous 1 mL concentrated cell suspension.
14. The suggested cell suspension dilution (0.38 v/v) was derived by empirical optimization using GFP reporter and may be altered based on individual optimization if desired. The volumetric ratio optimum is narrow, with lysate activity dropping rapidly with both less and more concentrated cell suspensions (unpublished data).
15. Take care not to place the nitrogen cavitation device exit tube adjacent to but not directly inside the receiver flask opening, to avoid flask breakage or loss of disrupted cells from the vessel via the nitrogen gas pressure wave.
16. For good lysate activity it is critical to avoid taking the unclarified zone just above the supernatant/pellet interface after the 30,000 × *g* spin, even if this requires removing less than 2/3 of the supernatant as usable cell lysate.
17. A typical small-scale reaction volume is 20 μL in a 384-well plate, comprising 14 μL supplemented lysate (i.e., 10 μL lysate plus 4 μL 5×FS), with 6 μL additional volume which will contain the DNA template for coupled transcription/translation.
18. Magnesium concentration affects lysate activity, although the optimum is fairly broad as determined using GFP as a reporter [4]. Hence, magnesium optimization did not appear necessary for routine lysate preparation in the *Leishmania* system. The magnesium concentration in the feed solution recipe (22.5 mM) represents a total concentration of 4.5 mM when diluted 1/5 into the final reaction (an additional 1.5 mM $Mg(OAc)_2$ is contributed by the gel-filtered lysate itself via the EB buffer). Typical optimum (final reaction) magnesium concentration is 4–5 mM for bioreactor-based lysates (using a temperature drop in the final production cultivation), as measured using unsupplemented lysate that was supplemented at the point of reaction with feeding solutions prepared as above but with altered magnesium content. The magnesium optimum

using multiple flask cultures (without temperature drop during cultivation) exhibited maximum activity at 5–7 mM Mg^{2+} [4]. Fine-tuning of the reaction magnesium concentration is possible by altering the $Mg(OAc)_2$ concentration in the concentrated feeding solution (as above) or by adding extra $Mg(OAc)_2$ in the final reaction makeup.

19. pLTE (3,437 bp) contains an Amp^R unit, T7 promoter, pBR322 origin of replication, lac promoter, and an ORF coding for enhanced GFP (eGFP) flanked by multicloning sites. Its sequence is shown in Table 4.

20. For C-terminal truncations, engineer Primer R1 to contain complement to the 3′-end of the desired gene followed by a stop codon in frame with the gene and random sequence 8 nucleotides in length. The random short sequence is required to prevent the stop codon from exonuclease digestion.

21. *Leishmania* cell-free lysate deteriorates rapidly after thawing, especially if warmed to room temperature or above. Deterioration occurs at a similar rate whether or not the translation reaction has been started. Lysate should not be thawed until as soon as practical before expression.

22. 20 nM represents the approximate saturation value for increasing GFP expression using pLTE plasmid as the vector backbone.

23. *Leishmania* extract contains endogenously biotinylated and fluorescent proteins that in some cases may be difficult to distinguish from translation products to be detected from fluorescence.

24. It is perhaps surprising that fluorophores such as eGFP and mCherry can survive SDS gel electrophoresis provided samples are not heated (98 °C 10 min completely inactivates both fluorophores, 70 °C for 10 min is indistinguishable from no heating at all). If using mCherry as a label and BODIPY®-FL to label all expressed bands, not heating samples allows gel scanning for both the mCherry and BODIPY®-FL labels simultaneously, as shown in Fig. 3. Unfortunately the absence of heating ensures that the size markers (based on linear DNA) are inaccurate. One solution is to run all samples as heated/unheated in separate lanes, with the heated samples (including BODIPY®-FL) used to accurately size expressed bands and the unheated samples used to locate fluorophores.

Table 4
Complete sequence of pLTE plasmid

GACGTCTAATACGACTCACTATAGGGACATCTTAAGTTTATTTTATTTTATTTTATTTTATTTTATTTTATTTT | ATTTTATTTTATTTTATTTTATTTAACCATGACA
GTAATGTATAAAGTCTGTAAAGACATTAAACACGTAAGTGAAACCATGGAGATCTCGAGCAAGGGCGAGGAGCTGTTCACCGGGGTGGTGCCCATCCTGGTCGAGCTG
GACGGCGACGTAAACGGCCACAAGTTCAGCGTGTCCGGCGAGGGCGAGGGCGATGCCACCTACGGCAAGCTGACCCTGAAGTTCATCTGCACCACCGGCAAGCTGCCC
GTGCCCTGGCCCACCCTCGTGACCACCCTGACCTACGGCGTGCAGTGCTTCAGCCGCTACCCCGACCACATGAAGCAGCACGACTTCTTCAAGTCCGCCATGCCCGAA
GGCTACGTCCAGGAGCGCACCATCTTCTTCAAGGACGACGGCAACTACAAGACCCGCGCCGAGGTGAAGTTCGAGGGCGACACCCTGGTGAACCGCATCGAGCTGAAG
GGCATCGACTTCAAGGAGGACGGCAACATCCTGGGGCACAAGCTGGAGTACAACTACAACAGCCACAACGTCTATATCATGGCCGACAAGCAGAAGAACGGCATCAAG
GTGAACTTCAAGATCCGCCACAACATCGAGGACGGCAGCGTGCAGCTCGCCGACCACTACCAGCAGAACACCCCCATCGGCGACGGCCCCGTGCTGCTGCCCGACAAC
CACTACCTGAGCACCCAGTCCGCCCTGAGCAAAGACCCCAACGAGAAGCGCGATCACATGGTCCTGCTGGAGTTCGTGACCGCCGCCGGGATCACTCTCGGCATGGAC
GAGCTATACAAGGAGCAGAAGCTGATCTCGGAGGAGGATCTGCAAGCTTGTCGACCTCTAGAGGATCCCCGGGGCTAAGCGGCCGCCCTCCTCCTCCTTTCTTGTTCC
TTTCACGTCGCCTTCTCGGTTGTAGCTGGCAGACGACGAGTCTTACTTTTACGTGTACTTCTCTATAGATGATGTATGATCTCTCTGCATGCGTGTTCGTGCATGTGT
CCGTGTGTTGGGTACGCGTGGTACCCTGCAGGAAGGAAGCTGAGTTGGCTGCTGCCACCGCTGAGCAATAACTAGTAATTACTAGCATAACCCCTTGGGGCCTCTAAA
CGGGTCTTGAGGGGTTTTTTGCTGAAAGGAGGACAGCTGATGATTGTCATGCTTGCCATCTGTTTTCTTGCAAGGTCAGAGGAATTCGTAATCATGGTCATAGCTGT
TTCCTGTGTGAAATTGTTATCCGCTCACAATTCCACACAACATACGAGCCGGAAGCATAAAGTGTAAAGCCTGGGGTGCCTAATGAGTGAGCTAACTCACATTAATTG
CGTTGCGCTCACTGCCCGCTTTCCAGTCGGGAAACCTGTCGTGCCAGCTGCATTAATGAATCGGCCAACGCGCGGGGAGAGGCGGTTTGCGTATTGGGCGCTCTTCCG
CTTCCTCGCTCACTGACTCGCTGCGCTCGGTCGTTCGGCTGCGGCGAGCGGTATCAGCTCACTCAAAGGCGGTAATACGGTTATCCACAGAATCAGGGGATAACGCAG
GAAAGAACATGTGAGCAAAAGGCCAGCAAAAGGCCAGGAACCGTAAAAAGGCCGCGTTGCTGGCGTTTTTCCATAGGCTCCGCCCCCCTGACGAGCATCACAAAAATC
GACGCTCAAGTCAGAGGTGGCGAAACCCGACAGGACTATAAAGATACCAGGCGTTTCCCCCTGGAAGCTCCCTCGTGCGCTCTCCTGTTCCGACCCTGCCGCTTACCG
GATACCTGTCCGCCTTTCTCCCTTCGGGAAGCGTGGCGCTTTCTCATAGCTCACGCTGTAGGTATCTCAGTTCGGTGTAGGTCGTTCGCTCCAAGCTGGGCTGTGTGC
ACGAACCCCCCGTTCAGCCCGACCGCTGCGCCTTATCCGGTAACTATCGTCTTGAGTCCAACCCGGTAAGACACGACTTATCGCCACTGGCAGCAGCCACTGGTAACA
GGATTAGCAGAGCGAGGTATGTAGGCGGTGCTACAGAGTTCTTGAAGTGGTGGCCTAACTACGGCTACACTAGAAGAACAGTATTTGGTATCTGCGCTCTGCTGAAGC
CAGTTACCTTCGGAAAAAGAGTTGGTAGCTCTTGATCCGGCAAACAAACCACCGCTGGTAGCGGTGGTTTTTTTGTTTGCAAGCAGCAGATTACGCGCAGAAAAAAAG
GATCTCAAGAGGATCCTTTGATCTTTTCTACGGGGTCTGACGCTCAGTGGAACGAAAACTCACGTTAAGGGATTTTGGTCATGAGATTATCAAAAAGGATCTTCACCT
AGATCCTTTTAAATTAAAAATGAAGTTTTAAATCAATCTAAAGTATATATGAGTAAACTTGGTCTGACAGTTACCAATGCTTAATCAGTGAGGCACCTATCTCAGCGA
TCTGTCTAGTTCGTTCATCCATAGTTGCCTGACTCCCCGTCGTGTAGATAACTACGATACGGGAGGGCTTACCATCTGGCCCCAGTGCTGCAATGATACCGCGAGACC
CACGCTCACCGGCTCCAGATTTATCAGCAATAAACCAGCCAGCCGGAAGGGCCGAGCGCAGAAGTGGTCCTGCAACTTTATCCGCCTCCATCCAGTCTATTAATTGTT
GCCGGGAAGCTAGAGTAAGTAGTTCGCCAGTTAATAGTTTGCGCAACGTTGTTGCCATTGCTACAGGCATCGTGGTGTCACGCTCGTCGTTTGGTATGGCTTCATTCA
GCTCCGGTTCCCAACGATCAAGGCGAGTTACATGATCCCCCATGTTGTGCAAAAAAGCGGTTAGCTCCTTCGGTCCTCCGATCGTTGTCAGAAGTAAGTTGGCCGCAG
TGTTATCACTCATGGTTATGGCAGCACTGCATAATTCTCTTACTGTCATGCCATCCGTAAGATGCTTTTCTGTGACTGGTGAGTACTCAACCAAGTCATTCTGAGAAT
AGTGTATGCGGCGACCGAGTTGCTCTTGCCCGGCGTCAATACGGGATAATACCGCGCCACATAGCAGAACTTTAAAAGTGCTCATCATTGGAAAACGTTCTTCGGGGC
GAAAACTCTCAAGGATCTTACCGCTGTTGAGATCCAGTTCGATGTAACCCACTCGTGCACCCAACTGATCTTCAGCATCTTTTACTTTCACCAGCGTTTCTGGGTGAG
CAAAAACAGGAAGGCAAAATGCCGCAAAAAAGGGAATAAGGGCGACACGGAAATGTTGAATACTCATACTCTTCCTTTTTCAATATTATTGAAGCATTTATCAGGGTT
ATTGTCTCATGAGCGGATACATATTTGAATGTATTTAGAAAAATAAACAAATAGGGGTTCCGCGCACATTTCCCCGAAAAGTGCCACCT

References

1. LeBowitz JH, Smith HQ, Rusche L et al (1993) Coupling of poly(A) site selection and trans-splicing in Leishmania. Genes Dev 7(6): 996–1007
2. Basile G, Peticca M (2009) Recombinant protein expression in Leishmania tarentolae. Mol Biotechnol 43(3):273–278
3. Mureev S, Kovtun O, Nguyen UT et al (2009) Species-independent translational leaders facilitate cell-free expression. Nat Biotechnol 27(8):747–752
4. Kovtun O, Mureev S, Jung W et al (2011) Leishmania cell-free protein expression system. Methods 55(1):58–64
5. Kovtun O, Mureev S, Johnston W et al (2010) Towards the construction of expressed proteomes using a Leishmania tarentolae based cell-free expression system. PLoS One 5(12):e14388
6. Yabuki T, Motoda Y, Hanada K et al (2007) A robust two-step PCR method of template DNA production for high-throughput cell-free protein synthesis. J Struct Funct Genomics 8(4):173–191

Chapter 2

Bioinformatics Analysis and Optimization of Cell-Free Protein Synthesis

Alexander A. Tokmakov, Atsushi Kurotani, Mikako Shirouzu, Yasuo Fukami, and Shigeyuki Yokoyama

Abstract

Cell-free protein synthesis offers substantial advantages over cell-based expression, allowing direct access to the protein synthetic reaction and meticulous control over the reaction conditions. Recently, we identified a number of statistically significant correlations between calculated and predicted properties of amino acid sequences and their amenability to heterologous cell-free expression. These correlations can be of practical use for predicting expression success and optimizing cell-free protein synthesis. In this chapter, we describe our approach and demonstrate how computational and predictive bioinformatics can be used to analyze and optimize cell-free protein expression.

Key words Cell-free protein synthesis, Heterologous expression, Rationalization, Optimization, Physicochemical and structural protein properties, Bioinformatics analysis

1 Introduction

Eukaryotic proteins and their domains are commonly expressed in recombinant form in *Escherichia coli* bacteria [1–3] and cell-free extracts [3–6]. However, obtaining the correct folding of eukaryotic proteins expressed in the bacterial host remains a great challenge. Inability of heterologous protein synthetic machinery to support correct protein folding is considered to be a major factor behind low expression yield and poor solubility of many recombinant proteins.

Various physicochemical properties of a polypeptide sequence have been correlated with soluble expression in bacteria [2, 7–10]. Recently we revealed a number of statistically significant correlations between the yield of heterologous cell-free protein synthesis and multiple calculated and predicted parameters of amino acid sequences [11]. They include protein length, hydrophobicity, p*I*, content of charged, nonpolar and aromatic residues, cysteine

Kirill Alexandrov and Wayne A. Johnston (eds.), *Cell-Free Protein Synthesis: Methods and Protocols*, Methods in Molecular Biology, vol. 1118, DOI 10.1007/978-1-62703-782-2_2,

content, solvent accessibility, presence of coiled coil, content of intrinsically disordered and structured (alpha-helix and beta-sheet) sequence, number of disulfide bonds and functional domains, presence of transmembrane regions, PEST motifs, and signaling sequences.

In addition, many eukaryotic proteins require multiple post-translational modifications (PTMs) to reach a native, biologically active conformation. However, the bacterial expression systems have only a limited capacity for PTMs. Most recently amenability of human polypeptide sequences to prokaryotic cell-free expression has been demonstrated to correlate with the presence of multiple bioinformatically predicted PTM sites [12].

Importantly, cell-free protein expression is well compatible with high-throughput protein production and optimization. It enables the use of PCR-generated linear DNA templates for programming protein synthesis without the need for their cloning into expression vectors. The protein synthesis in the cell-free system is fast and highly processive; its productivity reaches several milligrams of protein per milliliter of reaction mixture [13]; and it is amenable to efficient scaling. Coupling transcription and translation using DNA templates and bacteriophage RNA polymerases was shown to achieve the highest protein yields in a cell-free environment [14, 15]. The above features of cell-free protein synthesis allowed us to set up the protein expression pipeline described below. The developed protocol was used to screen for well-expressed and highly soluble polypeptide sequences from a large collection of candidate targets. This effort was carried out in the framework of the Japan's national structural genomic project "Protein 3000" launched in 2002 for the purpose of determining the structures of 3,000 proteins using X-ray and NMR methods [16–18].

This chapter does not cover the expression pipeline comprehensively, it is described in detail elsewhere [11, 19–21]. Instead the focus is set on the expression evaluation, data processing, bioinformatics analysis, and optimization of cell-free protein synthesis. In brief, the major steps of the screening-scale cell-free protein expression protocol are summarized in Fig. 1. They included linear template generation by the two-step PCR from the source human cDNA clones (*see* **Notes 1** and **2**), small-scale (20–50 μl) batch-mode coupled transcription/translation protein synthesis in the cell-free extract of *E. coli* (*see* **Note 3**), separation of soluble and insoluble reaction products by centrifugation, and estimation of protein yields and solubility by SDS PAGE and protein staining (*see* **Note 4**). DNA template generation and cell-free protein production were carried out in a 96-well format to allow simultaneous processing of multiple samples. The complete dataset of human proteins and their domains expressed in our project under the same uniform set of conditions according to the developed protocol comprised 3066 non-redundant amino acid sequences.

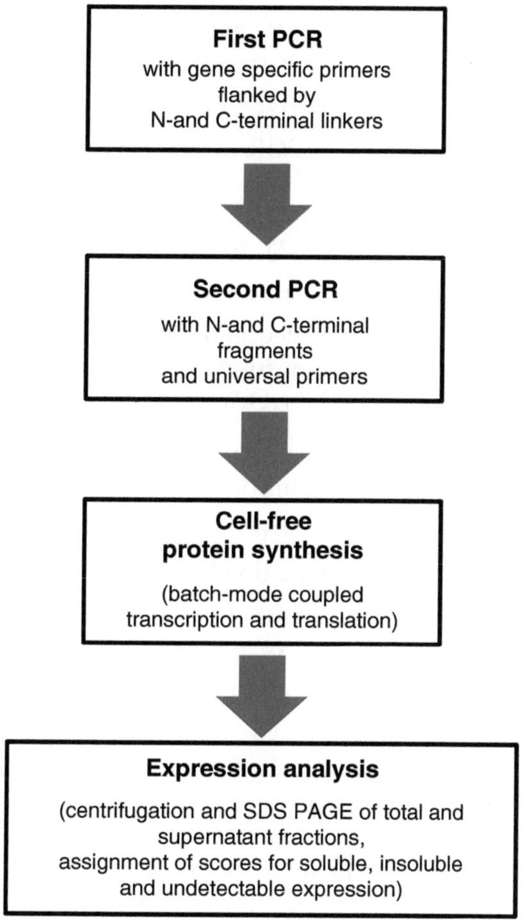

Fig. 1 Cell-free protein expression pipeline. Main steps of the small-scale protein production for the screening of well-expressed and highly soluble polypeptide sequences are presented

2 Materials

2.1 First-Step PCR

1. Expand High Fidelity PCR System (Roche, Basel, Switzerland).
2. dNTPs mixture (TOYOBO, Osaka, Japan).
3. Gene-specific forward and reverse primers (Invitrogen, Carlsbad, CA).
4. cDNA clones (*see* **Note 1**).
5. 96-Well PCR plates and strip caps.
6. A PCR thermal cycler.

2.2 Second-Step PCR

1. Expand High Fidelity PCR System (Roche, Basel, Switzerland).
2. dNTPs mixture (TOYOBO, Osaka, Japan).
3. T7 promoter and T7 terminator fragments (*see* **Note 2**).

4. Universal primer (*see* **Note 2**).
5. First-step PCR products.
6. 96-Well PCR plates and strip caps.
7. A PCR thermal cycler.

2.3 Cell-Free Reaction of Protein Synthesis

1. Bacterial cell-free S30 extract prepared as described previously ([22], *see* also **Note 3**).
2. Second-step PCR product.
3. Reaction buffer: HEPES–KOH, pH 7.5, containing PEG 8000, potassium glutamate, creatine phosphate, calcium folinate, NH_4OAc, cAMP, DTT, ATP, GTP, CTP, UTP.
4. Total *E. coli* tRNA (Roche, Basel, Switzerland).
5. Solution of $Mg(OAc)_2$.
6. Mixture of 20 amino acids in 10 mM DTT.
7. Creatine kinase (Roche, Basel, Switzerland).
8. T7 RNA polymerase prepared as reported previously [23–25].
9. 96-Well PCR plates and strip caps.
10. A thermostat.

3 Methods

The following methods are intended for analysis of output from an existing *E. coli*-based cell-free protein production pipeline. They include (1) categorical, rather than continuous quantification of protein expression; (2) identification of physicochemical and structural parameters of amino acid sequences and multiple PTM sites using Internet-based computational and predictive bioinformatics tools; (3) processing and presentation of correlation data for the continuous and discrete variable parameters; (4) statistical analysis of the observed correlations between calculated and predicted properties of proteins and their amenability to heterologous cell-free expression.

3.1 Estimation of Protein Yield and Solubility

After the completion of protein synthetic reaction, soluble and insoluble protein products are separated by centrifugation at 10,000 × *g* for 10 min (*see* **Note 4**). Five-microliter aliquots of total and supernatant fractions are subjected to SDS PAGE on 12.5 % gels and protein bands are visualized with Coomassie Blue staining. The expression yield and solubility are estimated by the intensities of specific bands in the total and supernatant fractions. The bands are quantified using image analyzing software, such as Image Gauge software (Fuji Film, Tokyo, Japan). For quantity calibrations, bovine serum albumin (0.2–2.0 μg/lane) can be used. Typically, proteins that are expressed at the levels of less than 0.1 mg/ml are difficult to reliably visualize on the Coomassie-stained gels, because the specific protein bands are masked by the endogenous *E. coli* proteins.

Based on the quantification results, the scores A, C, and N are assigned to all experimentally expressed proteins as follows: A, soluble proteins expressed at the levels of more than 0.1 mg/ml; C, expressed but insoluble proteins; and N, non-expressed proteins (expression level below 0.1 mg/ml) (*see* **Notes 5** and **6**). Analysis of protein expression in our dataset showed that the proteins of group A represented 25.7 %, the proteins of group C—46.7 %, and the proteins of group N—27.6 % of all proteins analyzed. Similar success rate of soluble expression has been reported for another subset of human proteins expressed in *E. coli* [26].

For the purpose of following bioinformatics analysis, it is important that all investigated sequences are initially expressed under the same uniform set of conditions, minimizing the influence of sequence-independent factors. The affinity purification tags, which represent the additions of a polypeptide fragment at C- or N-terminus, must be either avoided or should exert minimal effects on protein folding (*see* **Note 7**). Some tags, such as maltose-binding protein, glutathione-S-transferase, etc., are highly soluble, increasing overall solubility of the fused target proteins. This may hinder the analysis of expression correlations by diminishing the role of sequence-specific determinants of protein targets. In our dataset, all synthesized polypeptide products universally comprised the N-terminal poly-His tag to allow their purification at the step of large-scale production.

3.2 Calculation and Prediction of Multiple Parameters of Polypeptide Sequences

The physicochemical parameters of amino acid sequences, such as p*I*, charge, hydrophobicity, can be calculated using the free ProtParam tool available at the Expasy server (http://web.expasy.org/protparam/). Solvent accessibility is calculated with the ACCpro 4.0 software downloaded from the SCRATCH Protein Predictor server ([27], http://scratch.proteomics.ics.uci.edu/explanation.html) and content of secondary structure is calculated with the PREDATOR 2.1.2 tool [28] provided online (http://mobyle.pasteur.fr/cgi-bin/portal.py?#forms::predator). Content of disordered structure is predicted with the RONN software [29] available online (http://www.strubi.ox.ac.uk/RONN). Coiled coil structures, signal sequences, transmembrane domains, and PEST regions are predicted with the tools provided online (http://emboss.sourceforge.net/apps/cvs/emboss/apps/pepcoil.html, http://www.cbs.dtu.dk/services/SignalP/, http://bp.nuap.nagoya-u.ac.jp/sosui/sosuisignal/, http://emboss.bioinformatics.nl/cgi-bin/emboss/pestfind, respectively).

3.3 Prediction of Posttranslational Modifications

The sites of prenylation, asparagine glycosylation, phosphorylation, etc. can be predicted with the PROSITE scanning tool PS_SCAN provided online (http://www.hpa-bioinfotools.org.uk/cgi-bin/ps_scan/ps_scanCGI.pl). The sites of S-palmitoylation are predicted with the CSS-Palm tool [30] available online (http://csspalm.biocuckoo.org). Disulfide bonds are calculated with the DIpro tool [31] downloadable free for scientific use (http://download.igb.uci.edu/intro.html). Ubiquitination sites can be predicted using the predictor of protein ubiquitination UbPred [32] downloaded from http://ubpred.org/ and SUMOylation sites are predicted with the site-specific predictor SUMOsp 2.0 [33] freely downloadable for academic research (http://sumosp.biocuckoo.org/).

3.4 Data Processing and Presentation

At the step of bioinformatics analysis of the expression data, the variable parameters or features of the two types, continuous and discrete, are used to characterize physicochemical and structural properties of polypeptide sequences. Among them, the Yes/No type of discrete variables are the features that can be either absent from or present in proteins. Localization signals, single N- and C-terminal protein modifications and other single-event protein modifications are the examples of such features. To present the expression data associated with the Yes/No type variables, the bar graphs can be used, which show the percentage of protein targets in the expression groups A, C, and N. The graphs should be built for the two datasets of proteins (i.e., excluding and including the analyzed feature).

As an example of the Yes/No data type processing, a case of protein prenylation can be considered. Prenylation was found to be a low-abundant modification—only 16 proteins in the analyzed dataset have been predicted to contain potential prenylation sites, and only single sites of prenylation could be predicted in these proteins. This is largely consistent with the previous estimates that put the number of possible prenylated proteins in the mammalian proteome to less than 2 %, corresponding to total ~100–200 proteins being potential prenylation substrates [34, 35]. Relative rates of soluble (A), insoluble (B), and non-expressed (C) proteins with (+) or without (−) the predicted sites of prenylation are shown in Fig. 2. The total number of protein targets in the (+) and (−) subsets is indicated above the bars. Using this graph, it is easy to make a side-by-side comparison of the data for the two subsets of sequences and deduce the tendencies in the protein cell-free expression amenability associated with prenylation. The evaluation of statistical significance of the observed tendencies is described further (*see* Subheading 3.5).

To present the expression data associated with the continuous variables, such as sequence hydrophobicity, p*I*, solvent accessibility, etc., another type of data presentation is more convenient. In this case, the percentage of protein targets in the expression groups A,

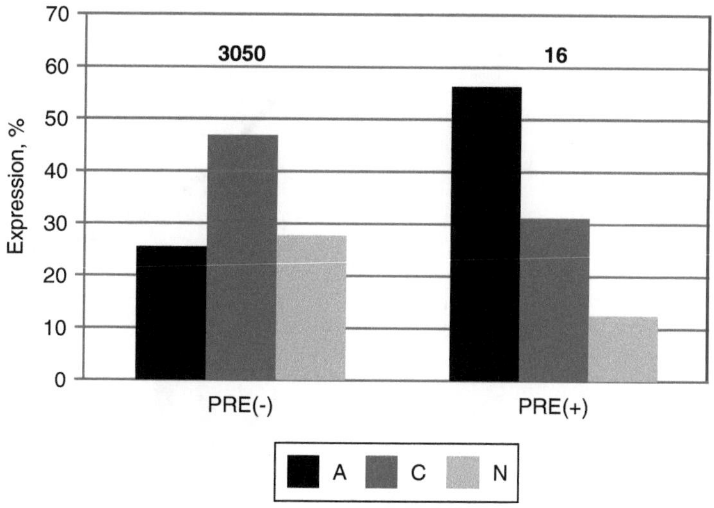

Fig. 2 Correlation between protein amenability to cell-free expression and predicted presence of prenylation. Relative rates of soluble (A), insoluble (B), and non-expressed (C) proteins with different probability of prenylation (+ or –) are shown

C, and N is plotted at different values of the studied parameter (Fig. 3a). The plot should cover the entire range of parameter values observed in the analyzed dataset. Curve smoothing is recommended, considering the continuous nature of these protein features. In addition, the distribution graph of dataset proteins according to parameter values should also be presented (Fig. 3b). The distribution graph provides the important information about the abundance of a studied protein feature in the analyzed dataset and its relation to expression amenability. For instance, distribution of the 3,066 dataset proteins according to their p*I* was found to be bimodal with the minimal representation of proteins at p*I* 7.0–7.5 (Fig. 3b). At the same time, the lowest rate of soluble expression and the highest rate of insoluble expression were observed for the proteins with p*I* 7.0–7.5 (Fig. 3a). These data propose that the proteins with neutral p*I* values may be underrepresented in the dataset due to their low solubility.

Data development for the discrete protein features repeatedly observed in the analyzed sequences, such as multiple S–S bonds, transmembrane regions, functional domains, abundant multi-site PTMs, is similar to the processing of the expression data associated with continuous variables. The graphs of relative soluble (A), insoluble (C), and undetectable expression (N), as well as the distribution graph should be provided in the complete range of discrete feature values. As an example, the correlations of cell-free expression amenability with the predicted number of disulfide bonds are shown in Fig. 4.

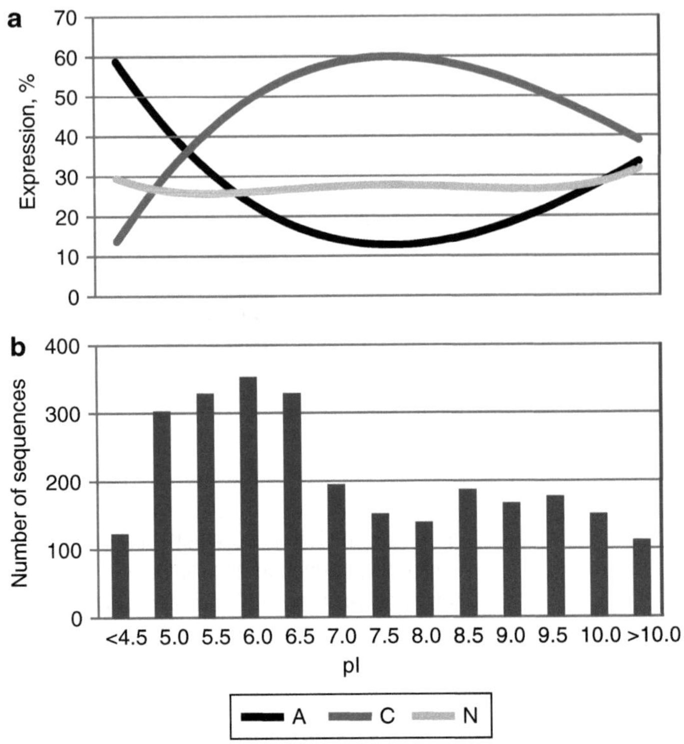

Fig. 3 Correlation between protein amenability to cell-free expression and p*I*. Relative rates of soluble (curve A), insoluble (curve C), and non-expressed (curve N) proteins with different p*I* values are presented in (**a**). Distribution of the dataset proteins according to their p*I* is shown in (**b**)

Finally, it is beneficial, for comprehensiveness of presentation, to summarize the observed correlations between expression amenability and multiple protein properties in one table. The major correlations revealed by the described analysis in our studies are presented in Table 2.

3.5 Statistical Significance of the Observed Correlations

At the stage of expression evaluation, all investigated proteins are categorized into three mutually exclusive classes—soluble (A), insoluble (C), and non-expressed (N) proteins, with the sum of these data (i.e., the total number of studied proteins) equaling 100 % (*see* **Note 8**). In other words, the data can only be placed in one class and not into another. Similarly, during the following bioinformatics analysis, the expressed protein targets are categorized into these three classes at different values of calculated or predicted parameters. Thus, the expression data represent, in essence, categorical datasets. To evaluate the statistical significance of the observed correlations, the categorical data analysis should be applied. This type of analysis is used to tell whether the difference between the sets of results is significant or not when the datasets contain the entries categorized into several mutually exclusive classes [36]. The estimation of statistical significance for each

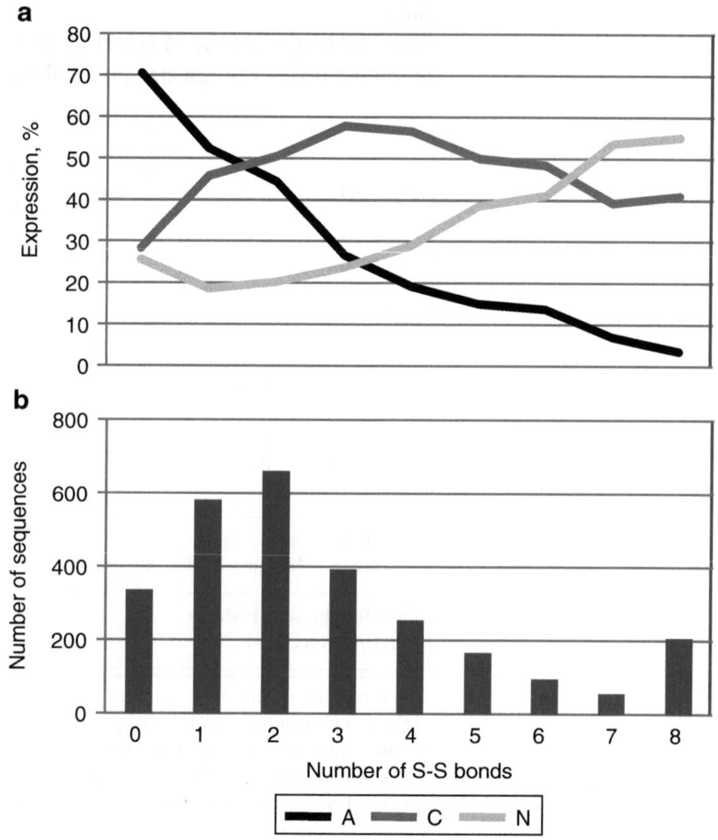

Fig. 4 Correlation between protein amenability to cell-free expression and predicted presence of disulfide bonds. Relative rates of soluble (curve A), insoluble (curve C), and non-expressed (curve N) proteins with different number of predicted S–S bonds are presented in (**a**). Distribution of the dataset proteins according to the number of S–S bonds is shown in (**b**)

expression group (A, C, and N) should be provided. In our project, to deduce the statistical differences between the expression datasets, the two-way contingency table test has been applied. The Fisher's exact *p*-values were computed using the tool available online (http://statpages.org/ctab2x2.html). A confidence level of 95 % was set up as the null hypothesis rejection threshold.

As an example, the evaluation of statistical significance of the data presented in Figs. 2, 3, and 4 is shown in Table 1. The categorical data analysis confirms a statistically significant difference in the ratios of soluble-expressed proteins in the two subsets, Pre (+) and Pre (−). On the other hand, although the tendencies towards the decrease in ratios of insoluble and undetectable protein expression can be observed for the sequences containing predicted sites of prenylation (Fig. 2), they cannot be validated by the statistical analysis (Table 1). Notably, the low statistical significance of these tendencies is certainly related to the small number of protein

Table 1
Statistical significance of correlations between calculated and predicted protein properties and cell-free expression amenability

Parameter	Expression		
	Soluble	Insoluble	Undetectable
Prenylation (number of sequences, +/–)	9/778	5/1427	2/845
p*I* (number of sequences, 5.5/7.5)	94/21	157/96	83/40
Disulfide bonds (number of sequences, +/–)	41/155	243/95	257/86
Prenylation (Fisher's exact *p*-value, +/–)	0.009	0.315	0.262
p*I* (Fisher's exact *p*-value, 5.5/7.5)	<0.001	0.004	0.911
Disulfide bonds (Fisher's exact *p*-value, +/–)	<0.001	<0.001	<0.001

The numbers of soluble, insoluble, and non-expressed polypeptide sequences with (+) or without (–) predicted prenylation sites and disulfide bonds, as well as the numbers of sequences with different values of the p*I* parameter ($5.0 < X < 5.5$ and $7.0 < X < 7.5$) are presented in the upper rows of the table. Analyzed subset of disulfide bond containing polypeptides comprised the sequences with more than five predicted S–S bonds

The Fisher's exact *p*-values calculated by the two-way contingency table analysis are presented in the lower rows

Boxes highlighted in *grey* denote the differences that are statistically significant at more than 95 % confidence level

sequences in the Pre (+) dataset because the confidence interval of categorical data is greatly affected by the sample size (*see* **Note 9**). The analysis of a more extended dataset of prenylated proteins is necessary to validate the observed tendencies. On the other hand, the lack of statistically significant difference demonstrated for the undetectable expression of proteins with $5.0 \le pI < 5.5$ and $7.0 \le pI < 7.5$ (Table 1) should reflect *de facto* absence of a tendency, considering the number of protein sequences in the analyzed datasets. Also, a high confidence level of the correlations between protein amenability to cell-free expression and predicted presence of disulfide bonds leans upon the large numbers of entries in the corresponding datasets (Table 1).

3.6 Optimizing the Conditions of Protein Synthetic Reaction

The purpose of cell-free synthesis in structural genomics and proteomics projects is to produce properly folded, functionally active proteins in the amounts sufficient for functional and structural studies. Thus, the optimization of cell-free protein synthesis concerns, as a rule, the yield of soluble (category A) expression. After the initial evaluation of protein expression using the described cell-free platform, it is often possible to optimize the reaction mixture for enhanced production of the selected proteins based on their individual physicochemical and structural properties. These properties can be calculated and predicted using bioinformatics algorithms and tools presented in Subheadings 3.2 and 3.3. Once the feature that may hinder soluble expression of a given protein is pinpointed, it becomes possible to tailor the conditions of cell-free protein synthesis specifically for this protein.

For instance, the high overall hydrophobicity of amino acid sequences is associated with worse protein amenability to soluble expression (Table 2). The poor solubility of hydrophobic proteins can be explained by their susceptibility to aggregation due to intermolecular hydrophobic interactions. It is often possible to neutralize these interactions and to increase protein solubility by using weak nonionic detergents that bind to the hydrophobic regions of proteins. This approach also allows the synthesis of membrane proteins that can be stabilized in the presence of detergent micelles. Importantly, the presence of mild nonionic detergents at the low solubilizing concentrations does not inactivate the protein synthetic activity of the bacterial extract. It was found that only nonionic detergents with low critical micelle concentration (c.m.c.) can be used in cell-free expression systems without decreasing protein yield [37]. In our project, G-protein-coupled receptors have been successfully expressed using the weak nonionic detergents Brij35 and digitonin [38].

Another example of reaction optimization concerns the synthesis of correctly folded eukaryotic proteins containing multiple disulfide bonds in the bacterial cell-free system. The predicted presence of disulfide bonds in a polypeptide negatively correlates with soluble protein expression and positively correlates with the insoluble expression (Fig. 4, Table 2). Among all protein features analyzed, the occurrence of disulfide bonds was found to be one of the most discriminative for expression propensity. Probability of soluble expression for the proteins that are predicted to bear more than six disulfide bonds per molecule falls below 10 % [11]. The formation of S–S bonds in eukaryotic proteins is greatly compromised in the bacterial extracts, as their reducing conditions differ from those in eukaryotic cells [39]. Pretreatment with iodoacetamine, which blocks the free sulfhydryl groups of endogenous bacterial proteins, was shown to abolish the disulfide-reducing activity of the extracts [40]. The use of the iodoacetamine-treated bacterial extracts complemented with a glutathione redox buffer and the DsbC disulfide isomerase allowed efficient production of complex mammalian proteins containing multiple disulfide bonds [41, 42].

To increase the rate of soluble expression of neutral proteins (i.e., the proteins with p*I* values close to 7.0), pH of the reaction

Table 2
Correlations of soluble, insoluble, and undetectable cell-free expression with calculated and predicted properties of proteins

Property	Expression		
	Soluble	Insoluble	Undetectable
Length	↓	—	↑
p*I*	↕	↕	—
Charge	↑	↕	↓
Hydrophobicity	↓	↑	—
Solvent accessibility	↑	↓	↑
Secondary structure	↑	↕	↓
Intrinsic disorder	↑	↓	↑
Disulfide bonds	↓	↑	↑
Coiled coil	↑	↓	↓
Transmembrane domains	↓	↓	↑
Localization signal sequences	↓	↑	—
PEST regions	↑	↓	↑
Prenylation	↑	—	—
Phosphorylation	↑	↓	↓
Asn glycosylation	↓	↑	—
Palmitoylation	↓	↕	↑
Ubiquitination	↑	↓	—
SUMOylation	↑	↓	↕

Upward and ***downward arrows*** indicate positive and negative correlations, respectively; ***straight horizontal lines*** denote the lack of correlation; and ***bidirectional arrows*** refer to the opposite tendencies of expression estimates at different values of calculated parameters

mixture can be optimized. The proteins with neutral p*I* values are not charged at the physiological pH of the original expression system and are prone to aggregation. Accordingly, the lowest rate of soluble expression has been observed for the proteins with p*I* 7.0–7.5 (Fig. 3). Shifting pH to more acidic or basic values will make these proteins charged, preventing their intermolecular interactions and resulting in better solubility. As a rule, setting the reaction mixture pH within the interval of 6.5–8.5 does not affect dramatically the total yield of protein synthetic reaction; however, often it can significantly increase the yield of soluble expression. Notably, pH also affects the oxidizing environment of the reaction mixture, because the thiol-disulfide exchange is inhibited at low pH (effectively, at pH below 8.0). It was found that the normal reaction pH was not high enough to allow the efficient formation and isomerization of disulfide bonds and both the oxidizing environment and an increased pH were required to produce the active protease domain of mammalian urokinase in a cell-free bacterial system [40].

Other conditions that can be optimized in the original reaction mixture for the enhanced production of the selected proteins include concentration of magnesium, DTT, DNA template, duration and temperature of protein synthesis, the energy substrate and amino acid composition. Additions of FAD, NAD, CoA, malic acid, 2-glutaric acid, succinic acid and introduction of chaperones have also been reported to stimulate cell-free protein synthesis. Usually, the positive effects of these compounds on the expression yields are not additive and it is quite difficult to explain their individual contributions in the particular cases.

3.7 Protein Engineering Aimed at Increasing Expression Success

In addition to easily mastered optimization of the reaction conditions, the cell-free gene expression technology combined with bioinformatics analysis allows feasible protein engineering and screening with the aim of increasing expression success. Several approaches have been developed to engineer the proteins that have enhanced amenability to heterologous cell-free expression.

For instance, the solubility of signal sequence-containing proteins can often be improved by the N-terminal truncation of signal sequence regions. These sequences direct proteins to certain cellular compartments and organelles, making their existence in the cytoplasm (i.e., in a free soluble form) implausible. Similarly, truncation of N-terminal or C-terminal membrane anchoring sequences in the single-pass transmembrane proteins can improve their solubility. The presence of the signal and anchoring (transmembrane) sequences was found to worsen protein amenability to cell-free soluble expression ([11], Table 2). As an example from our project, the human pyruvate dehydrogenase kinase 4 could be successfully expressed and crystallized for X-ray studies after truncating its mitochondrial targeting sequence [43]. The signal and anchoring sequences can be reliably identified in the analyzed proteins using the existing prediction algorithms (*see* Subheading 3.2).

Other common approach in the structural/functional protein analysis is to express the truncated forms of the complex multi-domain proteins. The worsening of protein expression with the number of functional domains and protein length has been demonstrated ([11], Table 2). The removal of domains is often able to improve protein propensity for soluble expression (*see* **Note 10**). The resolved protein structures deposited in the Protein Data Bank and 3D homology modeling can be used to define the functional domains in a protein. In the case when protein structure is largely unknown, it is possible to assign functional domains using prediction algorithms based on difference in the amino acid composition between domain and linker regions [44]. The domain truncation approach is exceptionally efficient when the domains to be expressed are well defined and display high intrinsic solubility.

Alternatively, fusing target proteins with other highly soluble polypeptides often helps to increase overall solubility of the fused protein products. The addition of the solubility tags, such as GST, MBP, SUMO, ubiquitin, at the N- or C-termini of highly hydrophobic sequences can be considered to increase the yield of their soluble expression. Interestingly, it was found that the presence of predicted sites of SUMOylation and ubiquitination in a protein directly correlates with its propensity for cell-free soluble expression ([12], Table 2).

Although, in general, single amino acid replacements cannot change significantly the integral characteristics of protein molecule related to its overall solubility, the dramatic global realignments of protein structure due to a single substitution event have been documented too. Thus, in some cases, introducing even a single mutation can significantly change the outcome of protein synthetic reaction, increasing the yield of soluble protein. Experimentally confirmed or bioinformatically predicted sites of PTMs may be the primary targets for site-directed mutagenesis. For example, phosphorylatable Ser, Thr, and Tyr residues in the amino acid sequences can be mutated into Glu or Asp residues to mimic the negative charge associated with the phospho group. It is not uncommon that the phosphorylated forms of proteins gain increased solubility. The presence of predicted phosphorylation sites in the polypeptide sequences was found to be associated with the increased production of properly folded soluble protein ([12], Table 2). Admittedly, the use of site-directed mutagenesis for the enhancement of cell-free protein expression is far from being established and it requires further development. A simple and widely used technique of alanine-scanning mutagenesis can be employed to determine the contribution of specific residues to the stability and solubility of a given protein.

4 Notes

1. The template human cDNA clones can be obtained from several commercial sources, such as Invitrogen, Carlsbad, CA, USA; OriGene Technologies, Rockville, MD, USA; Kazusa DNA Research Institute, Kisarazu, Chiba, Japan; Institute of Medical Science of Tokyo University, Tokyo, Japan; GeneCopoeia, Inc., Rockville, MD, USA; and Toyobo Engineering, Osaka, Japan.
2. The forward and reverse primers for the first PCR, the T7 promoter and T7 terminator fragments, and the universal primer for the second PCR have been constructed as described previously [19, 21].
3. To circumvent the low expression yields due to rare codons, the extracts prepared from the bacterial strains containing additional copies of the genes for low-abundant tRNAs such as BL21 codon plus (Stratagene, USA), BL21-Star (DE3), or Rosetta strains (Novagen, USA) should be used.
4. At the screening stage, the term "solubility" refers to the protein solubility in the cell-free extract that may differ from solubility of purified proteins. The steps should be taken to maintain an extract-soluble protein in the soluble state during the purification procedure normally following large-scale cell-free protein synthesis.
5. The score A provides the upper estimation of soluble protein expression, because the solubility status is evaluated after separation of soluble and insoluble protein products by centrifugation at $10{,}000 \times g$ for 10 min. This procedure cannot discriminate between small protein aggregates and truly soluble proteins.
6. Proteins that are expressed at a lower than expected molecular size should be considered as non-expressed, because they cannot attain proper structure and function when synthesized in the employed expression system.
7. The presence of a C- or N-terminal affinity tag makes impossible the analysis of correlations between the predicted modifications of the tagged terminus and cell-free protein synthesis. For instance, the presence of an N-terminal tag hinders the effects of the N-terminal protein modifications, such as N-terminal methionine excision, myristoilation, acetylation, etc., on the protein amenability to heterologous cell-free expression. Thus, the expression of untagged or C-terminally tagged protein targets should be considered in order to deduce the correlations between N-terminal modifications and heterologous protein synthesis.

8. Often, expressed proteins can be found in both soluble and insoluble fractions of the cell-free reaction mix. However, the applied categorical data analysis allows their classification into only one expression category. Usually, lane-to-lane comparison of total and supernatant fractions of the reaction in PAGE gels is sufficient to determine the preferential pattern of protein expression.
9. The confidence level of categorical data analysis increases dramatically with the number of sequences in the datasets of expressed proteins. To deduce statistically significant tendencies, rather small datasets may be used when analyzing robust correlations. Practically, for the discussed approach, the counts in every cell of the contingency table should be at least 5 [45].
10. It is important not to cut into domains when expressing truncated forms of proteins. The truncations should be made several residues upstream and downstream of the last major structural elements (α-helix or β-sheet) of a domain. It is recommended to produce several variants that are cut out at intervals of several residues at both N- and C-termini and examine their yield and solubility.

References

1. Yokoyama S (2003) Protein expression systems for structural genomics and proteomics. Curr Opin Chem Biol 7:39–43
2. Marsden RL, Orengo CA (2008) Target selection for structural genomics: an overview. Methods Mol Biol 426:3–25
3. Farokki N, Hrmova M, Burton RA et al (2009) Heterologous and cell-free protein expression systems. Methods Mol Biol 513:175–198
4. Spirin AS (2004) High-throughput cell-free systems for synthesis of functionally active proteins. Trends Biotechnol 22:538–545
5. Katzen F, Chang G, Kudlicki W (2005) The past, present and future of cell-free protein synthesis. Trends Biotechnol 23:150–156
6. He M (2008) Cell-free protein synthesis: applications in proteomics and biotechnology. Nat Biotechnol 23:126–132
7. Goh CS et al (2004) Mining the structural genomics pipeline: identification of protein properties that effect high-throughput experimental analysis. J Mol Biol 336:115–130
8. Bertone P, Kluger Y, Lan N et al (2001) SPINE: an integrated tracking database and data mining approach for identifying feasible targets in high-throughput structural proteomics. Nucleic Acids Res 29:2884–2898
9. Dyson MR, Shadbolt SP, Vincent KJ et al (2004) Production of soluble mammalian proteins in *Escherichia coli*: identification of protein features that correlate with successful expression. BMC Biotechnol 4:32
10. Idicula-Thomas S, Balaji P (2005) Understanding the relationships between the primary structure of proteins and its propensity to be soluble on overexpression in Escherichia coli. Protein Sci 14:582–592
11. Kurotani A, Takagi T, Toyama M et al (2010) Comprehensive bioinformatics analysis of cell-free protein synthesis: identification of multiple protein properties that correlate with successful expression. FASEB J 24:1095–1104
12. Tokmakov AA, Kurotani A, Takagi T et al (2012) Multiple post-translational modifications affect heterologous protein synthesis. J Biol Chem 287:27106–27116
13. Kigawa T, Yabuki T, Yoshida Y et al (1999) Cell-free production and stable-isotope labeling of milligram quantities of proteins. FEBS Lett 442:15–19
14. Kigawa T, Yokoyama S (1991) A continuous cell-free protein synthesis system for coupled transcription-translation. J Biochem 110:166–168
15. Kudlicki W, Kramer G, Hardesty B (1992) High efficiency cell-free synthesis of proteins: refinement of the coupled transcription/translation system. Anal Biochem 206:389–393
16. Yokoyama S, Hirota H, Kigawa T et al (2000) Structural genomics projects in Japan. Nat Struct Biol 7(Suppl):943–945

17. Yokoyama S (2005) Large-scale structural proteomics project at RIKEN: present and future. Tanpakushitsu Kakusan Koso 50:836–845
18. Yokoyama S, Kigawa T, Shirouzu M et al (2008) RIKEN structural genomics/proteomics initiative. Tanpakushitsu Kakusan Koso 53:632–637
19. Yabuki T, Motoda Y, Hanada K et al (2007) A robust two-step PCR method of template DNA production for high-throughput cell-free protein synthesis. J Struct Funct Genom 8:173–191
20. Kigawa T, Matsuda T, Yabuki T et al (2008) Bacterial cell-free system for highly efficient protein synthesis. In: Spirin AS, Swartz JR (eds) Cell-free protein synthesis. Wiley, Weinheim, pp 83–97
21. Kigawa T (2010) Analysis of protein functions through a bacterial cell-free protein expression system. Methods Mol Biol 607:53–62
22. Kigawa T, Yabuki T, Matsuda N et al (2004) Preparation of Escherichia coli extract for highly productive cell-free protein expression. J Struct Funct Genom 5:63–68
23. Davanloo P, Rosenberg AH, Dunn JJ et al (1984) Cloning and expression of the gene for bacteriophage T7 RNA polymerase. Proc Natl Acad Sci U S A 81:2035–2039
24. Grodberg J, Dunn JJ (1988) ompT encodes the Escherichia coli outer membrane protease that cleaves T7 RNA polymerase during purification. J Bacteriol 170:1245–1253
25. Zawadski V, Gross HJ (1991) Rapid and simple purification of T7 RNA polymerase. Nucleic Acids Res 19:1948
26. Ding HT, Ren H, Cheng Q et al (2002) Parallel cloning, expression, purification, and crystallization of human proteins for structural genomics. Acta Crystallogr D Biol Crystallogr 58:2102–2108
27. Cheng J, Randall AZ, Sweredoski MJ et al (2005) SCRATCH: a protein structure and structural feature prediction server. Nucleic Acids Res 33(Web Server issue):72–76
28. Frishman D, Argos P (1997) Seventy-five percent accuracy in protein secondary structure prediction. Proteins 27:329–335
29. Yang ZR, Thomson R, McMeil P et al (2005) RONN: the bio-basis function neural network technique applied to the detection of natively disordered regions in proteins. Bioinformatics 21:3369–3376
30. Ren J, Wen L, Gao X et al (2008) CSS-Palm 2.0: an updated software for palmitoylation sites prediction. Protein Eng Des Sel 21:639–644
31. Cheng J, Saigo H, Baldi P (2006) Large-scale prediction of disulfide bridges using kernel methods, two-dimensional recursive neural networks, and weighed graph matching. Proteins 62:617–629
32. Radivojac P, Vacic V, Haynes C et al (2010) Identification, analysis, and prediction of protein ubiquitination sites. Proteins 78:365–380
33. Ren J, Gao X, Jin C et al (2009) Systematic study of protein sumoylation: development of a site-specific predictor of SUMPsp 2.0. Proteomics 9:3409–3412
34. Gao J, Liao J, Yang GY (2009) CAAX-box protein, prenylation process and carcinogenesis. Am J Transl Res 25:312–325
35. Amaya M, Baranova A, van Hoek ML (2011) Protein prenylation: a new mode of host-pathogen reaction. Biochem Biophys Res Commun 416:1–6
36. Xu B, Feng X, Burdine RD (2010) Categorical data analysis in experimental biology. Dev Biol 348:3–11
37. Betton JM, Miot M (2008) Cell-free production of membrane proteins in the presence of detergents. In: Spirin AS, Swartz JR (eds) Cell-free protein synthesis: methods and protocols. Wiley, Weinheim, pp 165–178
38. Ishihara G, Goto M, Saeki M et al (2005) Expression of G-protein coupled receptors in a cell-free translational system using detergents and thioredoxin-fusion vectors. Protein Expr Purif 41:27–37
39. Tu BP, Weissman JS (2004) Oxidative protein folding in eukaryotes. J Cell Biol 164: 341–346
40. Kim DM, Swartz JR (2004) Efficient production of a bioactive, multiple disulfide-bonded protein using modified extracts of Escherichia coli. Biotechnol Bioeng 85:122–129
41. Yin G, Swartz JR (2004) Enhancing multiple disulfide bonded protein folding in a cell-free system. Biotechnol Bioeng 86:188–195
42. Yang J, Kanter G, Voloshin A et al (2004) Expression of active murine granulocyte-macrophage colony-stimulating factor in an Escherichia coli cell-free system. Biotechnol Prog 20:1689–1696
43. Kukimoto-Niino M, Tokmakov A, Terada T et al (2011) Inhibitor-bound structures of human pyruvate dehydrogenase kinase. Acta Crystallogr D Biol Crystallogr 67:763–773
44. Suyama M, Ohara O (2003) DomCut: prediction of inter-domain linker regions in amino acid sequences. Bioinformatics 19:673–674
45. Norman GR, Streiner DL (2000) Biostatistics: the bare essentials. B.C. Decker, Hamilton

Chapter 3

A Cell-Free Expression Screen to Identify Fusion Tags for Improved Protein Expression

Andrew Kralicek

Abstract

Cell-free protein synthesis can now be routinely used for the rapid screening of protein expression at the microliter level using PCR-amplified templates. However, identification of the optimal expression construct for a target protein can still be a problem. A rapid cell-free procedure is described here for the systematic assessment of a range of diverse fusion tags on the expression and solubility of any given target protein. Overlap/extension PCR is used to fuse a library of T7 promoter (T7p)-tag fragments with a gene-T7terminator (T7ter) fragment to produce cell-free expression templates encoding different fusion proteins. These constructs are then expressed in a series of small-scale (50 μL) *Escherichia coli* cell-free reactions and SDS-PAGE analysis is used to identify the optimal fusion tag(s). This screen is particularly useful for the identification of expression constructs for proteins that are normally poorly expressed or are insoluble.

Key words Cell-free protein synthesis, Overlap/extension PCR, Fusion tag, Expression screen

1 Introduction

Cell-free expression systems offer many advantages for protein production. Rapid screening of protein expression can be performed in a single day at the microliter level before scaling up to achieve milligram per milliliter quantities of target protein [1]. Molecules that aid protein folding can be added directly to the reaction mix [2–4], and cytotoxic proteins such as antimicrobial and membrane proteins can be synthesized because of the nonliving nature of the cell-free system [5–7]. Furthermore, expression templates generated by PCR can be used to explore the expression of previously unclonable genes and manipulated versions of proteins [8, 9].

Despite these advantages, successful cell-free protein synthesis is not guaranteed. Protein production yields can be improved by codon optimization [10], and N-terminal tagging with fusion tags which enhance translation and/or protein solubility [11]. But there is no way to predict which fusion tag is optimal for a

Kirill Alexandrov and Wayne A. Johnston (eds.), *Cell-Free Protein Synthesis: Methods and Protocols*, Methods in Molecular Biology, vol. 1118, DOI 10.1007/978-1-62703-782-2_3, © Springer Science+Business Media, LLC 2014

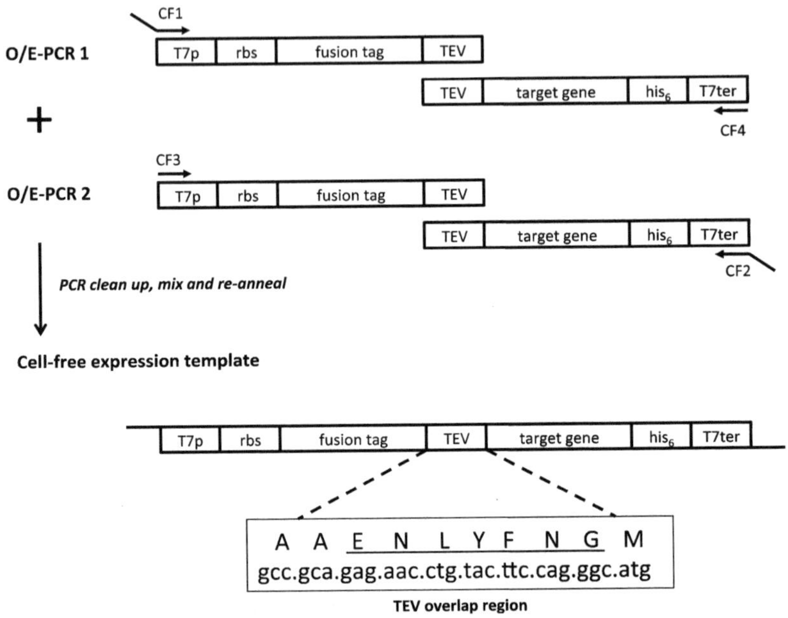

Fig. 1 Overlap/extension PCR using the TEV overlap region to produce a stable cell-free expression template encoding a fusion-tagged target protein. Two overlap/extension PCR reactions are performed with gel-purified T7p-tag-TEV and TEV-target gene-T7ter DNA fragments. In one reaction, the CF1/CF4 primer pair amplifies a PCR product with an 8 bp 5′ overhang sequence. In the other reaction, the CF3/CF2 primer pair amplifies a PCR product with a complementary 8 bp 3′ overhang sequence. Mixing and re-annealing of these products produces a DNA template with complementary overhangs that is believed to undergo ligation in the cell-free reaction to produce a stable circular template [8]. The sequence of the TEV overlap region is shown in the *box*; the TEV protease cleavage site is *underlined*. Figure is from Kralicek et al. [12], reproduced by permission of Elsevier

given target protein, so the conventional approach is to survey different tags empirically. The drawback of this approach is the time required to generate expression constructs encoding different tag-target protein constructs. This often causes researchers to abandon the characterization of proteins if commonly used tags fail to produce enough soluble protein.

In this chapter we describe a PCR-directed cell-free expression screen that can be used to identify the optimal N-terminal fusion tag for target protein production. An overlap/extension PCR approach is used to fuse an N-terminal sequence to a target gene in order to generate cell-free expression templates that can facilitate the rapid assessment of multiple tag-target protein fusions [12]. This requires the amplification of T7 promoter (T7p)-tag DNA fragments from appropriate vectors encoding the different fusion tags, as well as the amplification of a DNA fragment encoding the target gene with the T7 terminator (T7ter) (Fig. 1). Overlap/extension PCR of each T7p-tag fragment with the gene-T7ter fragment using a common linker sequence, the TEV protease sequence, enables the creation of a library of cell-free expression templates. Such templates can easily be prepared in a single day if a library of tag fragments is already at hand. These templates are then expressed on a small scale using an *E. coli* cell-free reaction system which is a modified version of the one previously described by Apponyi et al. [13]. SDS-PAGE analysis of the completed cell-free reactions determines the utility of the fusion tags.

2 Materials

2.1 Overlap/Extension PCR

1. Nuclease-free water.
2. Vectors encoding different Tag-TEV sequences under the control of a T7 promoter.
3. Vector encoding target gene sequence followed by the T7 terminator sequence. We have available the following nine **Tag-TEV** encoding vectors: pUC57-T7p-**CAT**-**TEV**, pETMCSI-**T7tag**-KcsArub, pUC57-T7p-**OmpAss**-**TEV**, pUC57-T7p-**Gb1**-**TEV**, pETMCSI-**Trx**-**TEV**-CIN1-his$_6$, pETMCSI-**Ppib**-**TEV**-KcsA-his$_6$, pETMCSI-**GST**-**TEV**-his$_6$, pETMCSI-**MBP**-**TEV**-EcOr 22a-his$_6$, pETMCSI-**NusA**-**TEV**-**his6** [12].
4. Primers required for PCR amplification of T7p-Tag-TEV and TEV-target gene-T7ter fragments, and the overlap/extension PCR reaction (Table 1).
5. Vent DNA polymerase, 10× ThermoPol Reaction buffer and 100 mM $MgSO_4$ (New England Biolabs).
6. 10 mM dNTPs.
7. DNA fragment Gel extraction kit.
8. PCR product purification kit.

2.2 S30 Extract Preparation

1. MilliQ (MQ) water.
2. 2 M glucose.
3. 1 mg/mL thiamine.
4. LB medium: 10 g/L tryptone, 5 g/L yeast extract, 5 g/L sodium chloride, and 2.5 mL/L 1 M sodium hydroxide.
5. LBT medium: LB, 0.01 mg/L thiamine.
6. 4× Z-medium: 165 mM potassium dihydrogen phosphate, 660 mM potassium phosphate dibasic, 40 g/L yeast extract.
7. 10× S30 buffer: 100 mM Tris acetate, pH 8.2, 160 mM potassium acetate, 140 mM magnesium acetate. Adjust pH to 8.3 with potassium hydroxide.
8. S30 buffer A: 1× S30 buffer containing 7.2 mM β-mercaptoethanol and complete EDTA-free protease inhibitor cocktail tablets (Roche).

Table 1
Primers required for PCR amplification of the cell-free expression template

Primer	Sequence	Design
CF1	5′-PO_4-TTA GCT GGT CGA TCC CGC GAA ATT AAT ACG	Targets the T7 promoter and adds on the 8 bp overhang (underlined)
CF2	5′-PO_4-CCA GCT AAC AAA AAA CCC CTC AAG ACC CG	Targets the T7 terminator and adds the complementary 8 bp overhang (underlined)
CF3	5′-PO_4-TCG ATC CCG CGA AAT TAA TAC G	Targets the T7 promoter
CF4	5′-PO_4-CAA AAA ACC CCT CAA GAC CCG	Targets T7 terminator
TEV4rev	CAT GCC CTG GAA GTA CAG GTT CTC TGC GGC	Targets the extended TEV sequence (Fig. 1)
TEV4-Tag-rev	CAT GCC CTG GAA GTA CAG GTT CTC TGC GGC—(25–30 nucleotides complementary to the 3′ end of the Tag)	Targets 3′ end of the Tag sequence and adds on the TEV overlap sequence
TEV4-Target-for	GCC GCA GAG AAC CTG TAC TTC CAG GGC—(25–30 nucleotides encoding the 5′ end of the Target gene)	Targets 5′ end of the Target gene sequence and adds on the TEV overlap sequence

9. S30 buffer B: 1× S30 buffer containing 1 mM PMSF and 1 mM DTT.
10. 50 % PEG8000 w/w: 50 % PEG8000 in 1× S30 buffer.
11. *E. coli* strain BL21 STAR (DE3) transformed with pRARE (isolated from or BL21 Rosetta (DE3) (Invitrogen).
12. Chloramphenicol.
13. 10 L Fermenter—authors use a Labfors® 3 (Infors).
14. Homogenizer—authors use the EmulsiFlex C5 homogenizer (Avestin).
15. Spectrapor #4 dialysis tubing, 12–14,000 MWCO (Spectrum Laboratories, Inc., Rancho Dominguez, CA, USA).

2.3 Cell-Free Protein Synthesis Reaction

1. MilliQ (MQ) water.
2. 2.0 M HEPES pH 7.5.
3. 0.5 M DTT.
4. 120 mM ATP.
5. 80 mM CTP/GTP/UTP.
6. 64 mM cyclic AMP.
7. 6.8 mM folinic acid.
8. 2.75 M ammonium acetate.

9. 4.16 M Potassium (L)-Glutamate.
10. 1.93 M magnesium acetate.
11. 1.6 M creatine phosphate.
12. 17.5 mg/mL *E. coli* MRE60 tRNA (Roche).
13. 10 mg/mL creatine kinase.
14. 40 U/μL RNasin (Promega).
15. 100 mM amino acid stock solutions: *Group A*—L-Aspartic acid, L-Glutamine, L-Threonine, L-Methionine, L-Tyrosine; *Group B*—L-Asparagine monohydrate, Glycine, L-Histidine, L-Leucine, L-Valine; *Group C*—L-Alanine, L-Arginine, L-Lysine hydrochloride, L-Proline, L-Serine; *Group D*—L-Isoleucine, L-Tryptophan, L-Cysteine-free base, L-Phenylalanine.
16. 1 μg/μL T7 RNA polymerase vector pKO1166.
17. Slide-A-Lyzer® MINI Dialysis Units (Pierce).

3 Methods

The following protocols are described: (Subheading 3.1) overlap/extension PCR to generate a range of cell-free expression templates encoding a target protein with different N-terminal fusion tags; (Subheading 3.2) preparation of the *E. coli* S30 extract required for cell-free protein synthesis; and (Subheading 3.3) how to set up the cell-free reaction to express the fusion proteins from their PCR generated templates.

3.1 Overlap/Extension PCR to Generate Cell-Free Expression Templates Encoding Different N-Terminally Tagged Fusion Proteins

Previously Wu et al. [8] developed a protocol to produce PCR-derived cell-free expression templates that are believed to form stable circular expression templates within the cell-free reaction. Their approach requires the generation of two versions of the PCR template, one with an 8 bp overhang at the 5′ end and the other with a complementary 8 bp overhang at the 3′ end. Mixing and re-annealing of these products results in half the linear PCR templates having complementary single-stranded overhangs at their 5′ and 3′ ends that are available for ligation by the endogenous ligase activity of the *E. coli* S30 extract. We have modified this protocol to include an overlap extension approach to produce a range of PCR templates encoding a target protein with different N-terminal fusion tags (Fig. 1) [12]. A 30-nucleotide GC-rich (60 %) sequence encoding the TEV protease cleavage is used as the overlap region to enable overlap/extension PCR.

As a prerequisite for the overlap/extension PCR reactions, PCR amplification of T7p-tag-TEV and TEV-target gene-T7ter fragments is required using a forward and reverse primer (Table 1). Each T7p-tag-TEV fragment is amplified from an appropriate vector encoding the tag sequence under the control of the T7 promoter,

Table 2
Composition of PCR reaction to amplify the TEV-target gene-T7ter fragment

Component	μL
20 ng/μL plasmid containing Target gene-T7ter	1
TEV4-Target-for primer	4
CF4 primer	4
10 mM dNTPs	2
10× ThermoPol reaction buffer	10
Vent DNA polymerase	0.5
100 mM magnesium sulfate	2
Nuclease-free water	76.5
Total	100

using the CF3 forward primer in combination with the TEV4rev reverse primer (if the vector encodes the tag followed by the TEV sequence) or otherwise a specific TEV4-Tag-rev primer (*see* **Note 1**). The TEV-target gene-T7ter fragment is amplified from an appropriate pET vector using the forward primer TEV4-Target-for in combination with the CF4 reverse primer. We usually incorporate a C-terminal his_6 tag into the design of the target gene (*see* **Note 2**).

Two separate overlap/extension PCR reactions are then carried out for each T7p-Tag-TEV and TEV-target gene-T7ter fragment combination to amplify each T7p-Tag-TEV-target gene-T7ter fusion construct (Fig. 1). In the first reaction, the CF1 forward primer that adds on the 8 bp overhang, and the CF4 reverse primer are used. In the second reaction, the CF3 forward primer and the CF2 reverse primer that adds the complementary 8 bp overhang are used. The resulting PCR products are purified, mixed, and re-annealed to produce each cell-free expression template (*see* **Note 3**).

3.1.1 PCR Amplification of the TEV-Target Gene-T7ter Fragment

1. Thaw the reagents listed in Table 2 on ice.
2. Prepare the PCR reaction according to Table 2, and mix by pipetting it up and down.
3. Dispense 100 μL of the mixed solution into a PCR tube and place on the heating block of the PCR machine.
4. Run the PCR program as follows: 94 °C denaturation step for 2 min; 32 cycles of 94 °C for 15 s (denaturation), 55 °C for 30 s (primer annealing), and 72 °C for 2.5 min (extension); 72 °C for 10 min (final elongation).
5. Load the complete reaction on a 1 % TAE agarose gel and perform gel electrophoresis for 1 h at 80 V.

Table 3
Composition of PCR reaction to amplify the T7p-Tag-TEV fragment

Component	μL
20 ng/μL plasmid containing T7p-Tag gene	1
CF4 primer	4
TEV4rev (or TEV4-Tag-rev)	4
10 mM dNTPs	2
10× ThermoPol reaction buffer	10
Vent DNA polymerase	0.5
100 mM magnesium sulfate	2
Nuclease-free water	76.5
Total	100

6. Use a UV illuminator to visualize the PCR products produced, excise the correct sized band, and then use a Gel Extraction Kit to purify the PCR fragment in a total volume of 50 μL elution buffer. Measure the absorbance at A260 to calculate the concentration of the PCR fragment.

3.1.2 PCR Amplification of the T7p-Tag-TEV Fragments

1. Thaw on ice the reagents listed in Table 3.
2. Prepare the PCR reaction according to Table 3, and mix by pipetting it up and down.
3. Follow **steps 3–6** in Subheading 3.1.1.

3.1.3 Overlap/Extension PCR to Produce the T7p-Tag-TEV-Target Gene-T7ter DNA Fragment

1. Thaw on ice the reagents listed in Table 4.
2. Prepare the two PCR Reactions, A and B, in separate tubes according to Table 4, and mix each reaction by sucking up and down with a pipette.
3. Dispense 100 μL of each solution into a separate PCR tube and place on the heating block of the PCR machine.
4. Run the PCR program as follows: 94 °C denaturation step for 2 min; 32 cycles of 94 °C for 15 s (denaturation), 60 °C for 30 s (primer annealing), and 72 °C for 3 min (extension); 72 °C for 10 min (final elongation).
5. Use a PCR purification kit to purify the PCR product from the A and B reactions separately in a total volume of 50 μL MilliQ water. Run 5 μL on a 1 % TAE gel to confirm correct product amplification. Measure the absorbance at A260 to calculate the concentration of the PCR fragment. A minimum concentration of 30 ng/μL is required.

Table 4
Composition of PCR reaction to amplify the T7p-Tag-TEV-target gene-T7ter fragment

Component	A (μL)	B (μL)
20 ng/μL T7p-Tag-TEV fragment	1	1
20 ng/μL TEV-target gene-T7ter fragment	1	1 μL
Forward primer	4 CF1	4 CF3
Reverse primer	4 CF4	4 CF2
10 mM dNTPs	2	2
10× ThermoPol reaction buffer	10	10
Vent DNA polymerase	0.5	0.5
100 mM magnesium sulfate	2	2
Nuclease-free water	75.5	75.5
Total	100	100

3.1.4 Re-annealing of the T7p-Tag-TEV-Target Protein-T7ter Fragments

1. Mix equimolar amounts of PCR product A with PCR product B in a microcentrifuge tube.
2. Heat the tube at 94 °C for 5 min and then leave to cool at room temperature for 5 min.
3. Store at −20 °C until further use.

Figure 2 is a picture of 1 % agarose gel showing examples of what the T7p-tag-TEV-target protein-T7ter fragments should look like. These DNA fragments were generated by overlap extension between a range of T7p-Tag-TEV fragments and a TEV-Adiponectin hypervariable domain (AHD)-his$_6$-T7ter fragment [12].

3.2 Preparation of the S30 Extract

In order to provide enough bulk extract for many cell-free screens, a culture of *E. coli* BL21 STAR (DE3)/pRARE is grown in rich media using a 10 L fermenter at 30 °C (*see* **Note 4**). The bacteria are then harvested during the exponential phase (OD600 = 3) to isolate cells containing high quantities of translating ribosomes. Cell lysis using a homogenizer is followed by centrifugation at 30,000 × *g* to yield the S30 extract. Retention of upper fraction is necessary to produce an S30 extract capable of synthesizing a target protein from a PCR-amplified template at comparable amounts to those achieved using plasmid templates [12]. A twofold concentration step is then employed by dialysis against PEG8000 to provide extra room in the final cell-free reaction for the addition of other components. Approximately 20 mL of concentrated S30 extract is obtained from a typical 10 L fermentation, which is

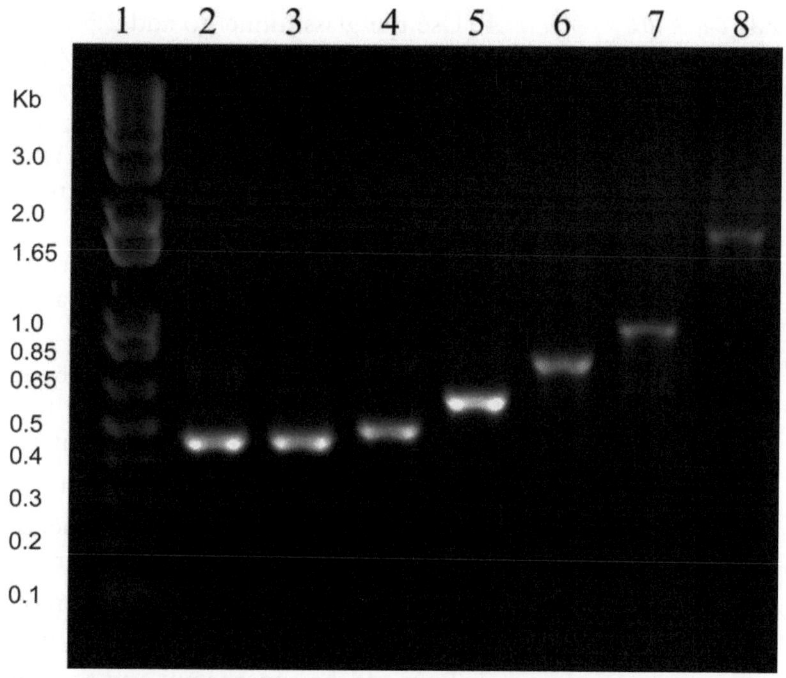

Fig. 2 Cell-free expression templates generated by overlap/extension PCR using the TEV overlap region. 1 % Agarose gel showing the final DNA fragments obtained from the overlap/extension PCR of a range of T7p-tag-TEV fragments with the TEV-AHD-his$_6$-T7ter fragment: *lane 1* is 1 kb + DNA standards (Life Technologies); *lanes 2–8* show the T7p-tag-TEV-AHD-his$_6$-T7ter fragments obtained for the CAT, T7tag, OmpAss, GB1, Trx, PpiB, and MBP tags. Figure is from Kralicek et al. [12], reproduced by permission of Elsevier

enough for 1,600 × 50 μL screening reactions (or 80 × 1 mL preparative reactions). The S30 extract is divided into aliquots, snap frozen in liquid nitrogen, and stored at −80 °C to ensure a lifespan of 1–2 years.

3.2.1 Day 1: Media and Buffer Preparation and Overnight 20 mL Culture

1. Prepare and autoclave the following solutions: 2 M glucose (200 mL), 1 mg/mL thiamine (200 mL), LB medium (500 mL), 4× Z-medium (2.5 L), 10× S30 buffer (2 L), 50 % PEG8000 (1 L), MilliQ water (1.79 L) (*see* **Note 5**). Also autoclave the 10 L glass fermenter vessel containing 5 L MilliQ water, as well as a 100-mL and a 1 L measuring cylinder, and a large glass funnel.
2. Grow overnight a 20 mL culture of BL21 star (DE3) pRARE in LBT containing 33 μg/mL of chloramphenicol at 30 °C with shaking at 200 rpm (*see* **Note 6**).

3.2.2 Day 2: Overnight 500 mL Culture

1. Inoculate 500 mL of LBT medium containing 33 μg/mL of chloramphenicol with the 20 mL overnight culture, and incubate overnight at 30 °C with shaking at 200 rpm.

3.2.3 Day 3: 10 L Fermenter Culture and Retrieval of Cell Pellets

1. Use the glass funnel to add 2.5 L of 4× Z-media, 112 mL of 2 M glucose, 100 mL of 1 mg/mL thiamine, 1.79 L autoclaved MilliQ water to 5 L of water already contained in the fermenter vessel, and mix using the internal rotor. Remove 1 mL of this mixture to be used as a blank for measuring the absorbance of the culture at 600 nm. Raise the temperature of the media to 30 °C.
2. Add the 500 mL overnight culture to the fermenter vessel and grow the resulting 10 L culture at 30 °C, with 100 % air and mixing at 500 rpm.
3. Remove 1 mL samples from the culture every 30 min, and measure their absorbance at 600 nm. After 5–6 h the culture should reach the desired endpoint absorbance of 3.
4. Prepare 400 mL of 1× S30 buffer A and place on ice.
5. Drain the culture from the fermenter vessel into 5-L conical flasks and place at 4 °C in a cold room.
6. Harvest the cells by centrifugation of the culture for 10 min at 10,000 × *g* at 4 °C using pre-weighed 0.5 L centrifuge tubes.
7. Pour off the supernatant and then do a further 5 min spin to compact the pellets. Use a pipette to remove any residual supernatant.
8. Resuspend the cell pellets in 200 mL 1× S30 buffer A, using glass rods to break the pellets up, and by pipetting up and down with a 5 mL pipette.
9. Centrifuge this solution for 10 min at 10,000 × *g* at 4 °C and then pour off the supernatant and perform a further 5 min spin to compact the pellets.
10. The cell pellets are then resuspended again in 200 mL 1× S30 buffer A, centrifuged for 10 min at 10,000 × *g* at 4 °C, the supernatant is decanted, and the pellets are compacted by a further 5 min spin.
11. Snap freeze the pellets by immersing the centrifuge tubes in liquid nitrogen and store at −80 °C.

3.2.4 Day 4: S30 Extract Preparation

1. Prepare 100 mL of 1× S30 buffer A and place on ice. Prepare three lots of 3 L each of 1× S30 buffer B (*see* **Note 7**), place these and the 50 % PEG8000 solution (1 L) at 4 °C.
2. Remove the centrifuge tubes from the freezer and defrost for 20 min at room temperature. Weigh the centrifuge tubes to determine the cell pellet weight (a typical 10 L fermentation will yield ~60 g) and then place the tubes on ice and resuspend the pellets in 1× S30 buffer A at a ratio of 1.3 mL/g of cells.
3. Pass the suspension through an Emulsiflex C5 homogenizer in a single pass at a pressure of ~15,000 psi (*see* **Note 8**).

4. Centrifuge the lysed cells at 5,200 × *g* at 4 °C for 15 min to remove any cell debris and measure the volume of the supernatant.
5. Centrifuge the supernatant at 30,000 × *g* at 4 °C for 30 min. Remove the upper two-thirds (by calculated volume) of the supernatant with a 5 mL pipette. This is the first S30 fraction.
6. Centrifuge this S30 fraction again at 30,000 × *g* at 4 °C for 30 min and remove the upper two-thirds of the supernatant. This is the second S30 fraction.
7. Dialyze the second S30 fraction in Spectropor #4 dialysis tubing (12–14 MWCO) against three consecutive volumes of 3 L of 1× S30 buffer B at 4 °C for 1 h each time.
8. Transfer the 50 % PEG8000 solution to a plastic box and add DTT to a final concentration of 1 mM. Place the dialysis tubing in this mixture and leave the box at 4 °C on a rocker at low speed. Leave this rocking until the volume inside the dialysis tubing has been reduced by half (*see* **Note 9**). The volume is measured by comparison with dialysis tubing containing the expected volume of water.
9. Dialyze the condensed S30 extract in Spectropor #4 dialysis tubing (12–14 MWCO) against 3 L of 1× S30 buffer B at 4 °C for 1 h.
10. Aliquot 50 μL of the S30 extract into 0.5-mL microcentrifuge tubes and immediately snap freeze in liquid nitrogen and store at −80 °C.

Each new batch of S30 extract must be tested to identify the optimal volume that yields maximum protein production. This volume is usually 12 μL of S30 extract per 50 μL reaction volume. To confirm similar productivity for a new S30 extract, synthesis reactions are performed at 6, 9, 12, 15, and 18 μL of S30 extract per 50 μL reaction volume (*see* **Note 10**).

3.3 How to Set Up the Cell-Free Reaction Screen

The complexity of the cell-free reaction requires that appropriate stock solutions and mixes are prepared in order to significantly reduce the reaction set-up time. There are three main steps: preparation of the stock solutions, the master mix, and finally the setting up of the cell-free reaction screen. Preparation of stock solutions and the 10× Master mix should be done on a different day than the cell-free reaction screen. The preparation of sufficient amounts of stock solutions and 10× Master mix to cater for 500 × 50 μL reactions is described below. The cell-free reaction is based on a continuous exchange cell-free (CECF) approach in which 50 μL of each cell-free reaction is added to a microdialysis unit placed in a feeding solution to prolong the reaction by providing a continuous supply of substrates and the removal of inhibitory by-products. The pKO1166 vector is used in the cell-free reaction to supply fresh T7 RNA polymerase continuously for mRNA production [4].

Table 5
Stock solutions required for the 10× Premix, 10× Master mix, and 1.1× Reaction mix

Reagent	Stock concentration	MW	Dry weight (g)	Stock volume (mL)	Final concentration in reaction mixture	Supplier	Storage (°C)
HEPES pH 7.5[a]	2.0 M	238.3	7.15	15	55 mM	BDH	+4
DTT	0.17 M	154.3	0.144	5.5	1.7 mM	Sigma	−80
ATP	120 mM	551.1	0.364	5.5	1.2 mM	Sigma	−80
C/G/UTP[b]	80 mM CTP	527.1	0.232	5.5	0.8 mM	Sigma	−80
	80 mM GTP	523.2	0.230		0.8 mM	Sigma	
	80 mM UTP	484.1	0.213		0.8 mM	Sigma	
Cyclic AMP	64 mM	369.2	0.130	5.5	0.64 mM	Sigma	−80
Folinic acid	6.8 mM	511.5	0.0191	5.5	68 μM	Sigma	−80
Ammonium acetate	2.75 M	77.1	1.166	5.5	27.5 mM	BDH	−80
Potassium (L)-glutamate[c]	4.16 M	203.2	25.36	30	208 mM	Sigma	−80
Magnesium acetate	1.93 M	214.5	2.28	5.5	19.3 mM	Fluka	+4
Creatine phosphate	1.6 M	327.2	15.71	30	80 mM	Roche	−80
E. coli MRE60 tRNA[d]	17.5 mg/mL		0.0175	1	0.175 mg/mL	Roche	−80
Creatine kinase[d]	10 mg/mL	18 kDa	0.01	1	250 μg/mL	Roche	−80
RNasin	40 U/μL	50 kDa			500 U/mL	Promega	−20
pKO1166	1 μg/μL				32 μg/mL		−20
S30 extract					240 μL/mL		−80
PCR template	>30 ng/μL				10 μg/mL		−80

[a]pH using 10 M KOH
[b]CTP, GTP, and UTP are added to the same tube
[c]Add slowly small amounts to 2/5 of final volume. Vortexing and heating to 50 °C may aid solubilization. Once dissolved, make up to correct volume with MilliQ water
[d]Divide into 5-μL aliquots and snap freeze in liquid nitrogen

3.3.1 Preparation of the Stock Solutions

Table 5 details the preparation of stock solutions required for the preparation of the 10× Premix, 10× Master mix, and the 1.1× Reaction mix; and Table 6 details the preparation of the four 100 mM amino acid stock solutions. Depending on the final volume,

Table 6
100 mM Group A–D amino stock solutions

Amino acid	Reagent name	MW	Weight (g)
Group A (acidic)—dissolve amino acids together in 10 mL of in 0.7 M HCl (0.67 mL 32 % HCl + 9.33 mL H_2O)			
Asp	L-Aspartic acid	133.1	0.1331
Gln	L-Glutamine	146.15	0.1462
Thr	L-Threonine	119.1	0.1191
Met	L-Methionine	149.2	0.1492
Tyr	L-Tyrosine	181.2	0.1812
Group B (slightly acidic)—dissolve amino acids together in 10 mL of in 0.11 M HCl (0.105 mL 32 % HCl + 9.89 mL H_2O)			
Asn	L-Asparagine monohydrate	150.1	0.1501
Gly	Glycine	75.07	0.0751
His	L-Histidine	155.16	0.1552
Leu	L-Leucine	131.2	0.1312
Val	L-Valine	117.1	0.1171
Group C (water soluble)—dissolve amino acids together in 10 mL of MilliQ water			
Ala	L-Alanine	89.09	0.0891
Arg	L-Arginine	210.7	0.2107
Lys	L-Lysine hydrochloride	182.06	0.1821
Pro	L-Proline	115.1	0.1151
Ser	L-Serine	105.1	0.1051
Group D (basic)—dissolve amino acids together in 10 mL of 1.09 M KOH (0.612 g KOH tablets/10 mL)			
Ile	L-Isoleucine	131.7	0.1317
Trp	L-Tryptophan	204.2	0.2042
Cys	L-Cysteine-free base	121.6	0.1216
Phe	L-Phenylalanine	165.2	0.1652

we routinely place the weighed solutes in either 15 or 50 mL polypropylene tubes, and then make the solutions up in MilliQ water to the required volume. Most of the stocks to be stored at −80 °C are first snap frozen in liquid nitrogen and can be refrozen multiple times. Exceptions to this rule are the creatine kinase and *E. coli* tRNA stocks, which must be aliquotted before being snap frozen (refreezing of these after thawing is not recommended).

3.3.2 Preparation of the 10× Master Mix

1. Remove from the −80 °C freezer stock solutions required for the 10× Premix (Table 7), the 10× Master mix (Table 8), and the four 100 mM amino acid stock solutions (Table 9) and allow to defrost at room temperature (*see* **Note 11**). Stock solutions stored at −20 °C are placed directly on ice.
2. Prepare the 10× Premix according to Table 7.
3. Mix all the components of the 10× Master mix (Table 8) together, except for the 15 mM amino acid solution.

Table 7
10× Premix solution for 500 cell-free screening reactions

Stock solution	mL
2 M HEPES pH 7.5	14.71
0.17 M DTT	5.35
120 mM ATP	5.35
80 mM C/G/U TP	5.35
64 mM cyclic AMP	5.35
6.8 mM folinic acid	5.35
2.75 M ammonium acetate	5.35
MilliQ water	6.69
Total	53.48

Table 8
Master mix solution for 500 cell-free screening reactions

Stock solution	mL
10× Premix	50.42
4.16 M potassium (L)-glutamate	25.21
1.93 M magnesium acetate	50.42
1.6 M creatine phosphate	25.21
15 mM amino acids[a]	33.61
Total	139.50

[a]Prepare this freshly once the other components of the 1.1× master mix have been mixed together

4. Prepare the 15 mM amino acids solution according to Table 9. Add the four 100 mM amino acid stock solutions in the order specified, and vortex after each addition to the mixture.
5. Immediately add the required volume of 15 mM amino acids to the 10× Master mix solution and vortex (*see* **Note 12**).
6. Aliquot the 10× Master mix in 570 μL lots (enough for 2×50 μL reactions/950 μL feeding solution), snap freeze in liquid nitrogen, and store at −80 °C.

Table 9
15 mM amino acid solution for 500 cell-free screening reactions

100 mM stock solution	mL
Group A	5.40
Group B	5.40
Group C	5.40
Group D	5.40
MilliQ H_2O	14.40
Total	36.00

Table 10
1.1× Reaction mix solution and feeding solution for one cell-free reactions

Stock solution	1.1× Reaction mix (μL)	Feeding solution (μL)
Master mix	15.2	262.8
10 mg/mL creatine kinase	1.4	–
40 U/μL Rnasin	0.7	–
17.5 mg/mL *E. coli* total tRNA	0.3	–
S30 fraction[a]	13.2	–
pKO1166 (1 μg/μL)	1.8	–
Water[a]	2.8	687.2
Total	35.3	950

[a]Adjust depending on potency of S30 extract

3.3.3 Setting Up the Cell-Free Reaction (See ***Note 13****)*

1. For each reaction, soak a Slide-A-Lyzer® MINI Dialysis Units (Pierce) in 200 mL of MilliQ water for 30 min to wet their membranes.
2. Working on ice, mix the components needed for the 1.1× reaction mixture required for a 50 μL reaction (Table 10) in a 1.5 mL microcentrifuge tube.
3. Prepare the feeding solution (Table 10) in a 50 mL polypropylene tube, and adjust the pH of the solution to 7.5 using 1 M KOH.
4. Prepare each of the individual cell-free reactions according to Table 11 in a 1.5 mL microcentrifuge tube using the appropriate PCR-amplified template.

Table 11
50 μL cell-free reaction

Stock solution	Final reaction (μL)
1.1× Reaction mix	32.1
0.05 μg/μL PCR template[a]	10
Water[a]	7.9
Total	50

[a]Adjust depending on final concentration of each PCR template

5. Place 950 μL of Feeding solution into nine 1 mL Nunc cryogenic tubes.
6. For each reaction, pipette 50 μL of reaction mixture on top of the membrane of a Slide-A-Lyzer® MINI Dialysis Unit, and then cap it.
7. Place each dialysis unit into the Feeding solution in a cryogenic tube by holding the tube on a 45° angle and sliding in the dialysis unit in a manner that ensures no air bubbles are trapped under the membrane (some of the Feeding solution will spill out in the process).
8. Hold the dialysis unit firmly in the cryogenic tube and then wrap Parafilm® around both to secure them together and prevent evaporation.
9. Place the cell-free reactions in a stable polystyrene holder and incubate at 30 °C in a thermomixer shaking at 300 rpm overnight for 16–20 h.

Only 1 μL of each completed reaction is needed for loading on an SDS-PAGE gel for analysis by Coomassie staining. High speed centrifugation of each reaction beforehand will determine whether any expressed proteins are soluble. Figure 3 shows the results of cell-free screen performed for the AHD, a peptide sequence that could not previously be expressed by cell-free protein synthesis [12]. Nine different fusion-tagged AHD-his_6 templates were cell-free expressed. The PpiB tag was found to be the best fusion tag for cell-free synthesis of this peptide. To scale up the synthesis of PpiB-TEV-AHD-his6, we then cloned this construct into the pETMCSI vector. Preparative scale synthesis was performed using 1 mL cell-free reaction against 10 mL of feeding solution, with the vector template present at 16 μg/mL. The PpiB tag was removed by adding TEV protease directly to the completed cell-free reaction, and then the resulting AHD-his_6 peptide was purified by IMAC chromatography.

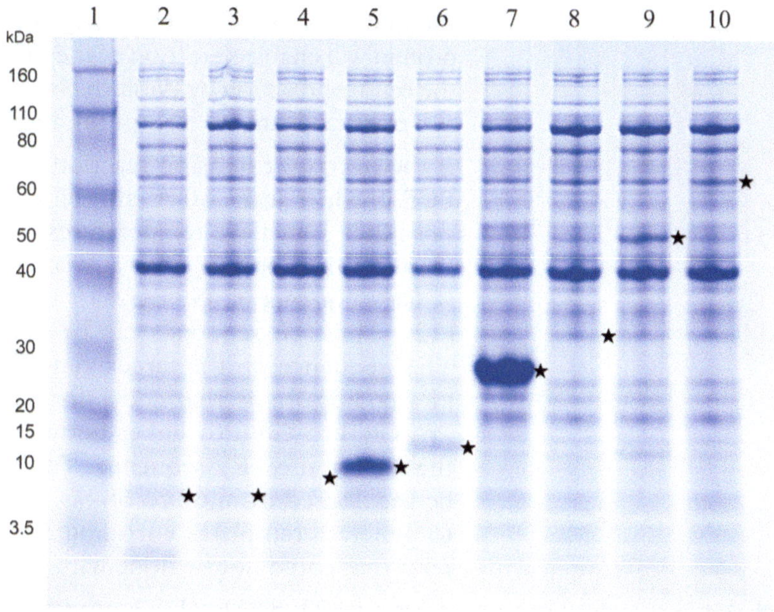

Fig. 3 SDS-PAGE analysis of the cell-free protein synthesis reactions for the different fusion-tagged AHD-his$_6$ constructs. A 4–12 % SDS-PAGE gel showing the cell-free expression reactions for the different fusion-tagged AHD-his$_6$ constructs. *Lane 1* represents Novex Sharp protein standards (Life Technologies); *lanes 2–10* are the cell-free expressions for the different tags: CAT-TEV, T7-TEV, OmpAss-TEV, GB1-TEV, Trx-TEV, PpiB-TEV, GST-CGr-TEV, MBP-TEV, and NusA-CGr-TEV. Expected size of each fusion protein product is indicated by a *black star*

4 Notes

1. Commercially available vectors encoding a tag under the control of the T7 promoter can be used as a template to generate the T7p-tag-TEV fragment if a reverse primer specific to the tag is used to add on the TEV overlap sequence (*see* Table 1, TEV4-Tag-rev). However, our strategy was either to clone large tags into a our standard cell-free expression vector, pET-MCSI, with a TEV sequence following the tag, or in the case of short peptide tags, the complete T7p-tag-TEV sequence was synthetically made by Genscript and supplied in the pUC57 vector. This made it easy to amplify the library of T7p-tag-TEV fragments in parallel using only the CF3 forward primer and the universal TEV4rev reverse primer.

   Most T7p-tag-TEV fragments can be easily gel purified because of their large size. However for short peptide tags, the T7p-tag-tev fragment generated by the CF3/TEV4rev primer combination is often too small for gel purification; a solution is to replace the CF3 primer with a primer for a sequence upstream from the T7 promoter, which will yield a PCR product at least 300 nucleotides in length.

2. Incorporation of a C-terminal his_6 tag in the design of the target protein will enable detection of the full-length protein product in the cell-free reaction by Western blot using an anti-his6 antibody, as well as facilitate the later purification of the target protein.
3. If the overlap/extension reaction fails to produce the desired PCR product at the 60 °C annealing temperature, one can survey other annealing temperatures. If there is still no success, then another option is to use a 15-nucleotide C_3G_3 repeat (CGr) sequence $C_3G_3C_3G_3C_3$ [14] as the overlap sequence [12]. The CGr sequence can be incorporated in front of the TEV sequence on the target gen-T7ter fragment, and placed in a coding frame that codes for an *E. coli* codon-optimized Pro-Gly-Pro-Gly-Pro linker between the end of the tag and the start of the TEV sequence. The annealing temperature can be then raised from 60 to 68 °C. This strategy has enabled us to generate templates encoding NusA and GST-tagged proteins that could not be done with just the TEV overlap region.
4. *E. coli* BL21 STAR (DE3) contains a mutation in the gene encoding RNase E which knocks out the activity of this enzyme which is a major source of RNA degradation. Seki et al. [15] have demonstrated that lowering the temperature of induction to between 20 and 34 °C greatly improves the resulting amount of cell-free synthesis from linear DNA templates, as it reduces the degree of exonuclease activity in the S30 extract.
5. The 50 % PEG8000 solution (1 L) is prepared by adding 500 g of PEG8000 to 500 mL of 1× S30-buffer which is left mixing overnight on a magnetic stirrer at room temperature to dissolve, and then autoclaved.
6. BL21 star (DE3) needs first to be transformed with the pRARE plasmid (isolated from the *E. coli* Rosetta™ strain (Novagen)). Chloramphenicol is present to maintain the pRARE plasmid in the culture. Perform a 1/1,000 dilution from a 33 mg/mL stock made in ethanol.
7. The PMSF and DTT are added to the 1× S30 buffer B solution just before use from a 100 mM PMSF (in isopropanol) and a 0.5 M DTT stock solution, respectively. Note PMSF is toxic, so wear safety glasses, a mask and gloves when weighing it out.
8. An alternative to the homogenizer is to use a single pass through a French press at 6,000 psi [13].
9. This takes 30 min to 1 h to complete and it pays to check every 5–10 min to prevent extract precipitation.
10. Others recommend also optimizing the concentration of magnesium acetate by surveying concentrations between 15 and 25 mM [13]. We have not found this necessary.
11. After defrosting each of the 100 mM amino acid stock solutions at room temperature, check for precipitation. If precipitation is

observed, then vortex the mixture for 5 min to resolubilize the precipitant; sometimes heating at 50 °C for 5 min can also help.

12. Failure to add the 15 mM amino acid solution immediately will result in its precipitation within 2–5 min at room temperature. The precipitated amino acids cannot be resolubilized.
13. The protocol described is only for one cell-free reaction and feeding solution. More than likely you will wish to do multiple reactions to test the effect of a number of different tags. It is therefore advisable to prepare in one volume more than enough 1.1× Reaction mix to cover all the reactions.

Acknowledgments

This work has been supported by New Zealand Government grants from Plant & Food Research (Blue Skies fund) to A.V.K., the New Zealand Foundation for Research Science and Technology (CO6X0701) to A.V.K. Thanks goes to Colm Carraher (Plant & Food Research, Auckland, New Zealand) for critical reading of this manuscript.

References

1. Kralicek AV, Ozawa K (2011) Advances in cell-free protein synthesis for NMR sample preparation. In: Dingley AJ, Pascal S (eds) Biomolecular NMR spectroscopy, vol 3, Advances in biomedical spectroscopy. Ios Press, Amsterdam, pp 3–19
2. Matsuda T, Kigawa T, Koshiba S et al (2006) Cell-free synthesis of zinc-binding proteins. J Struct Funct Genomics 7(2):93–100
3. Endo S, Tomimoto Y, Shimizu H et al (2006) Effects of *E. coli* chaperones on the solubility of human receptors in an in vitro expression system. Mol Biotechnol 33(3):199–209
4. Ozawa K, Jergic S, Crowther JA et al (2005) Cell-free protein synthesis in an autoinduction system for NMR studies of protein–protein interactions. J Biomol NMR 32(3): 235–241
5. Schwarz D, Dotsch V, Bernhard F (2008) Production of membrane proteins using cell-free expression systems. Proteomics 8(19): 3933–3946
6. Xu Z, Chen H, Yin X et al (2005) High-level expression of soluble human beta-defensin-2 fused with green fluorescent protein in *Escherichia coli* cell-free system. Appl Biochem Biotechnol 127(1):53–62
7. Xun Y, Tremouilhac P, Carraher C et al (2009) Cell-free synthesis and combinatorial selective *15*N-labeling of the cytotoxic protein amoebapore A from *Entamoeba histolytica*. Protein Expr Purif 68(1):22–27
8. Wu PS, Ozawa K, Lim SP et al (2007) Cell-free transcription/translation from PCR-amplified DNA for high-throughput NMR studies. Angew Chem Int Ed Engl 46(18):3356–3358
9. Yabuki T, Motoda Y, Hanada K et al (2007) A robust two-step PCR method of template DNA production for high-throughput cell-free protein synthesis. J Struct Funct Genomics 8(4):173–191
10. Chumpolkulwong N, Sakamoto K, Hayashi A et al (2006) Translation of 'rare' codons in a cell-free protein synthesis system from *Escherichia coli*. J Struct Funct Genomics 7(1): 31–36
11. Esposito D, Chatterjee DK (2006) Enhancement of soluble protein expression through the use of fusion tags. Curr Opin Biotechnol 17(4):353–358
12. Kralicek AV, Radjainia M, Mohamad Ali NA et al (2011) A PCR-directed cell-free approach to optimize protein expression using diverse fusion tags. Protein Expr Purif 80(1):117–124

13. Apponyi MA, Ozawa K, Dixon NE et al (2008) Cell-free protein synthesis for analysis by NMR spectroscopy. Methods Mol Biol 426:257–268
14. Cha-aim K, Fukunaga T, Hoshida H et al (2009) Reliable fusion PCR mediated by GC-rich overlap sequences. Gene 434(1–2):43–49
15. Seki E, Matsuda N, Yokoyama S et al (2008) Cell-free protein synthesis system from *Escherichia coli* cells cultured at decreased temperatures improves productivity by decreasing DNA template degradation. Anal Biochem 377(2):156–161

Chapter 4

One-Pot, Microscale Cell-Free Enzyme Expression and Screening

Aarthi Chandrasekaran and Anup K. Singh

Abstract

The lack of high-throughput approaches for expression and screening of large enzyme libraries remains a major bottleneck for current enzyme engineering efforts. To address this need, we have developed a high-throughput, fluorescence-based approach for rapid one-pot, microscale expression, and screening of industrial enzymes. In this chapter, we present the protocol for integration of cell-free protein expression with activity screening of enzymes in two formats: (1) a 96-well plate format and (2) a microscale-array format. Our one-pot method is ideally suited for rapid, first pass screening of enzymes and can also be used to perform detailed mechanistic analysis such as measurement of kinetics, determination of optimum temperature, and to study enzyme inhibition.

Key words One-pot method, Enzyme engineering, Cell-free enzyme expression, Activity screening, Microscale-array platform

1 Introduction

Enzymes have a wide variety of applications in medicine and biotechnology. A number of enzymes have been used as drugs (e.g., digestive enzymes), as catalysts for synthesis of novel materials and chemicals (e.g., detergents, plastics, textiles, paper), in the food industry and for production of lignocellulosic biofuels (e.g., cellulases that break down biomass, enzymes involved in synthesis of biofuels) [1–4]. The quest for generating enzymes with improved catalytic efficiencies and better tolerance for the harsh industrial process conditions (such as extreme temperatures, pressure, and pH) has resulted in the development of enzyme engineering approaches to tailor-make enzymes for specific applications [5–8]. Enzyme engineering approaches such as directed evolution [9–12] and rational design [6, 7] generate large DNA libraries which need to be expressed and analyzed. Therefore, there is a critical need for the development of high-throughput approaches for rapid enzyme expression and activity screening.

Kirill Alexandrov and Wayne A. Johnston (eds.), *Cell-Free Protein Synthesis: Methods and Protocols*, Methods in Molecular Biology, vol. 1118, DOI 10.1007/978-1-62703-782-2_4,

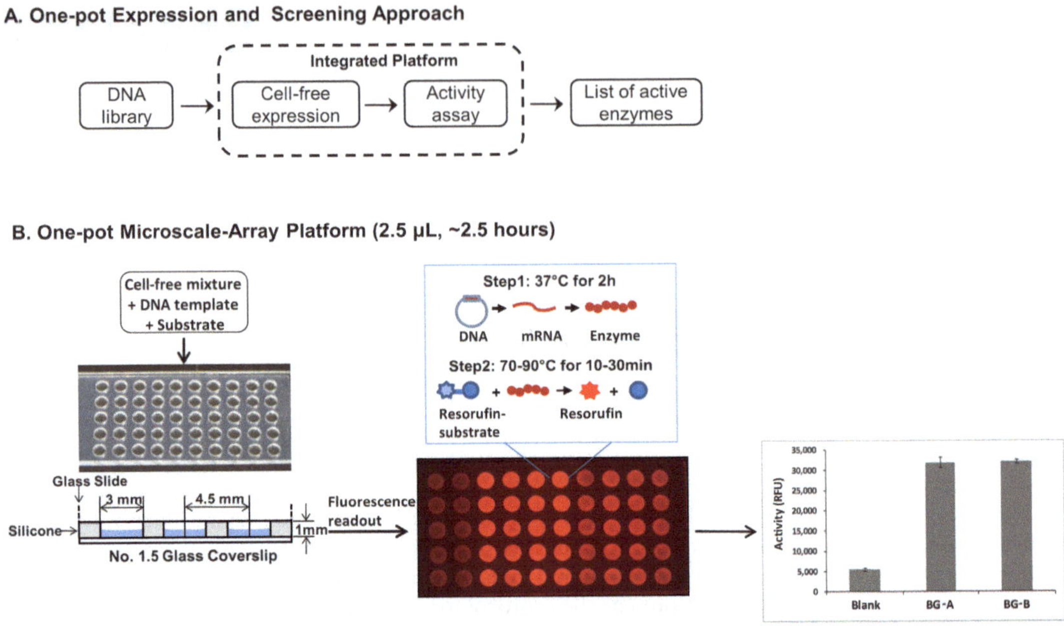

Fig. 1 (**a**) One-pot approach for integrated expression and activity screening of enzymes. (**b**) Microscale-array platform for rapid one-pot expression and activity screening. Starting from DNA libraries, this platform can be used to express and screens enzymes in 2.5 μL in less than 3 h. Fluorescent substrates are used for activity screening and quantification of the fluorescent product is a direct measure of enzyme activity. The platform can also be used for enzyme kinetics measurements and for studying enzyme inhibition [14]

Typically, the functional screening of large libraries of enzymes is a multistep process: (1) expression in heterologous host cells (such as *E. coli*, mammalian or yeast cells), (2) protein purification and quantification, followed by (3) activity screening of enzymes. These approaches are time consuming (take days to weeks to screen a large library), labor intensive, use large amounts of reagents and are not easily adaptable to high-throughput formats [13]. In order to overcome these limitations, we have developed a one-pot, microscale, fluorescence-based method for rapid expression and screening of enzymes (Fig. 1a) [14]. Our method utilizes a commercial *E. coli*-based cell-free protein expression system (Expressway cell-free expression system, Invitrogen, Carlsbad, CA) that affords rapid conversion of DNA templates to functional enzymes [15]. In comparison with the cell-based approaches, cell-free expression can be achieved within hours in microliter volumes. Further, the open and flexible format of cell-free protein synthesis allows for expression of proteins that are otherwise difficult to express in host cells (such as toxic or insoluble proteins) [15–17]. By combining cell-free expression with screening using fluorescent substrates, we achieve rapid, sensitive screening of enzyme libraries in one-pot experiments.

In this chapter, we describe the integration of enzyme expression and activity screening in two different formats. The first is a tube/96-well plate format in which we mix and incubate DNA templates, cell-free reaction mixture, and the fluorescent substrates all in one-pot in 50–100 μL volumes. This format uses standard laboratory equipment and can be easily adapted to high-throughput liquid handling robotic systems. Starting with DNA libraries, functional enzymes can be expressed and screened in less than 3 h. This method therefore has significant advantages over the conventional approaches. Despite all these advantages, the high cost of cell-free protein expression systems remains a major bottleneck for using this format for high-throughput analyses. For instance, a typical prokaryotic cell-free protein expression kit (Expressway; *see* Subheading 2.2) has enough reagents to set up 20 cell-free reactions in 100 μL volumes in the 96-well plate format. This results in a cost of over $20/reaction. Therefore, screening large libraries of enzymes would quickly become cost prohibitive. Further, while this format is well suited for screening mesophilic enzymes (due to minimal evaporation at 37 °C), it is not ideal for screening enzymes with higher optimal temperatures due to issues with increased evaporation at elevated temperatures.

To address these limitations, we have adapted the one-pot expression and screening of enzymes to a commercial microwell array platform (Electron Microscopy Sciences, Hatfield, PA, Subheading 2.2) [14]. This format consists of a multiwell silicone gasket that has an array of 10 × 5 wells with each well having a minimum reaction volume of 2.5 μL (Subheading 3.4). This format uses 20–40-fold lower volumes as compared with the 96-well plate format and therefore results in a significant reduction in reagent usage and costs. For example, one Expressway cell-free kit can be used to express over 800 proteins in 2.5 μL reaction volumes and the cost dramatically decreases to about $0.5/reaction. Further, this format is fully sealed and therefore evaporation is minimized, making it very attractive to screen both mesophilic and thermophilic enzymes. Additionally, we achieve enzyme expression and screening in this platform in about 2–2.5 h. In 1 day, it is therefore possible to screen hundreds of enzymes. Finally, we can also use this platform to quantify enzyme kinetics, estimate optimum temperatures for enzyme activities, and study enzyme inhibition (or to screen inhibitor libraries).

In order to demonstrate the utility of our one-pot method, we tested a panel of mesophilic and thermophilic enzymes. We expressed and screened a beta-glucuronidase (from *Arabidopsis thaliana*) and four cellulases (two beta-glucosidases and two cellobiohydrolases) from thermophilic bacteria and archaea (*see* Table 1 for a list of enzymes that have been screened using this method). Resorufin-β-D-glucopyranoside, resorufin-cellobioside and 4-methylumbelliferyl β-D-glucuronide (4MUGlcU) serve as excellent substrates for fluorometric monitoring of the activities of beta-glucosidase, cellobiohydrolase, and beta-glucuronidase enzymes, respectively [18, 19]. Enzymatic activity hydrolyzes the glycosidic linkage between the dye

Table 1
List of enzymes used to validate one-pot expression and screening in microscale array [14]

Name	Uniprot ID	Source organism (activity)	Substrate used for activity screening	Optimal assay conditions	Detection of product
BG-A	A5IL97	*Thermotoga petrophila* (*Beta-glucosidase*)	Resorufin-β-D-glucopyranoside	75 °C for 10 min	Ex: 571 nm Em: 585 nm
BG-B	A5IL43	*Thermotoga petrophila* (*Beta-glucosidase*)	Resorufin-β-D-glucopyranoside	75 °C for 10 min	Ex: 571 nm Em: 585 nm
CBH-A	O58925	*Pyrococcus horikoshii* (*Cellobiohydrolase*)	Resorufin-cellobioside	80–95 °C for 30 min	Ex: 571 nm Em: 585 nm
CBH-B	A4XIF7	*Caldicellulosiruptor saccharolyticus* (*Cellobiohydrolase*)	Resorufin-cellobioside	70 °C for 30 min	Ex: 571 nm Em: 585 nm
GUS	NA	*Arabidopsis thaliana* (*Beta-glucuronidase*)	4-Methylumbelliferyl β-D-glucuronide (4MUGlcU)	37 °C for 10 min	Ex: 360 nm Em: 465 nm

and the sugar thereby releasing the highly fluorescent resorufin or 4-methylumbelliferone (4-MU) dye. The quantification of this fluorescent product therefore provides a direct measure of the enzyme activity. Several such synthetic substrates are commercially available for fluorometric analysis of activities of other families of enzymes including, but not limited to, cytochrome P450, glutathione transferases, luciferase, proteases, esterases, and lipases. Furthermore, our method is independent of the type of cell-free expression system and can be easily adapted to use with the wheat germ, rabbit reticulocyte, and insect cell-based in vitro protein translational systems [20–22].

2 Materials

2.1 Gateway Cloning

1. High-Fidelity PCR master mix with DNA polymerase such as Phusion® (New England Biolabs, Ipswich, MA).
2. pENTR-gus (Invitrogen, Carlsbad, CA).
3. Nuclease-free water.
4. Nuclease-free microcentrifuge tubes.
5. Gel extraction kit such as QIAquick (Qiagen, Valencia, CA).
6. Gateway BP clonase II reaction enzyme mix (Invitrogen, Carlsbad, CA).
7. Chemically competent *E. coli* such as One Shot® TOP10 cells (contains cells and SOC medium) (Invitrogen, Carlsbad, CA).

Store cells at −80 °C and thaw individual vials of cells as required on ice. SOC medium can be thawed and stored at room temperature after the first use.

8. LB medium: Dissolve 25 g of LB medium powder in 1 L water. Autoclave to sterilize.
9. LB agar plates: Dissolve 6 g of agar in 400 ml of LB medium. Autoclave to sterilize.
10. 14 ml Polypropylene round bottom test tubes.
11. Antibiotics: Ampicillin, Kanamycin (Invitrogen, Carlsbad, CA).
12. Plasmid purification kits such as Qiaprep Spin Miniprep kit (Qiagen, Valencia, CA).
13. Spectrophotometer such as Nanodrop (Thermo Scientific, Wilmington, DE).
14. pDONR™ 221 vector (Invitrogen, Carlsbad, CA). Store at −20 °C.
15. pEXP4-DEST vector (C-terminal fusion vector with a C-terminal 6X-his tag and Lumio tag) (Invitrogen, Carlsbad, CA). Store at −20 °C.
16. pEXP3-DEST vector (N-terminal fusion vector with a N-terminal 6X-his tag and Lumio tag) (Invitrogen, Carlsbad, CA). Store at −20 °C.
17. LR clonase II enzyme mix (Kit contains Gateway LR clonase II and 2 µg/µL Proteinase K) (Invitrogen, Carlsbad, CA). Store at −20 °C.
18. TE buffer: 10 mM Tris–HCl, 1 mM EDTA (pH 8.0).
19. Codon optimization tools and DNA synthesis services (GenScript, Piscataway, NJ).
20. DNA sequencing services (Quintara BioSciences, Albany, CA).

2.2 One-Pot Cell-Free Enzyme Expression and Screening

1. Expressway™ Lumio™ Expression and Detection System or Expressway™ Mini Cell-Free Expression System (Kit contains following components: *E. coli* Extract, IVPS reaction buffer, feed buffer, 19 amino acid mix, methionine, DNase/RNase-free water, T7 Enzyme Mix and a positive expression control vector) (Invitrogen, Carlsbad, CA). Store the 19 amino acid mix, methionine and control vector at −20 °C. Store all the other reagents of the cell-free kit at −80 °C.
2. Dimethyl sulfoxide (DMSO for molecular biology).
3. Fluorescent substrates (e.g., resorufin β-D-glucopyranoside, resorufin-cellobioside, 4-methylumbelliferyl β-D-glucuronide (4MUGlcU)) (Marker Gene Technologies, Eugene, OR). Make stock solution for each substrate at a concentration of 50 mM in DMSO. Aliquot and store at −20 °C. Light sensitive, store in the dark.

4. 1.7 ml Sterile, DNAse/RNAse-free microcentrifuge tubes.
5. 96-Well black clear bottom plates.
6. Eppendorf thermomixer (Eppendorf, Hauppauge, NY).
7. Fluorescence microplate reader such as Spectramax M2 (Molecular Devices, Sunnyvale, CA).
8. CultureWell™ chambered coverglass (Electron Microscopy Sciences, Hatfield, PA). In the rest of the chapter, this commercial chambered coverglass consisting of a silicone multiwell gasket attached to a glass coverslip is referred to as the microscale-array platform.
9. Clear glass slides.
10. Oven (VWR, Randor, PA).
11. Digital imaging system such as FluorChem Q system (Protein Simple, Santa Clara, CA).
12. Data analysis and graphing software such as SigmaPlot (Systat software, Inc., Sunnyvale, CA).

3 Methods

In order to generate the entry and expression clones we used Gateway® cloning procedures as outlined in Subheadings 3.1 and 3.2.

3.1 Generation of Entry Clones

1. Obtain the amino acid sequences of the enzymes that need to be expressed and assayed.
2. Codon-optimize the nucleotide sequences of the different enzymes and have the DNA synthesized for protein expression in *E. coli* (*see* **Note 1**).
3. Set up a PCR reaction to incorporate *att*B1 and *att*B2 overhangs into the functional domains of the template DNA. Towards this, design and order primers for the PCR reaction (as per the instructions in the Invitrogen manual for Gateway cloning). Make stock solutions of the *att*B1 and *att*B2 primers in nuclease-free water at a final concentration of 10 μM.

   The following are the sequences used for standard Gateway primers for generating C-terminal fusion proteins (see **Note 2**):

   *att*B1 forward primer: 5′-GGGGACAAGTTTGTACAAAAAA-GCAGGCTTCGAAGGAGATAGAACCATG-(18–25 bp gene-specific sequence)-3′

   *att*B2 reverse primer: 5′-GGGGACCACTTTGTACAAGAA-AGCTGGGTC- (18–25 bp gene-specific sequence)-3′
4. In a nuclease-free tube, set up a PCR reaction using a high-fidelity PCR master mix such as Phusion. Follow the protocol

(for the reaction set up and thermocycling conditions) as per the manufacturer's guidelines for routine PCR with phusion DNA polymerase.

5. Gel-purify the PCR product using a gel extraction kit such as the QIAquick kit following the manufacturer's instructions.
6. In order to generate the entry clone, set up a BP reaction by mixing the following reagents in a PCR tube on ice: *att*B purified PCR product: 20–80 ng, pDONR™ 221 entry vector: 60 ng, Gateway BP clonase II enzyme mix: 0.5 μL. Make up the final volume to 2.5 μL using TE buffer (if required).
7. Mix and incubate for 2 h at room temperature.
8. Use 1 μL of each BP reaction mix to transform chemically competent *E. coli* such as One Shot TOP10 cells. Thaw vials of cells on ice, add the BP reaction mix, and incubate for 20–30 min on ice. Then heat shock the cells at 42 °C for 45 s followed by incubation on ice for 2 min. Add 250 μL of SOC medium to each vial of cells and shake the vials on an incubator shaker for an hour at 37 °C.
9. Spread 50–100 μL of the cells on an agar plate with kanamycin as the antibiotic. Incubate the plates overnight in a 37 °C warm room.
10. Pick colonies (at least two colonies from each plate) into 14 ml round bottom tubes with 5 ml of LB medium with kanamycin. Incubate these liquid cultures on a shaker overnight at 37 °C.
11. Pellet the cells and isolate and purify the DNA using a plasmid purification kit such as QIAprep Spin Miniprep kit following the manufacturer's instructions. Elute the entry clone in 50 μL of nuclease-free water.
12. Quantify the DNA using a nanodrop spectrophotometer.

3.2 Generation of Expression Clones (Cell-Free DNA Template)

1. In a nuclease-free microcentrifuge tube, set up an LR recombination reaction to generate expression clones for the different enzymes. Add the following reagents to the tube: Purified entry clones (in pDONR221 vector): 150 ng, pEXP3-DEST or pEXP4-DEST (destination vector): 150 ng, TE buffer (pH 8.0): up to 8 μL.
2. To each reaction mixture, add 2 μL of the thawed LR clonase II enzyme mix. Vortex and incubate the tubes for 1 h at room temperature.
3. Stop the reaction by adding 1 μL of proteinase K enzyme (provided in the kit) and incubate for 10–15 min at 37 °C.
4. Use 1 μL of each LR reaction mix to transform chemically competent *E. coli* such as One Shot TOP10 cells. Thaw vials of cells on ice, add the LR reaction mix, and incubate for 20–30 min on ice. Then heat shock the cells at 42 °C for 45 s

followed by incubation on ice for 2 min. Add 250 μL of SOC medium to each vial of cells and shake the vials on an incubator shaker for an hour at 37 °C.

5. Spread 50–100 μL of the cells on an agar plate with ampicillin as the antibiotic. Incubate the plates overnight in a 37 °C warm room.
6. Pick colonies (at least two colonies from each plate) into 14 ml round bottom tubes with 5 ml of LB medium with ampicillin. Incubate these liquid cultures on a shaker overnight at 37 °C.
7. Pellet the cells and isolate and purify the DNA using a plasmid purification kit such as the QIAprep Spin Miniprep kit following the manufacturer's instructions.
8. Elute the expression clones in 50 μL of nuclease-free water.
9. Quantify the DNA using a nanodrop spectrophotometer.
10. Confirm the expression clones by sequencing the DNA using T7 forward and reverse sequencing primers.

3.3 One-Pot Enzyme Expression and Screening in the Tube/96-Well Plate Format

1. Thaw the different components of the Expressway cell-free expression kit on ice (*see* **Note 3**).
2. Thaw the appropriate fluorescent substrate previously reconstituted in DMSO (Subheading 2.2). Dilute this DMSO stock 1:10 in nuclease-free water to a final concentration of 5 mM.
3. In a nuclease-free microcentrifuge tube (*see* **Note 4**), prepare a 50 μL batch of cell-free reaction mix by mixing the different cell-free components:

   10 μL *E. coli* lysate + 10 μL reaction buffer + 12.5 μL feed buffer + 2.5 μL amino acids + 1.5 μL methionine + 0.5 μL T7 RNA polymerase + 5 μL of the substrate solution (5 mM solution as in **step 2**) + 0.5 μg of the expression clone. Make up the final volume to 50 μL using nuclease-free water.
4. Separately, set up a control reaction with all the components of the cell-free reaction mix including the substrate (as mentioned in **step 3**) but without the expression clone (*see* **Note 5**). Make up the final reaction volume to 50 μL using nuclease-free water.
5. Mix all enzyme and control reactions with a pipette.
6. Incubate the tubes in the dark with shaking on an Eppendorf thermomixer for 2 h at 37 °C at 1,250 rpm to allow for transcription and translation (*see* **Notes 6** and **7**).
7. Further incubate the tubes at appropriate optimum temperatures and times for the activity assay (Table 1) (*see* **Note 8**).
8. Transfer the reagents to a 96-well black clear bottom plate for analysis (*see* **Note 7**).
9. Read the plate on a fluorescence microplate reader such as the Spectramax M2 at the appropriate excitation and emission wavelengths for the fluorescent product.

3.4 One-Pot Enzyme Expression and Screening in Microscale-Array Format

The protocol below describes one-pot enzyme expression and screening in 2.5 μL volumes. This method uses a microscale-array platform that consists of a multiwell silicone gasket attached to a 24 × 50 mm glass coverslip (Electron Microscopy Sciences, Hatfield, PA, Subheading 2.2). The gasket used in this study comprises of an array of 50 wells (10 rows × 5 columns). The dimensions of each well are 3 mm (diameter) and 1 mm (depth) with a well to well spacing of 4.5 mm (*see* Fig. 1b).

1. Set up the microscale platform with the silicone gasket (multiwell array) facing upwards.
2. Thaw the different components of the Expressway cell-free expression kit on ice (*see* **Note 3**).
3. Thaw the appropriate fluorescent substrates previously reconstituted in DMSO (Subheading 2.2) at room temperature. Dilute the substrate stocks 1:10 in nuclease-free water to a concentration of 5 mM.
4. Determine the number of reactions that need to be set up for each family of enzymes. Include one negative control for each family of enzymes to be tested per array.
5. In a nuclease-free microcentrifuge tube, prepare a batch of cell-free reagents by mixing the different components of the cell-free kit. The volume of this master mix (and individual components) will depend on the total number of reactions that need to be set up. For example, to screen five beta-glucosidases, set up a master mix for 8 reactions (5 enzymes + 1 negative control + 2 extra reactions) (*see* **Note 9**) in a tube as follows:

   4 μL *E. coli* lysate + 4 μL reaction buffer + 5 μL feed buffer + 1 μL amino acids + 0.6 μL methionine + 0.2 μL T7 RNA polymerase + 2 μL of the appropriate diluted substrate (5 mM stock concentration from **step 3**) (*see* **Note 10**).
6. Mix by pipetting carefully. Do not vortex.
7. Separately, in microcentrifuge tubes, dilute each DNA template (expression clones in pEXP3/4-DEST) in nuclease-free water to prepare a stock solution of 62.5 ng/μL.
8. Add 2.1 μL of the master mix to each well in the array platform.
9. Then add 0.4 μL of the template DNA (expression clone) to the well. The final DNA amount in each well should be 25 ng.
10. Mix reagents in the well carefully by pipetting without generating any bubbles (*see* **Note 11**).
11. In another well, set up a control reaction by adding 2.1 μL of the master mix. To this mixture add only 0.4 μL of the nuclease-free water without any DNA template (*see* **Note 5**). Pipette carefully to mix.

12. Gently press down a clear glass slide on the top of the silicone gasket to seal the array (*see* **Note 12**).
13. Incubate the sealed array platforms in a warm room or an incubator at 37 °C for 2 h to allow for in vitro transcription and translation (*see* **Note 6**).
14. The array is then incubated at appropriate temperatures and times for the activity assay in an oven. *See* Table 1 for a list enzymes (and corresponding reaction conditions) that have been tested using this platform (*see* **Note 13**).
15. Remove the array from the oven and let it cool briefly.
16. Scan the array using the digital imaging system such as FluorChem Q using appropriate filters and measure the fluorescence in each well by densitometry (*see* **Note 14**).

3.5 Enzyme Kinetics and Enzyme Inhibition Measurements in Microscale-Array Format

Once the active enzymes have been identified from the initial library (*see* **Note 15**), the array format can be used to estimate enzyme kinetics and quantify kinetic parameters such as K_m and V_{max} for individual active enzymes. The microwell array format can also be used to estimate enzyme inhibition or to screen a library of inhibitors. The following protocol can be used to estimate enzyme kinetic parameters and enzyme inhibition for a single enzyme in the microwell array.

1. Set up a separate master mix of cell-free reagents for each substrate/inhibitor concentration.
2. For enzyme kinetics: Dilute the fluorescent substrates in nuclease-free water over an entire range of concentrations (0.5–10 mM). In each tube, mix 4 μL *E. coli* lysate + 4 μL reaction buffer + 5 μL feed buffer + 1 μL amino acids + 0.6 μL methionine + 0.2 μL T7 RNA polymerase + 0.2 μg of the expression plasmid + 2 μL of substrate at a specific concentration. Make up the solution to 20 μL using nuclease-free water.
3. For enzyme inhibition studies: Dilute the inhibitor to be tested over an entire concentration range in nuclease-free water. In a tube, add 4 μL *E. coli* lysate + 4 μL reaction buffer + 5 μL feed buffer + 1 μL amino acids + 0.6 μL methionine + 0.2 μL T7 RNA polymerase + 2 μL of fluorescent substrate (at 5 mM concentration) + 0.2 μg of the expression plasmid + 2 μL of the inhibitor at a specific concentration. Make up the solution to 20 μL using nuclease-free water.
4. For both measurements, set up a control reaction with all components of the cell-free reaction mix (with substrate and inhibitor as appropriate) but without the DNA template.
5. Add 2.5 μL/well of the cell-free mixture for each enzyme and control reaction to five wells in the array to generate replicates (*see* **Note 16**).

6. Seal and incubate the array as mentioned as per Subheading 3.4.
7. Scan the array using the digital imaging system with appropriate filter sets and measure the fluorescence in each well using densitometry.
8. For enzyme kinetics measurements, plot the average fluorescence values vs. the substrate concentration and estimate the kinetic parameters using a data analysis and graphing software such as SigmaPlot. For enzyme inhibition measurements, plot the average fluorescence values vs. the inhibitor concentration using SigmaPlot.

4 Notes

1. In this study, a commercial, proprietary algorithm from GenScript (OptimumGene codon optimization technology) was used for codon optimization for protein expression in *E. coli*. The majority of gene synthesis companies now offer codon optimization services in addition to custom gene synthesis for expression in multiple host organisms.
2. While designing the standard Gateway primers (to include the *att*B sites), it is important to take into account the following considerations. Since, we are interested in expressing C-terminal fusion enzymes, it is important for the *att*B1 forward primer to include a start codon sequence to initiate transcription and a Shine-Dalgarno and/or Kozak consensus sequence for initiation of translation in *E. coli* or eukaryotic cells, respectively. While designing the *att*B2 reverse primer, it is important to exclude or remove any stop codons. Following the BP reactions, the cellulase genes in the *attB-PCR* products are transferred to the entry vector pDONR™ 221. An additional LR reaction transfers the genes to an expression vector pEXP4-DEST which includes a C-terminal 6X-His fusion tag, a C-terminal Lumio fusion tag (CCPGCC) and a stop codon. In this study, we obtained the entry clone for the beta-glucuronidase (pENTR-GUS) directly from Invitrogen. The pENTR-gus entry clone does not contain a start codon or a Shine-Dalgarno ribosome binding site sequence. For entry clones without these required regulatory elements, it is important to recombine them with the N-terminal fusion vector, pEXP3-DEST, which contains a start codon, Shine-Dalgarno sequence, N-terminal 6X-His tag, and N-terminal Lumio tag (CCPGCC).
3. Thaw out all the reagents in the cell-free kit, except the nuclease-free water, on ice. Keep the reagents on ice throughout the experiment. Soon after setting up the experiment, return the reagents to the appropriate storage temperatures. The nuclease-free water can be stored at room temperature.

4. For optimal enzyme expression, it is important to use sterile, nuclease-free tubes, nuclease-free tips, and nuclease-free water for all mixing and incubation steps. Any DNAse or RNAse in the reaction mix will degrade the DNA template or the transcript and therefore reduce or eliminate protein expression.
5. It is important to set up control or blank reactions for each family of enzymes tested. These control samples contain all the components of the cell-free system plus 0.5 mM of the appropriate substrate but no expression plasmid. These reactions give us an estimate of the background fluorescence signals of the different substrates.
6. The cell-free expression system used in this study is a T7-based coupled transcription–translation system that uses a T7 RNA polymerase for transcription. Starting from the DNA template, mRNA transcripts, and functional proteins are generated within 2 h. In this study, we use plasmids as DNA templates but it also possible to directly use linear PCR products with required regulatory elements (T7 promoter, ribosome binding site, T7 terminator) as the DNA template.
7. For the tube/96-well plate format, it is important to continuously shake the Eppendorf tubes throughout the incubation steps to ensure optimal protein/enzyme yields. This can be achieved using an Eppendorf thermomixer that allows for temperature control while shaking. Protein yields are significantly lower without shaking. The thermomixer supports multiple formats and has exchangeable thermoblocks for tubes and 96/384-well plates. If user has access to the thermoblock for 96-well plates, it is possible to set up the cell-free reactions directly in a 96-well plate (as opposed to setting them up in a microcentrifuge tube as mentioned in Subheading 3.3). The plate will have to be sealed with a plate sealer before incubation. After the incubation step, the plate sealer is peeled off and the plate is read using a fluorescence microplate reader at the appropriate excitation and emission wavelengths.
8. While screening mesophilic enzymes, the additional incubation step for activity screening (following the incubation step for transcription/translation) can be skipped. Since the optimum temperature for these enzymes is around 37 °C, activity can be observed as and when the enzymes are being translated. For example, with the beta-glucuronidase, enzyme activity can be detected as early as 20–30 min after setting up the one-pot reaction.
9. While preparing the master mix, include enough reagents for two extra reactions (apart from the reactions to be tested and the negative control). Since the method uses microscale volumes, it is important to have this extra volume to account for any losses during sample mixing/handling.

10. For the microscale-array platform, the volumes for the individual components of the cell-free master mix can be scaled up depending on the number of reactions to be set up. While scaling up, it is important to keep the ratios of the individual components of the master mix the same as described in Subheading 3.4, **step 5**.
11. While adding reagents for each cell-free reaction or control to the wells in the array, it is important to evenly spread out the solution over the entire bottom surface of the well. Further, make sure that there are no bubbles while adding reagents to the wells. If there are bubbles, gently remove them with a pipette tip. Failure to do so will complicate analysis of fluorescence in each well by densitometry.
12. For the microscale-array platform, it is critical to tightly seal the platform before the incubation steps. Failure to do so can result in evaporation of samples from the wells particularly at high temperatures and can cause significant well-to-well variation. We use a clean glass slide for sealing because silicone sticks strongly to glass. Further, the elastomeric property of silicone enables a good seal even at high temperature. This prevents evaporation making this platform attractive for screening not only mesophilic but also thermophilic enzymes.
13. It is possible to use our array platform to estimate the temperature dependence of enzyme activities or to estimate thermostability of enzymes. Set up the array as described in Subheading 3.4 and seal the array. Set up one sealed array for every temperature that needs to be tested. A typical temperature range would be 25–95 °C (in 10 °C increments). Incubate each sealed array platform at desired temperature for appropriate times and then scan the array as described in Subheading 3.4.
14. The fluorescent wavelengths of 4-MU are 365 nm (Excitation) and 455 nm (Emission) and for resorufin are 571 nm (Excitation) and 585 nm (Emission). In this study, we used the FluorChem Q digital imager to scan the array platform for quantification of fluorescent product. While this imager has filters for detection of free resorufin dye, it does not have appropriate filters for detecting fluorescence of 4-MU product formed from beta-glucuronidase activity. Hence, in this case, we used a standard epifluorescence microscope fitted with an appropriate filter set to measure fluorescence in each well.
15. The primary goal of our platform is rapid first pass screening of activities of large cellulase mutant libraries. However, for applications in enzyme engineering, it is essential to estimate and compare the specific activities of the libraries of mutant and wild-type enzymes. This would require quantification of protein expression levels for different DNA mutants and normalization of their activities. In order to get an estimate of protein

expression levels in these microscale volumes, we used EGFP as a model protein and quantified EGFP fluorescence to get an estimate of the protein expression levels. We set up EGFP cell-free reactions as described in Subheading 3.4 using EGFP-pEXP4-DEST template DNA. We also included serial dilutions of a commercial purified recombinant EGFP protein to generate a standard curve. After incubating the platform for 2 h to allow for protein expression, the array platform was incubated at 4 °C overnight to allow for proper folding and maturation of EGFP. Then the platform was scanned using the digital imaging system with FITC filters, and fluorescence was quantified by densitometry.

16. For the enzyme kinetics and inhibition experiments, we set up our assays to account for well-to-well variation. Therefore, we prepare a master mix that combines cell-free reagents with DNA, substrate, and inhibitor (if present). We then add the reaction mixture to five wells in the array. The data from the five wells are averaged and the standard error is calculated. The error bars are an indication of well-to-well variation. This estimation of variation is particularly important when incubating the platform at elevated temperatures and for accurate quantification of enzyme kinetic parameters and inhibition. If measurement of well-to-well variation is not critical, the user can choose to set up the master mix with the substrate separately, add it to each well and then add the DNA and inhibitor (as appropriate) to each well (as indicated in Subheading 3.4).

Acknowledgments

The authors would like to thank Dr. Joshua I. Park for assistance with Gateway cloning techniques and Dr. Chung-Yan Koh for critical reading of the manuscript. This work was part of the DOE Joint BioEnergy Institute (http://www.jbei.org) supported by the US Department of Energy, Office of Science, Office of Biological and Environmental Research, through contract DE-AC02-05CH11231 between Lawrence Berkeley National Laboratory and the US Department of Energy. Sandia is a multiprogram laboratory operated by Sandia Corp., a Lockheed Martin Co., for the US Department of Energy under Contract DE-AC04-94AL85000.

References

1. Kirk O, Borchert TV, Fuglsang CC (2002) Industrial enzyme applications. Curr Opin Biotechnol 13:345–351
2. Polaina J, MacCabe AP (eds) (2007) Industrial enzymes structure, function and applications, vol XII. Springer, Dordrecht
3. Schäfer T, Kirk O, Borchert TV et al (2005) Enzymes for technical applications. Biopolymers online. Wiley, Weinheim
4. Schoemaker HE, Mink D, Wubbolts MG (2003) Dispelling the myths—biocatalysis in industrial synthesis. Science 299:1694–1697

5. Schulein M (2000) Protein engineering of cellulases. Biochim Biophys Acta 1543: 239–252
6. Dalby PA (2007) Engineering enzymes for biocatalysis. Recent Pat Biotechnol 1:1–9
7. Bornscheuer UT, Pohl M (2001) Improved biocatalysts by directed evolution and rational protein design. Curr Opin Chem Biol 5: 137–143
8. Zhang YHP, Himmel ME, Mielenz JR (2006) Outlook for cellulase improvement: screening and selection strategies. Biotechnol Adv 24: 452–481
9. Bloom JD, Meyer MM, Meinhold P et al (2005) Evolving strategies for enzyme engineering. Curr Opin Struct Biol 15:447–452
10. Cherry JR, Fidantsef AL (2003) Directed evolution of industrial enzymes: an update. Curr Opin Biotechnol 14:438–443
11. Dalby PA (2011) Strategy and success for the directed evolution of enzymes. Curr Opin Struct Biol 21:473–480
12. Turner NJ (2009) Directed evolution drives the next generation of biocatalysts. Nat Chem Biol 5:567–573
13. He M (2008) Cell-free protein synthesis: applications in proteomics and biotechnology. N Biotechnol 25:126–132
14. Chandrasekaran A, Bharadwaj R, Park JI et al (2010) A microscale platform for integrated cell-free expression and activity screening of cellulases. J Proteome Res 9:5677–5683
15. Katzen F, Chang G, Kudlicki W (2005) The past, present and future of cell-free protein synthesis. Trends Biotechnol 23:150–156
16. Jackson AM, Boutell J, Cooley N et al (2004) Cell-free protein synthesis for proteomics. Brief Funct Genomic Proteomic 2:308–319
17. Spirin AS (2004) High-throughput cell-free systems for synthesis of functionally active proteins. Trends Biotechnol 22:538–545
18. Coleman DJ, Studler MJ, Naleway JJ (2007) A long-wavelength fluorescent substrate for continuous fluorometric determination of cellulase activity: resorufin-beta-D-cellobioside. Anal Biochem 371:146–153
19. Jefferson RA, Kavanagh TA, Bevan MW (1987) GUS fusions: beta-glucuronidase as a sensitive and versatile gene fusion marker in higher plants. EMBO J 6:3901–3907
20. Endo Y, Sawasaki T (2006) Cell-free expression systems for eukaryotic protein production. Curr Opin Biotechnol 17:373–380
21. Litterer L (2009) Expressing mammalian proteins using insect and rabbit cell-free lysates. Promega Notes. www.promega.com
22. Katzen F, Kudlicki W (2006) Efficient generation of insect-based cell-free translation extracts active in glycosylation and signal sequence processing. J Biotechnol 125:194–197

Chapter 5

Cell-Free Translation of Biofuel Enzymes

Taichi E. Takasuka, Johnnie A. Walker, Lai F. Bergeman, Kirk A. Vander Meulen, Shin-ichi Makino, Nathaniel L. Elsen, and Brian G. Fox

Abstract

In nature, bacteria and fungi are able to utilize recalcitrant plant materials by secreting a diverse set of enzymes. While genomic sequencing efforts offer exhaustive lists of genes annotated as potential polysaccharide-degrading enzymes, biochemical and functional characterizations of the encoded proteins are still needed to realize the full potential of this natural genomic diversity. This chapter outlines an application of wheat germ cell-free translation to the study of biofuel enzymes using genes from *Clostridium thermocellum*, a model cellulolytic organism. Since wheat germ extract lacks enzymatic activities that can hydrolyze insoluble polysaccharide substrates and is likewise devoid of enzymes that consume the soluble sugar products, the cell-free translation reactions provide a clean background for production and study of the reactions of biofuel enzymes. Examples of assays performed with individual enzymes or with small sets of enzymes obtained directly from cell-free translation are provided.

Key words Wheat germ extract, Transcription, Translation, Glycoside hydrolase, Cellulase, Xylanase, Mannanase, Functional screen

1 Introduction

The availability of cheap energy has been a major factor supporting economic development worldwide during the past century [1, 2]. Among the many available energy sources, liquid fuels, which are primarily used for transportation, are the most valuable but are also increasingly problematic, with major perturbations in their supply and price occurring in the mid-1980s and again after 2000. Given this recurring state of affairs, efforts to secure new sources of liquid fuels have intensified worldwide, with ongoing research in academic, industrial, and government settings.

The Great Lakes Bioenergy Research Center is a multi-institution collaborative effort funded by the US Department of Energy, with a mission to perform basic research needed to generate and improve technology to convert cellulosic biomass to ethanol

Kirill Alexandrov and Wayne A. Johnston (eds.), *Cell-Free Protein Synthesis: Methods and Protocols*, Methods in Molecular Biology, vol. 1118, DOI 10.1007/978-1-62703-782-2_5, © Springer Science+Business Media, LLC 2014

and advanced biofuels. Methods described in this chapter have been developed and carried out in the Great Lakes Bioenergy Research Center in order to evaluate the genomic contents of new isolates from highly cellulolytic environmental niches and to better understand the capabilities of model cellulolytic organisms. This approach is also applicable to engineering of multi-protein mixtures for improved deconstruction of biomass.

1.1 Overview of Biomass Deconstruction Needed for Biofuel Production

The complexities of biomass, particularly the diversity of cellulosic and hemicellulosic substructures that make up a large portion of plants, make the economic utilization of cellulosic biomass a difficult challenge. Currently, biomass is chemically pretreated to achieve a partial decomposition or extraction of lignin, which exposes the entrained polysaccharide substructures and also alters its physical state [3]. After chemical pretreatment, the biomass is subjected to enzymatic hydrolysis to produce soluble sugar solutions suitable for fermentation by microbes into biofuels. Because of its complex structure, many different enzymes are necessary for efficient, high yield deconstruction of biomass [4].

1.2 Enzymes Needed to Deconstruct Cellulose

Natural cellulolytic organisms are known to deconstruct biomass materials into simple sugars [5]. To better understand this natural capability, the DOE Joint Genome Institute and other researchers have sequenced over 5,000 natural organisms and metagenomes, including an increasing number of proven cellulolytic microbes and fungi. Bioinformatics approaches have sorted the protein-coding sequences of these genomes into many putative cellulolytic functions. As one leading example, the Carbohydrate-Active Enzymes database (CAZy, http://www.cazy.org [6]) provides an intensively curated list that is separated into glycoside hydrolase (GH), glycoside transferase (GT), carbohydrate esterase (CE), polysaccharide lyase (PL), and carbohydrate-binding module (CBM) families.

Efficient hydrolysis of pure cellulose requires concerted activities provided by several different members of the GH family [7]. Endocellulases or endoglucanases attack crystalline regions in the interior of cellulose molecules, while exoglucanases or cellobiohydrolases degrade cellulose from either the reducing or nonreducing ends of a glucan strand to produce cellobiose. Cellobiosidases then hydrolyze cellobiose to glucose. Both endo- and exocellulases have associated CBM domains, which increase the efficiency of binding the catalytic GH domain to the insoluble polysaccharide. Related cascades of enzymes react with polysaccharides such as xylan, mannan, and pectin, which collectively represent the hemicellulose fractions. Deconstruction of biomass is potentiated when enzymes capable of reaction with the cellulose and hemicellulose fractions are simultaneously present [8].

Among cellulolytic microorganisms, *Clostridium thermocellum*, an anaerobic, thermophilic, Gram-positive bacterium, has demonstrated one of the fastest growth rates on crystalline cellulose [9]. Since it can also ferment sugars to ethanol, *C. thermocellum* has become an important model organism for biofuel research. *C. thermocellum* secretes dozens of cellulases, xylanases, mannanases, and other polysaccharide-degrading enzymes that allow the efficient conversion of polysaccharides into soluble oligosaccharides. Biochemical [10, 11], transcriptomic [12], and proteomic [13] studies have helped to establish the identities of these enzymes, and many have been purified and studied for their catalytic function and 3-dimensional structures. Additionally, *C. thermocellum* produces a unique extracellular assembly termed the cellulosome, a multienzyme complex in which many different enzymes are recruited to a protein called scaffoldin through tight-binding interactions of dockerin and cohesin domains [14]. The formation of this large complex is believed to provide high efficiency in polysaccharide degradation by increasing local enzyme concentration. However, because the scaffoldin has a repetitive multi-domain structure of ~200 kDa, obtaining pure preparations of it through heterologous expression has been a significant challenge. Consequently, protein engineering has been used to develop more experimentally tractable substitutes for the scaffoldin and various assemblies of enzymes from *C. thermocellum* [15, 16].

1.3 Overview of Experimental Strategy and Method

In this chapter, we describe an approach for producing engineered fusions of GH families with CBM domains. Figures 1 and 2 show an overview of the plasmid constructs and cloning methods that were applied to genes from *C. thermocellum* to produce forms of the enzymes that efficiently react with insoluble substrates without a need for the scaffoldin. These general principles are applicable to the study of the catalytic and polysaccharide binding domains from other organisms.

Identification of candidate genes includes bioinformatics evaluation and, if available, correlation with experimental results such as transcriptomics or proteomics. Since Flexi Vector cloning uses SgfI and PmeI restriction sites [17], candidate genes containing these restriction sites will require mutagenesis (or gene synthesis) to eliminate them. Fortunately, most genomes contain a low frequency of these restriction sites [17]. Furthermore, since most biofuel enzymes are secreted from the natural host, they contain hydrophobic signal sequences at the N-terminus that should be removed in the design for cell-free translation. This simple excision can dramatically improve the solubility of translated proteins. Bioinformatics tools such as SignalP (http://www.cbs.dtu.dk/services/SignalP/) are useful to identify possible positions for truncation of the sequence encoding a signal peptide from the native gene. In this work, we also removed nucleotide sequences encoding the C-terminal dockerin domains in order to facilitate our studies of catalytic potential outside of the cellulosomal format.

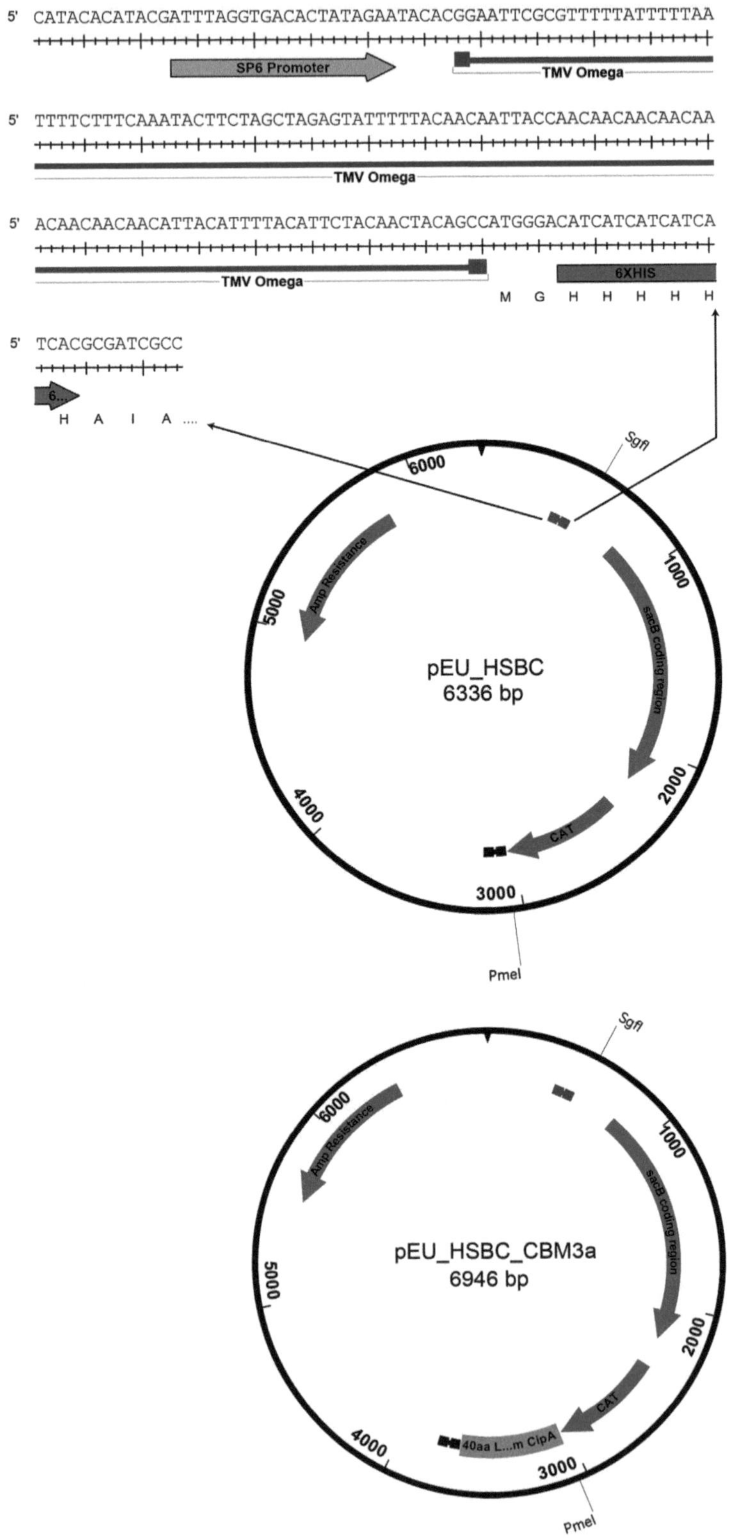

Fig. 1 Maps of two cloning vectors used in this study. The *top portion* shows the nucleotide sequence of the SP6 promoter, TMV-omega sequence, the start codon (ATG) and sequence encoding a His6 tag, and the Sgfl restriction site. Circular maps of pEU_HSBC and pEU_HSBC_CBM3a are shown below. In both plasmids,

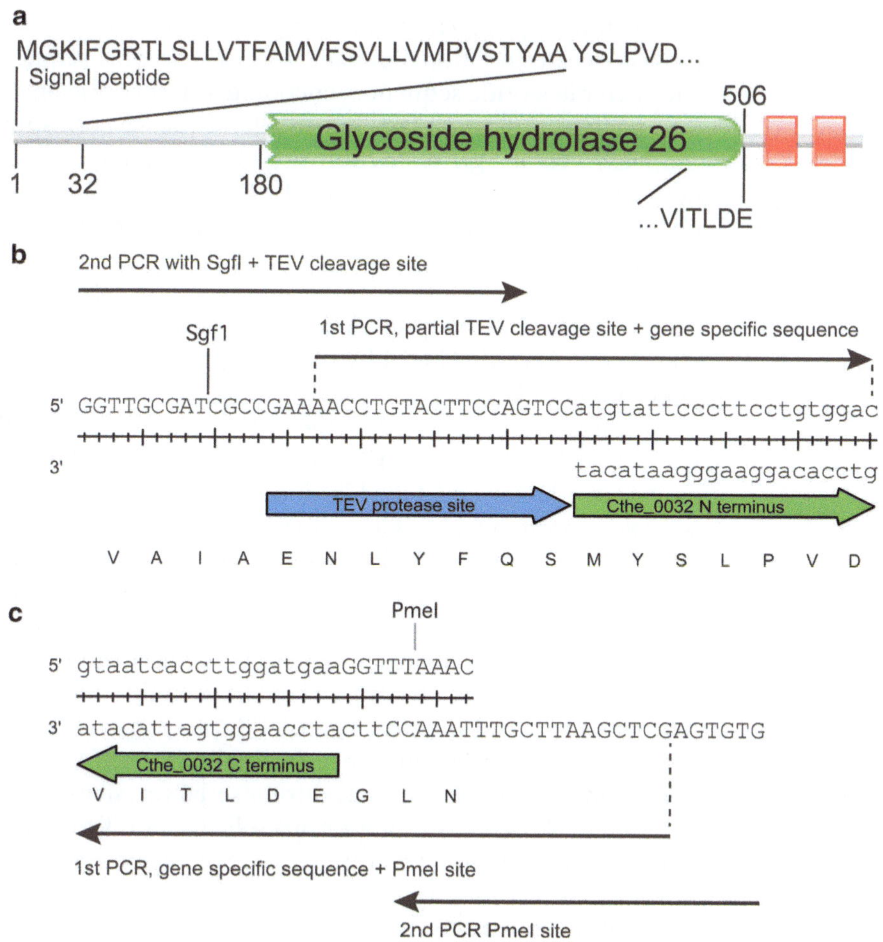

Fig. 2 Design of primer pairs and a schematic of the cloning method. (**a**) The Cthe_0032 (ManA) gene is used as an example. The PFAM image of the domain structure [39] shows the predicted amino acid residue positions for the signal peptide (residues 1–32), the GH26 domain (residues 180–506, *green*), and the two dockerin domains (*red*). The forward primer (**b**) and reverse primer design (**c**) for PCR needed for two-step Flexi Vector transfer are shown with nucleotide sequences and corresponding amino acid sequences

Endo, Sawasaki, and colleagues at Ehime University, Matsuyama, Japan, created an optimized wheat germ cell-free translation plasmid, pEU [18]. For the protocols presented here, the parental plasmid was modified by the UW Center for Eukaryotic Structural Genomics to create pEU-HSBC (Fig. 1). This Flexi Vector cloning-compatible plasmid contains an SP6 promoter at the 5′ end of the

Fig. 1 (continued) ampicillin resistance is used as the transformation selection marker. During cloning, the sacB-CAT cassette is replaced by the gene of interest, which can permit positive selection by growth on agar medium containing sucrose if desired. In pEU_HSBC_CBM3a, the 40-amino acid linker and CBM3a coding sequence is placed in frame at the 3′ end of sacB-CAT cassette so that the target protein is expressed as a fusion with the additional linker and CBM3a domain

critical transcription/translation region, followed by the tobacco mosaic virus translational enhancer (TMV-omega) sequence [19], and then nucleotide sequences encoding a start codon and Gly residue, a His6 purification tag, and an AIA sequence corresponding to the Flexi Vector SgfI restriction site [17]. The empty vector contains a toxic cassette consisting of the SacB gene [20] and chloramphenicol acetyl transferase genes. At the 3′ end, pEU-HSBC provides a stop codon by use of the Flexi Vector PmeI site, which thus yields a candidate protein with the N-terminal tag MGHHHHHHAIA– provided by the vector and the remaining coding sequence produced by the PCR primer design and targeted gene amplification.

Another vector created by the Great Lakes Bioenergy Research Center, pEU-HSBC-CBM3a, creates a C-terminal fusion to a 40 amino acid linker and CBM3a domain from the *C. thermocellum* scaffoldin protein CipA (Cthe_3077). Upon translation, candidate genes cloned into pEU-HSBC-CBM3a yield an unnatural fusion protein with an N-terminal tag MGHHHHHHAIA–, the candidate domains, and then a fusion to CBM3a, which is known to bind to crystalline cellulose. Genes cloned according to the Flexi Vector method are easily transferred between different Flexi Vector plasmids [17], and the presence of a 3′ pF1K homology sequence increases the transfer efficiency by inhibiting self-ligation of the two plasmid vector backbones during serial transfer of genes.

Primer design has an important role in this effort. By consideration of the domain structure of a gene of interest, domains of interest can be excised from the complete gene to assemble better performing constructs, to test specific aspects of the catalytic capabilities of multi-domain enzymes, or to obtain engineered assemblies. Figure 2 schematizes this primer design process as applied to Cthe_0032, a GH26 protein. In this work, the catalytic domain of the protein was cloned, expressed, and assayed. By use of the 2-stage PCR method, it is possible to build additional functionality at the 5′ and 3′ regions of the cloned gene. For example, inclusion of the coding sequence for a protease recognition site immediately after the SgfI site but before the coding sequence of the gene of interest is useful for removal of N-terminal tags from expressed proteins for structure determination studies [21]. Alternatively, omission of a stop codon from the coding sequence of the gene of interest can be used to create C-terminal fusion proteins such as some of the examples provided here (Table 3, CBM3a fusions).

Plasmid DNA purified according to the protocols given is the essential substrate for a transcription reaction, which provides an mRNA that becomes the substrate for cell-free translation. Both the transcription and translation reactions are carried out by robots [22], and the individual translation reactions are then suitable for assay with a wide variety of biofuel relevant substrates, can be combined into higher-order assemblies of enzymes, or can be subjected to additional purification if desired.

Table 1
Summary of the substrates used in enzyme assays

Substrate (abbreviation)	Assay type	Activity detected	Source
4-Methylumbelliferyl-β-D-glucopyranoside (MUG)	Fluorescence	β-Glucosidase	Sigma-Aldrich (M3633)
4-Methylumbelliferyl-β-D-cellobioside (MUC)	Fluorescence	Cellobiohydrolase	Sigma-Aldrich (M6018)
4-Methyumbelliferyl-β-D-mannopyranoside (MUM)	Fluorescence	β-Mannosidase	Sigma-Aldrich (M0905)
4-Methylumbelliferyl-β-D-xylopyranoside (MUX)	Fluorescence	β-Xylosidase	Sigma-Aldrich (M7008)
Filter paper (FP)	DNS	Cellulase	GE Healthcare (Whatman 10001-185)
Phosphoric acid-swollen cellulose (PASC)	DNS	Cellulase	Prepared as described in Subheading 3.6; also *see* reference [38]
Birch xylan (xylan)	DNS	Xylanase	Sigma-Aldrich (X0502)
1, 4-β-D-Mannan (mannan)	DNS	Mannanase	Megazyme (90302b)
AFEX-treated switchgrass (AFEX-SG)	DNS	Biomass deconstruction	Great Lakes Bioenergy Research Center
Ionic liquid-treated switchgrass (IL-SG)	DNS	Biomass deconstruction	Joint Bioenergy Institute

1.4 Protein Functional Assays

During our initial work on cell-free translation of biofuel enzymes, we discovered that wheat germ extract had no endogenous capability for hydrolysis of the insoluble substrates cellulose, xylan, or mannan and likewise no competing enzymatic reactions that would consume soluble sugars or contain other products such as NADH that would be produced from coupled enzyme assays. Moreover, a low-level soluble glucosidase activity present in the wheat germ extract was easily inactivated at temperatures higher than 45 °C. Thus many proteins translated in wheat germ extract can be assayed for their various cellulolytic activities without a need to obtain purified preparations of target polypeptides.

Table 1 summarizes the substrates used in this work. Assays with small molecule fluorogenic analogs, purified polysaccharides, and biomass materials have all been carried out with enzymes expressed in cell-free translation. In many cases, assays of single enzymes are inadequate for some discovery purposes, particularly when synergistic enzyme activities are required to disassemble the complicated ultrastructure of biomass. With robotic cell-free translation, new samples of enzymes can be rapidly obtained for combinatorial tests of synergistic activity or as replacements for individual enzymes in mixtures with established reactivity. We briefly introduce examples of this capability in the chapter.

The methylumbelliferyl glycosides are small molecule fluorogenic analogs used in enzyme screening to deduce the potential reactivity of biofuel enzymes with oligomeric substrates. Hydrolysis of the glycosidic bond releases the fluorophore 4-methylumbelliferone [23, 24]. The high sensitivity of the fluorescence detection, the short reaction time, and the small amounts of enzyme needed contribute to a broad glycosidase diagnostic utility of the small molecule analogs.

Insoluble substrates are most realistic for biofuel enzymology but also provide increasing experimental challenges [25]. While individual enzymes can often be reliably assayed using purified polysaccharides, studies of biomass most often require the combination of chemical pretreatment and mixtures of synergistic enzymes to provide sufficient reactivity. The determination of enzyme function with insoluble polysaccharides can be accomplished by a variety of methods [26]. While samples from cell-free translation are acceptable for most of these, we focus here on use of the dinitrosalicylic acid (DNS) assay [27] because of its broad applicability to a variety of products, relatively simple methodology, sufficient sensitivity, and low cost. When an enzyme hydrolyzes a polysaccharide, the reducing-end sugar product can react with the DNS reagent, producing a colored adduct with an absorbance maximum of ~540 nm. The color formation permits the use of optical spectroscopy for detection, and the method is well suited for high throughput because the sample handling process can be reduced to relatively few liquid handling steps. In addition, we have adapted a simple protein pull-down assay to assess the function of CBM domains [28] and show examples of this qualitative evaluation using pure cellulose as a binding substrate.

2 Materials

2.1 Equipment

1. The Protemist DT II benchtop cell-free translation robot (CellFree Sciences Co., Ltd., Yokohama, Japan) is used for preparative protein production. This robot uses a bilayer cell-free protein synthesis method [29].
2. The Protemist DT II software (CellFree Sciences Co., Ltd., Yokohama, Japan) is used to program and execute the robotic translation experiment.
3. A spectrometer is used to measure DNA concentrations in plasmid preparations.
4. The Criterion stain-free gel imaging system (Bio-Rad Laboratories, Inc., USA) is used for qualitative and quantitative measurement of protein production (*see* **Note 1**).

Criterion Precast 4–20 % gels (#345-0426, Bio-Rad Laboratories, Inc., USA) are used for protein electrophoresis.

5. Image Lab software version 3.0 (Bio-Rad Laboratories, Inc., USA) is used to visualize and quantitate proteins separated by the Criterion imaging system.
6. A shaking device (suggested Microtube Rotisserie (LABQUAKE®, Barnstead Thermolyne, NH, USA)) is used for enzyme assays.
7. A microplate reader is used for making fluorescence and optical density measurements of samples in either 96-well or 384-well microtiter plate formats (suggested instrument Infinite® M1000 Pro, TECAN Group Ltd., Switzerland).

2.2 Cloning Reagents

1. 18 MΩ water is used to prepare all buffers and other reagents.
2. The pEU plasmid is the parental plasmid ([18], CellFree Sciences Co., Ltd., Yokohama, Japan). The wheat germ cell-free translation plasmid pEU-HSBC, shown in Fig. 1, was created from this parent.
3. *C. thermocellum* ATCC 27405 genomic DNA was provided by Dr. Paul Weimer (US Department of Agriculture Dairy Forage Research Center, Madison, WI).
4. Herculase® II Fusion Enzyme with dNTP Combo (Agilent Technologies, Santa Clara, CA, USA) is used for PCR reactions.
5. 10× Flexi Enzyme Blend (Promega, Madison, WI, USA) containing optimized amounts of SgfI and PmeI endonucleases is used for restriction enzyme digestions.
6. T4 DNA Ligase (New England Biolabs Ltd., UK) is used for ligation reactions.
7. Chemically competent cells (suggested strain *Escherichia coli* 10G, Lucigen, Middleton, WI, USA) are used for transformation.
8. BigDye® v2.0 (Applied Biosystems, CA, USA) is used for DNA sequencing reactions.

2.3 Transcription and Translation Reagents

1. A Qiagen Miniprep kit (Qiagen, Germany) is used for plasmid DNA preparations.
2. Proteinase K (Sigma-Aldrich, St. Louis, MO, USA) is used to eliminate residual RNase from plasmid DNA preparations. The 10× Proteinase K buffer consists of 100 mM Tris–HCl, pH 8.0, containing 50 mM EDTA and 1 % (w/v) SDS.
3. Transcription buffer (TB) consists of 400 mM HEPES–KOH, pH 7.8, 100 mM magnesium acetate, 10 mM spermidine hydrochloride, and 50 mM DTT. This buffer is used in cell-free translation reactions carried out in the Protemist DT II benchtop robot.

4. Nucleotide solution (NTPs) consists of 25 mM each of ATP, GTP, CTP, and UTP (Sigma-Aldrich, St. Louis, MO, USA) dissolved in Milli-Q water. This solution is used in the transcription stage of the cell-free translation reaction.
5. SP6 RNA polymerase and RNase inhibitor (Promega, Madison, WI, USA) are used in transcription reactions.
6. Translation buffer (40× solutions 1–4; CellFree Sciences Co., Ltd., Yokohama, Japan) is used with the Protemist DT II robot.
7. The amino acid mixture consists of 2 mM of each of the 20 L-amino acids.
8. Wheat germ cell-free extract, either WEPRO2240 or WEPRO2240H (CellFree Sciences Co., Ltd., Yokohama, Japan), is used in the Protemist DT II robot (*see* **Note 2**).
9. A 50 mg/mL solution of creatine kinase (Roche Applied Sciences, Indianapolis, IN, USA) is prepared in Milli-Q water and used in translation reactions.

2.4 Protein Purification Reagents

1. Ni-Sepharose high-performance chromatography resin (suggested resin is GE Healthcare, Piscataway, NJ, USA) is used for purification of His-tagged proteins from the wheat germ cell-free translation reactions.
2. Purification loading buffer (100 mM MOPS, pH 7.4, containing 300 mM NaCl, 2 mM $CaCl_2$, and 25 mM imidazole) is used to load sample to the resin.
3. Purification washing buffer (100 mM MOPS, pH 7.4, containing 300 mM NaCl, 2 mM $CaCl_2$, and 50 mM imidazole) is used to reduce nonspecific binding of proteins to the resin.
4. Purification elution buffer (100 mM MOPS, pH 7.4, containing 300 mM NaCl, 2 mM $CaCl_2$, and 250 mM imidazole) is used to elute protein from the resin.
5. Laemmli sample buffer (#161-0737, Bio-Rad Laboratories, Inc., USA) is the 2× protein sample buffer.
6. Protein molecular weight size markers (suggested markers are Precision Plus Protein Unstained Standards, Bio-Rad Laboratories, Inc., USA) are used for internal standards in electrophoresis experiments.
7. VIVASPIN 500, 10,000 MWCO PES (Sartorius Stedim Biotech, Goettingen, Germany) is used for concentrating protein samples after purification (*see* **Note 3**).
8. U-bottom 96-well plates are from Greiner Bio-One (Monroe, NC).

2.5 Enzyme Assay Substrates

1. The fluorogenic substrates 4-methylumbelliferyl-β-D-glucopyranoside (MUG), 4-methylumbelliferyl-β-D-cellobioside (MUC), 4-methylumbelliferyl-β-D-mannopyranoside (MUM),

and 4-me-thylumbelliferyl-β-D-xylopyranoside (MUX) (Sigma-Aldrich, St. Louis, MO, USA) are used in small molecule analog assays. These substrates are prepared fresh for each experiment by dissolving an appropriate mass in a 50 mL solution of 0.1 M sodium phosphate buffer, pH 8.0, in order to achieve 0.2 mM substrate concentration.

2. Whatman filter paper 1001-185 (GE Healthcare, UK) is used as a representative substrate for cellulase assays. A standard paper hole punch tool is used to create ~0.7 cm diameter disks, corresponding to a weight of 3.0 mg.
3. Phosphoric acid-swollen cellulose (PASC) is used as a representative substrate for cellulase assays.
4. Birch xylan (Sigma-Aldrich, St. Louis, MO, USA) is used as a substrate for xylanase assays.
5. Mannan (Sigma-Aldrich, St. Louis, MO, USA) is used as a substrate for mannanase assay and mannan pull-down assay.
6. Ammonia fiber expansion (AFEX) switchgrass ground to 1 mm particle size and washed ionic liquid-treated switchgrass are used as biomass substrates (*see* **Notes 4** and **5**).
7. Sigmacell 20 (Sigma-Aldrich, St. Louis, MO, USA) is used as a substrate for the protein pull-down assay.

2.6 Enzyme Assay Reagents

1. A 50 mM sodium phosphate buffer, pH 8.0, is used for all reactions.
2. The colorimetric reagent 3,5-dinitrosalicylic acid (Sigma-Aldrich, St. Louis, MO, USA) is used in reducing sugar determinations. To prepare the stock reagent solution, 10.6 g of 3,5-dinitrosalicylic acid (Sigma Aldrich, St. Louis, MO) and 19.8 g of NaOH are dissolved in 1 L of Milli-Q water. Then 306 g of Rochelle salts (Sigma-Aldrich, St. Louis, MO, USA), 7.6 mL of phenol, and 8.3 g of sodium metabisulfite (Sigma-Aldrich, St. Louis, MO, USA) are added. The final volume of the reagent solution is adjusted to 1.5 L by addition of Milli-Q water. The reagent stock is stored at 4 °C in foil-covered bottles until use.
3. For reducing sugar assays with reactions containing cellulose, β-glucosidase (Lucigen, Middleton, WI) is added to convert all soluble oligosaccharides to glucose. Although not essential, this addition allows the quantitation of total monomer sugar equivalents produced.

2.7 Polyacrylamide Gel Silver Staining Reagents

1. A silver staining method was used to detect protein–polysaccharide interactions in polyacrylamide gels. Reagents used are sodium thiosulfate (Sigma-Aldrich, St. Louis, MO, USA), silver nitrate, $AgNO_3$ (Sigma-Aldrich, St. Louis, MO, USA), and formaldehyde (Fisher Scientific, Hampton, NH, USA).

3 Methods

3.1 Design of First-Step PCR Primers for Candidate Genes

Figure 2 summarizes how PCR primers are designed by using the *C. thermocellum* Cthe_0032 gene, Uniprot A3DBE4, as an example (*see* **Note 6**).

1. The boundaries of the translocation signal peptide (residues 1–32) are used to define the position for truncation at the 5′ end of the gene. This truncation removes the signal peptide, which is not necessary for cell-free translation. According to this criterion, the gene-specific portion of the forward primer is 5′-atgtattcccttcctgtggac-3′ (Fig. 2b).
2. The first-step forward PCR primer is then completed by appending a nucleotide sequence encoding a portion of the tobacco etch virus protease recognition sequence to the 5′ end of the sequence identified in **step 1**. With this addition, the complete sequence of the first-step forward PCR primer is 5′-AACCTGTACTTCCAGTCCatgtattcccttcctgtggac-3′, indicated as "1st PCR, partial TEV cleavage site + gene-specific sequence" in Fig. 2b.
3. The end of the GH26 domain (residue 506) is used to define the position for truncation at the 3′ end of the gene. This truncation removes the dockerin domain. According to this criterion, the first-step reverse complement primer for the gene-specific sequence is 5′-ttcatccaaggtgattacata-3′.
4. For insertion of a stop codon at the end of the gene-specific sequence, the first-step reverse PCR primer is then completed by appending a nucleotide sequence encoding a stop codon and the PmeI restriction sequence to the 5′ end of the sequence identified in **step 3**. With this addition, the complete sequence of the first-step reverse PCR primer is 5′-GCTCGAATTCGTTTAAACTAttcatccaaggtgattacata-3′, indicated as "1st PCR, gene-specific sequence + PmeI site" in Fig. 2c. This primer is also shown in Table 2, Cthe_0032R.
5. For creation of a C-terminal fusion protein, the first-step reverse PCR primer is then completed by appending a nucleotide sequence encoding a PmeI restriction sequence to the 5′ end of the sequence identified in **step 1**. With this addition, the complete sequence of the first-step reverse PCR primer is 5′-GCTCGAATTCGTTTAAACCttcatccaaggtgattacata-3′. This primer is also shown in Table 2, Cthe_0032R_CBM3a.

3.2 PCR Cloning and Plasmid Preparation

1. A two-step PCR method is used to amplify genes for this cloning method (*see* **Note 7**). The first-step PCR is performed using the forward and reverse primer pairs indicated in Table 2, with a reverse primer designed to create either a catalytic domain alone or a fusion protein. Each primer possesses ~20

Table 2
Oligonucleotides used for cloning

Primer name	Sequence[a]	Direction
Cthe_0032F	5′-AACCTGTACTTCCAGTCCatgtattcccttcctgtggac-3′	Forward
Cthe_0032R	5′-GCTCGAATTCGTTTAAACTAttcatccaaggtgattacata-3′	Reverse
Cthe_0032R_CBM3a	5′-GCTCGAATTCGTTTAAACCttcatccaaggtgattacata-3′	Reverse
Cthe_0212F	5′-AACCTGTACTTCCAGTCCatgtcaaagataactttccc-3′	Forward
Cthe_0212R	5′-GCTCGAATTCGTTTAAACTAttaaaaaccgttgtttttgatta-3′	Reverse
Cthe_0269F	5′-AACCTGTACTTCCAGTCCatggtgccttttaacacaaa-3′	Forward
Cthe_0269R_CBM3a	5′-GCTCGAATTCGTTTAAACctcctgttatgtacaacaaagtg-3′	Reverse
Cthe_0270F	5′-AACCTGTACTTCCAGTCCatgaaaaaaataccgttactta-3′	Forward
Cthe_0270R	5′-GCTCGAATTCGTTTAAACTAtcaatcatcaacaggtatattgt-3′	Reverse
Cthe_0405F	5′-AACCTGTACTTCCAGTCCatggatccgaacaatgacgactg-3′	Forward
Cthe_0405R_CBM3a	5′-GCTCGAATTCGTTTAAACtccatttgaaccaagaggtatc-3′	Reverse
Cthe_0412F	5′-AACCTGTACTTCCAGTCCatgggacatcatcatcatcatca-3′	Forward
Cthe_0412R_CBM3a	5′-GCTCGAATTCGTTTAAACagctgtaacccatgcaaacg-3′	Reverse
Cthe_0536F	5′-AACCTGTACTTCCAGTCCatgggacatcatcatcatcatca-3′	Forward
Cthe_0536R	5′-GCTCGAATTCGTTTAAACTAtttatacggcaactcactta-3′	Reverse
Cthe_0543F	5′-AACCTGTACTTCCAGTCCatgggacatcatcatcatcatca-3′	Forward
Cthe_0543R	5′-GCTCGAATTCGTTTAAACTAttactgttcagccgggaattttt-3′	Reverse
Cthe_0578F	5′-AACCTGTACTTCCAGTCCatggactataactatggagaag-3′	Forward
Cthe_0578R_CBM3a	5′-GCTCGAATTCGTTTAAACggtaccattgggttctacacc-3′	Reverse
Cthe_0625F	5′-AACCTGTACTTCCAGTCCatgggacatcatcatcatcatca-3′	Forward
Cthe_0625R	5′-GCTCGAATTCGTTTAAACTActattctaccggaaatttatcta-3′	Reverse
Cthe_0797F	5′-AACCTGTACTTCCAGTCCatgggacatcatcatcatcatca-3′	Forward
Cthe_0797R	5′-GCTCGAATTCGTTTAAACTAtcttcaacgccggcacctctc-3′	Reverse
Cthe_0912F	5′-AACCTGTACTTCCAGTCCatgggacatcatcatcatcatca-3′	Forward
Cthe_0912R	5′-GCTCGAATTCGTTTAAACTActccttcgattacagttcc-3′	Reverse
Cthe_1256F	5′-AACCTGTACTTCCAGTCCatggcggtagatatcaagaaaat-3′	Forward
Cthe_1256R	5′-GCTCGAATTCGTTTAAACTAttattccacgttgtttattttgt-3′	Reverse
Cthe_1838F	5′-AACCTGTACTTCCAGTCCatgggacatcatcatcatcatca-3′	Forward
Cthe_1838R_CBM3a	5′-GCTCGAATTCGTTTAAACgttaactatagcataaaatgc-3′	Reverse
Cthe_2872F	5′-AACCTGTACTTCCAGTCCatgggacatcatcatcatcatca-3′	Forward
Cthe_2872R	5′-GCTCGAATTCGTTTAAACTAtccgtagtactcgcccagggaaa-3′	Reverse
2nd universal primerF	5′-GGTTGCGATCGCCGAAAACCTGTACTTCCAG-3′	Forward
2nd universal primerR	5′-GTGTGAGCTCGAATTCGTTTAAACC-3′	Reverse

[a] *Lowercase letters* indicate the gene-specific sequence

nucleotides of gene-specific sequence and ~20 nucleotide overhang, respectively. The *C. thermocellum* genomic DNA is used as the amplification template. The PCR reaction contains 0.5 μL of 1 μg/μL *C. thermocellum* genomic DNA, 1.25 μL of 10 μM primer pair, 0.5 μL of 100 mM dNTP, 10 μL of 5× PCR buffer, and 1 μL of Herculase DNA polymerase in a total volume of 50 μL. The first-step PCR reaction cycle consists of 95 °C for 5 min followed by 95 °C for 30 s, a ramp from 50 to 60 °C for 30 s, and 72 °C for *X* min for 35 cycles, where *X* is increased by 1 min for each 1,000 base pairs in the template, and then the reaction cycle is ended by 72 °C for 5 min.

2. The second-step PCR is performed using the first PCR product as a DNA template and the forward and reverse universal primers (Table 2). The forward universal primer is indicated in Fig. 2b as "2nd PCR with SgfI + TEV cleavage site," while the reverse universal primer is indicated as "2nd PCR PmeI site."
3. The PCR reaction contains 5 μL of 1st PCR reaction product, 2.5 μL of 10 μM primer pair, 2 μL of 100 mM dNTP, 20 μL of 5× PCR buffer, and 2 μL of Herculase DNA polymerase in a total volume of 100 μL. The second-step PCR reaction cycle consists of 95 °C for 5 min followed by 95 °C for 30 s, ramp from 50 to 60 °C for 30 s, and 72 °C for *X* min for 35 cycles, where *X* is increased by 1 min for each 1,000 base pairs in the template, and then the reaction cycle is ended by 72 °C for 5 min.
4. Plasmid vector DNA (pEU-HSBC-CBM3a, *see* **Note 8**) and PCR product are digested using the Flexi Enzyme Blend. A typical reaction contains 3.5 μL of the DNA product, 0.5 μL of the Flexi Enzyme Blend, and 1 μL of buffer in a total volume of 5 μL. The reaction is carried out at 37 °C for 60 min and followed by 65 °C for 20 min.
5. Ligation of the digested PCR product and the plasmid vector DNA (pEU-HSBC-CBM3a) is carried out using T4 DNA Ligase with 1 mM dATP in 1× ligation buffer for 12 h at 16 °C.
6. The ligated sample is transformed into chemically competent *E. coli* 10G cells by a standard heat shock transformation protocol. Transformants are selected on Luria Bertani agar plates containing 100 μg/mL of ampicillin and 5 % (w/v) of sucrose.
7. Single colonies are screened by colony PCR methods to identify transformants that contain an insert of the approximately correct size. Transformants with the correct insert size are then submitted for DNA sequencing, in this case at the University of Wisconsin Biotechnology Center. The nucleotide sequences are analyzed by software comparison of predicted and desired sequences (suggested DNASTAR software, v.10, DNASTAR, Madison, WI, USA).
8. A sequence-verified plasmid carrying the gene of interest is prepared for transcription by using a Qiagen Miniprep kit.

The purified plasmid DNA is treated with proteinase K in 1× Proteinase K buffer at 37 °C for 1 h to eliminate residual RNase. After the digestion, the plasmid preparation is treated by phenol/chloroform extraction. The plasmid preparation is recovered from the aqueous phase of the phenol/chloroform extraction by ethanol precipitation with a 2.5× volume addition of absolute ethanol supplemented with a 1/10 volume of 3 M ammonium acetate. The precipitation reaction is incubated at −80 °C for 1 h. The precipitated DNA is recovered by centrifugation at 16,100 × *g* for 15 min. The amount of DNA is measured by spectrometry using A_{260} measurement. Based on this measurement, the concentration of plasmid DNA is adjusted to 1 μg/μL.

3.3 Protein Cell-Free Translation

1. A 5 μL aliquot of a 1 μg/μL preparation of purified plasmid DNA is used for each transcription reaction carried out in the Protemist DT II.
2. The transcription mixture (45.25 μL total volume per reaction) consists of 29 μL of water, 10 μL of 5× of transcription buffer (CellFree Sciences Co., Ltd., Japan), 5 μL of 25 mM NTPs, 0.625 μL of SP6 RNA polymerase, and 0.625 μL of RNase inhibitor enzymes (80,000 units/mL).
3. The translation mixture (59.2 μL total volume per reaction) consists of 56 μL of wheat germ extract (CellFree Sciences Co., Ltd., Japan), 3 μL of 2 mM amino acid mix, and 0.2 μL of a 20 μg/μL solution of creatine kinase (Roche, USA).
4. The 1× translation buffer (1.1 mL per reaction) containing ATP, GTP, and 20 amino acids is prepared from a SUB-AMIX (CFS-SUB, CellFree Sciences Co., Ltd., Japan).
5. The plasmid DNA, transcription mixture, translation mixture, and translation buffer are placed onto the appropriate positions of the deck of the Protemist™ DT II robot as described in the instrumental manual. The Protemist™ DT II carries out fully automated transcription and bilayer mode translation reactions in 28 h duty cycle [29].

3.4 Verification of Translated Protein in Wheat Germ Extract

1. Translated proteins are analyzed by using denaturing gel electrophoresis (suggested Bio-Rad Criterion system, *see* **Notes 1, 12, 13**). A 5 μL sample of the translation reaction and an equal volume of 2× protein sample buffer are mixed and heated at 95 °C for 10 min. Electrophoresis is performed in constant voltage mode at 220 V for 50 min in 1× Criterion buffer system.

3.5 Purification of His-Tagged Proteins

1. Ni-Sepharose resin is equilibrated with water to make an ~50 % (v/v) slurry.
2. For each translation reaction from Protemist DT II robot, place 50 μL of the Ni-Sepharose slurry into each well of the filter plate, retained in the U-bottom 96-well plate. Add

100 μL of the Ni binding/washing buffer to each well. This combination of a filter plate and the U-bottom plates is the purification plate.

3. Add 100 μL of the supernatant from the soluble translation product plate to each well of the purification plate.
4. Mix the solution in a plate shaker for 10 min at room temperature, taking caution to not spill or cross-contaminate the wells.
5. Centrifuge the plate (suggested JS 5.9 rotor and Avanti J-30I centrifuge for 1 min at 3,640 rpm, 2,500×*g*, and ambient temperature).
6. Add 150 μL of Ni binding/washing buffer to wash out nonspecifically bound proteins. Gently mix the solution using a plate shaker for 10 s at room temperature. Centrifuge the plate. Repeat this step three times. The filtrate should not contain the translated protein, but can be retained for analysis of binding efficiency and product yield if needed.
7. Place the filter plate onto a new U-bottom 96-well plate. Add 50 μL of the Ni elution buffer into each well of the purification plate. Use a plate shaker to gently mix the solution for 1 min at room temperature to elute the bound proteins.
8. Centrifuge the plate and save the filtrate for analysis by denaturing gel electrophoresis (Fig. 3) and enzymatic assays.

3.6 Preparation of Phosphoric Acid-Swollen Cellulose

1. Sigmacell 50 microcrystalline cellulose (20 g, Sigma, St. Louis, MO) is swollen in 800 g of cold (0 °C) 80 % phosphoric acid, with rapid stirring with a plastic rod for ~1 h in an ice bath.
2. The cellulose is diluted with 2 L of cold water, thoroughly mixed, and allowed to settle, and the overlying liquid is removed by siphoning. This washing and settling procedure is repeated several times to reduce the acid content.
3. The cellulose slurry is neutralized with solid $NaHCO_3$, rinsed, decanted as above, and then secured inside bags formed from nylon-reinforced paper towelling. These bags are filled with ~1 L of distilled water, and the excess liquid is squeezed off; this process is repeated 20 times. The bags are then sealed, suspended in buckets containing 5 L of cold deionized water, and dialyzed for 10 days, with frequent changes of water. Prior to each change of water, the bags are tightly hand squeezed to facilitate removal of the equilibrated solutions.
4. After completion of dialysis (when the phosphate content reaches < 1 μg/L) the cellulose is lyophilized (*see* **Note 9**).

3.7 Fluorogenic Substrate Hydrolysis Assay

5. A 10 μL aliquot of cell-free translation reaction containing a translated enzyme is mixed with the fluorogenic substrate and assay buffer to give a total reaction volume of 100 μL. These

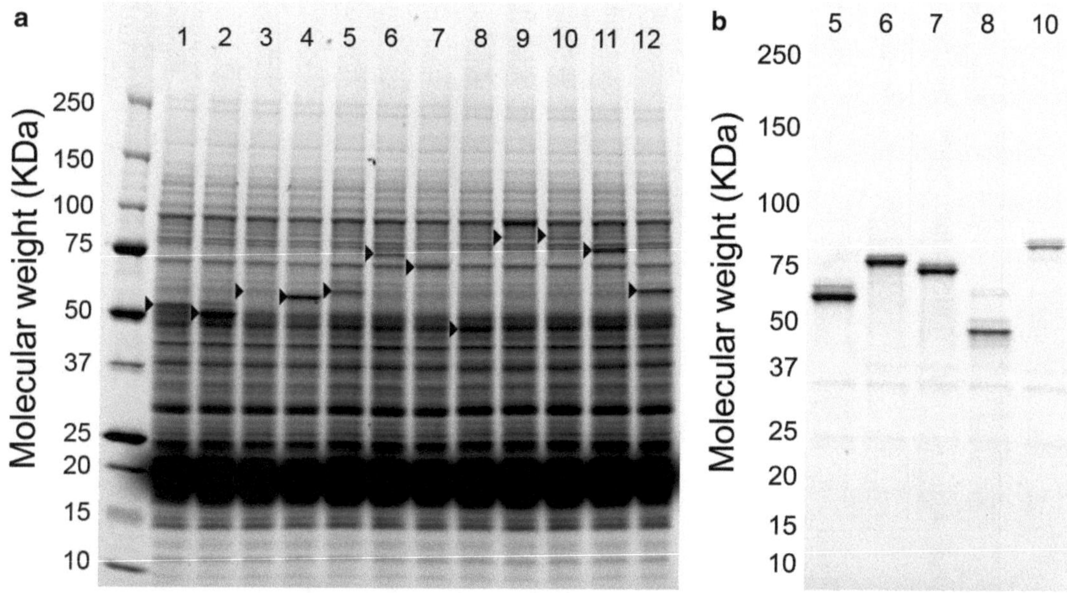

Fig. 3 SDS-PAGE analysis (**a**) of the wheat germ cell-free translation of *C. thermocellum* proteins and (**b**) purification of proteins containing an N-terminal His-tag. Lane 1, Cthe_0032; lane 2, Cthe_0212; lane 3, Cthe_0269; lane 4, Cthe_0270; lane 5, Cthe_0536; lane 6, Cthe_0543; lane 7, Cthe_0625; lane 8, Cthe_0797; lane 9, Cthe_0912; lane 10, Cthe_1256; lane 11, Cthe_1838; lane 12, Cthe_2872. The domains included in the translated proteins and results of catalytic assays on these samples are provided in Table 3

reactions are conveniently prepared in a 96-well plate and include technical replicates.

6. The fluorogenic reaction is carried out at 37 °C for 15 min in plate reader. The fluorescence measurement is performed using excitation at 360 nm and emission detection at 460 nm. A negative control is made using a blank wheat germ translation reaction.
7. Results from assays with a panel of fluorogenic substrates are summarized in Table 3. Enzymes that give a positive fluorescence increase after subtraction of the background fluorescence in the wheat germ extract control with the fluorogenic substrate are assigned to have positive reactivity with the indicated substrate (*see* **Note 10**).

3.8 Reducing Sugar Assays

1. A 50 μL aliquot of a cell-free translation reaction containing a translated enzyme is mixed with 50 μL of a 20 mg/mL suspension of pure polysaccharide substrate in 50 mM phosphate buffer, pH 8.0 (*see* **Note 10**). Pure polysaccharides suitable for use with this method include PASC, xylan, and mannan. These samples are conveniently prepared in a 96-well microtiter plate and sealed with TempPlate sealing film (USA Scientific, Ocala, FL).
2. The reaction is carried out at 60 °C for 20 h. A 30 μL aliquot of the reaction and a 60 μL aliquot of DNS reagent are com-

Table 3
Summary of assay results obtained for individual *C. thermocellum* enzymes

Gene locus	Gene name	Domains produced in translation reaction	His6 added[a]	CBM3a added[b]	Uniprot ID	FP	PASC	Xylan	Mannan	MUG	MUC	MUX	MUM	Measured function(s)
Cthe_0032		GH26			A3DBE4	–	–	–	+	–	–	–	–	Mannanase
Cthe_0040	celI	GH9 CBM3			Q02934	+	+	–	–	–	–	–	–	Endocellulase
Cthe_0212	bglA	GH1			P26208	–	–	–	–	+	–	–	–	β-Glucosidase
Cthe_0269	celA	GH8		Yes	A3DC29	+	+	–	–	–	–	–	–	Endocellulase
Cthe_0270	chiA	GH18			A3DC30	–	–	–	–	–	–	–	–	None detected
Cthe_0405	celL	GH5		Yes	A3DCG4	+	+	–	–	–	+	–	–	Exocellulase
Cthe_0412	celK	CBM_4_9 CelDN GH9		Yes	A3DCH1	–	+	+	–	+	+	–	–	Exocellulase
Cthe_0536	celB	GH5	Yes		P04956	+	+	–	–	–	+	–	–	Exocellulase
Cthe_0543	celF	GH9 CBM3	Yes		P26224	+	+	–	–	–	–	–	–	Endocellulase
Cthe_0578	celR	GH9 CBM3		Yes	A3DCY5	+	–	–	–	+	–	–	–	Endocellulase
Cthe_0625	celQ	GH9 CBM3	Yes		A3DD31	+	+	–	–	–	–	–	–	Endocellulase
Cthe_0797	celE	GH5 Lipase_GDSL	Yes		A3DDK3	+	+	+	+	–	+	–	–	Exocellulase, mannanase, xylanase
Cthe_0912	xynY	CBM_4_9 GH10 CBM_4_9			A3DDW7	–	–	+	–	–	–	–	–	Xylanase
Cthe_1256	bglB	GH3	Yes		P14002	–	–	–	–	+	+	+	–	β-Glucosidase, β-xylosidase
Cthe_1838	xynC	CBM_4_9 GH10		Yes	A3DGI0	–	–	+	–	–	–	–	–	Xylanase
Cthe_2872	celG	GH5			Q05332	+	+	–	–	–	+	–	–	Exocellulase

[a]Proteins produced using pEU-His-FV, which adds an N-terminal MGHHHHHHAIA tag
[b]Protein produced using pEU-His-FV-CBM3a, which gives a native N-terminus and adds fusion of a 40-amino acid linker followed by the CBM3a domain from Cthe_3077 to the C-terminus

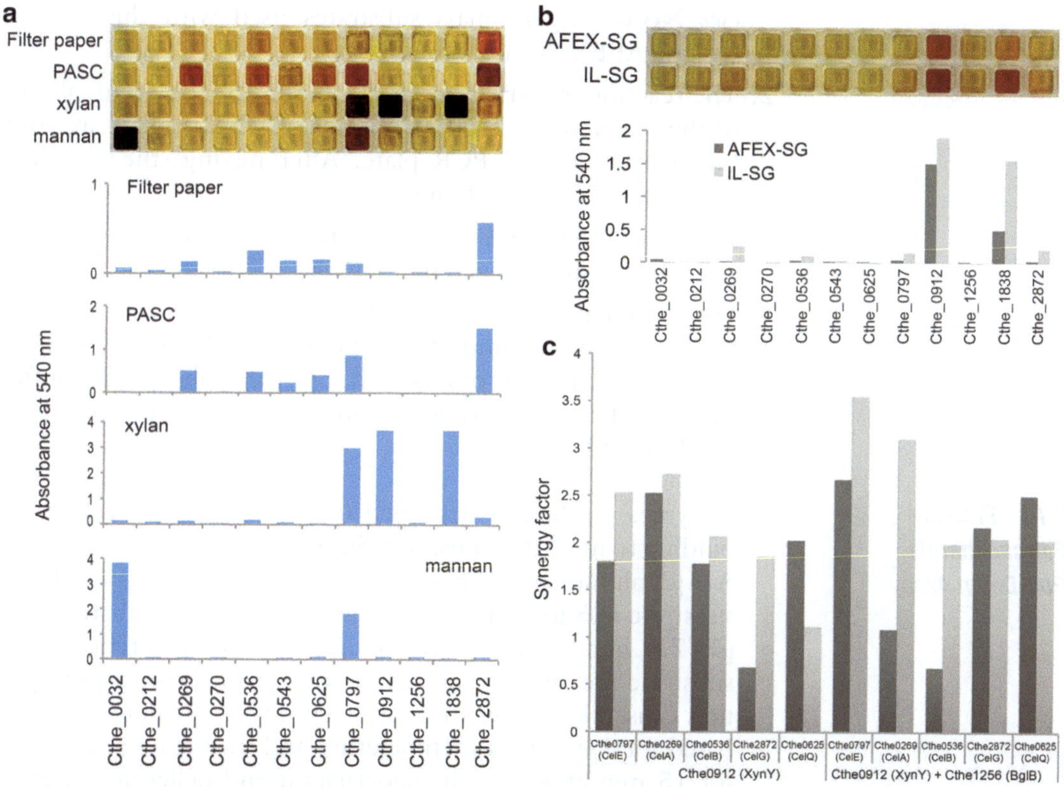

Fig. 4 Representative endpoint assays of reducing sugar formation from *C. thermocellum* enzymes prepared by robotic cell-free translation and assayed without further purification. The reactions were carried out in a 384-well microtiter plate. The substrates used are indicated, and detection was by use of the DNS reagent, which reacts with reducing-end sugars, yielding a change in color from *pale yellow* to *red* (increase in absorbance at 540 nm). (**a**) Reactions of a set of individual cellulase and xylanase enzymes with pure polysaccharides, where each row contains the results for the panel of enzymes indicated with a different substrate. (**b**) The same set of single enzymes was tested for reaction with AFEX-SG and IL-SG. Table 3 gives a summary of the assay results. (**c**) Synergistic improvement in biomass deconstruction obtained from binary combinations of a cellulase with a xylanase. The influence of β-glucosidase in alleviation of product inhibition is also demonstrated

bined in a 96-well PCR plate. After mixing, the sample is heated at 95 °C for 10 min.

3. The absorbance at 540 nm is used to determine the amount of reducing sugar present (Fig. 4a). A water blank containing DNS reagent is used for the background subtraction. The amount of reducing sugar produced is determined by comparison to standard curves generated using the corresponding pure monomer sugar, i.e., glucose, xylose, or mannose.
4. Results of the DNS assay with pure polysaccharides are shown in Fig. 4a and summarized in Table 3.

3.9 Biomass Assays

1. A 50 μL aliquot of cell-free translation reaction containing a translated enzyme is mixed with 50 μL of a 20 mg/mL suspension of pretreated biomass in 50 mM phosphate buffer, pH 8.0

(*see* **Note 10**). Biomass substrates used with this method include AFEX-SG and IL-SG.

2. The reaction is carried out at 60 °C for 20 h. A 30 μL aliquot of the reaction and a 60 μL aliquot of DNS reagent are combined in a 96-well PCR plate. After mixing, the sample is heated at 95 °C for 10 min.
3. Results of the DNS assay with pretreated biomass are shown in Fig. 4b and summarized in Table 3.
4. For combinatorial assays, 25 μL of two separate cell-free translation reactions are mixed with 50 μL of a 20 mg/mL suspension of pretreated biomass in 50 mM phosphate buffer, pH 8.0. The reaction is carried out as described in **step 2** above (*see* **Note 11**).

3.10 Protein–Polysaccharide Pull-Down Assay

1. Enzymes purified as described in Fig. 3 are used to test for binding to insoluble substrates Sigmacell 20 and PASC (Fig. 5). A 5 μL aliquot of purified enzyme is mixed with 1 mg of each substrate in 5 μL of total reaction in 50 mM phosphate buffer, pH 7.0, in a 96-well PCR plate. Reactions of protein without substrate are also prepared as a control for protein precipitation caused by aggregation or denaturation. Samples are incubated for 1 h at room temperature and then spun at 5,000 × *g* for 15 min at 4 °C. The supernatant and pellet are carefully separated, and 10 μL of 2× sample buffer is added to the supernatant, while the pellet is suspended in 5 μL of 2× sample buffer and an equal volume of water. The samples are heated for 5 min at 95 °C.
2. Denatured samples are separated using polyacrylamide gel electrophoresis and analyzed as described above.
3. The gels are placed into fixing solution (50 % methanol, 12 % acetic acid, 0.5 mL/L 37 % formaldehyde) for 1 h.
4. The gels are washed in 50 % ethanol three times for 5 min and pretreated with a 0.2 g/L solution of $Na_2S_2O_3{\cdot}5H_2O$ for 1 min.
5. The gels are rinsed with Milli-Q water 3× for 20 s and stained in a solution containing 2 g/L $AgNO_3$ and 0.75 mL/L of 37 % formaldehyde for 30 min.
6. The gels are rinsed again with Milli-Q water 3× for 20 s; developed in a solution containing 60 g/L Na_2CO_3, 0.5 mL/L of 37 % formaldehyde, and 4 mg/L of $Na_2S_2O_3{\cdot}5H_2O$ for 5–10 min; and rinsed with Milli-Q water 2× for 5 s.
7. The development is stopped with 50 % methanol (v/v) and 12 % (w/v) acetic acid for 10 min. The gel is washed in 50 % (v/v) methanol for 20 min and imaged.

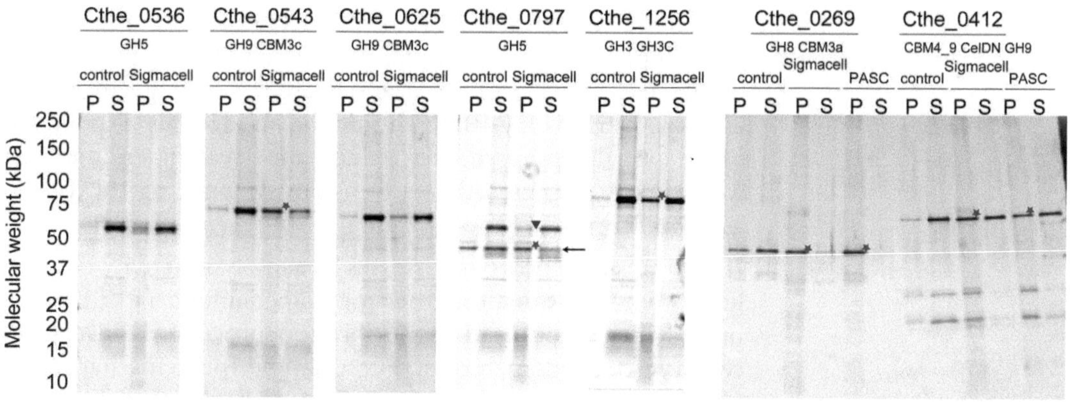

Fig. 5 Pull-down assays of enzymes using Sigmacell 20 as the binding substrate. Five *C. thermocellum* proteins that have reactivity with insoluble cellulose were produced in cell-free translation reactions, purified as described, and tested for the ability to bind to cellulose. After introduction of the cellulose to the translation mixture and incubation, the pull-down sample was separated into pellet (P) and supernatant (S) fractions by centrifugation. Proteins with cellulose binding capacity were enriched in the pellet fraction after centrifugation (*marked by stars*). Soluble and pellet fractions were treated with denaturing buffer, separated by gel electrophoresis, and visualized by silver staining. For each protein, the control lane provides information on translated protein that pelleted in the absence of cellulose. Cthe_0536 lacks a CBM domain and did not bind to cellulose, while Cthe_0543 contained a natural CBM3c domain and was bound. Cthe_0625 also contained a natural CBM3c domain, but did not bind to cellulose. This second CBM3c domain has a different primary sequence that the similarly named domain in Cthe_0543. Surprisingly, Cthe_0797 formed a soluble complex with an endogenous wheat germ protein (*marked by triangle*); this interaction was broken when Cthe_0797 bound to cellulose. The multi-domain protein Cthe_1256 also exhibited binding to cellulose. In this figure, Cthe_0269 and Cthe_0412 were obtained by recombinant *E. coli* protein expression system. The Cthe_0269 is engineered such that additional CBM3a domain derived from Cthe_3077 scaffoldin is present at C-terminus. A native form of Cthe_0412 possesses a CBM4_9 domain at N-terminus. Pull-down assays are carried out for testing their binding capacity to two different forms of cellulose, Sigmacell 20 and PASC. Results indicate that both enzymes bind to two forms of cellulose. Apparently, Cthe_0269 binds more tightly to substrates than Cthe_0412, indicating that CBM3a binds to cellulose material more effectively than CBM4_9 domain

4 Notes

1. The Criterion stain-free gel imaging system is used to obtain a gel image, and the presence of translated proteins is determined by visual inspection of lanes containing control extract and translation reactions. In the stain-free analysis, tryptophan residues in the separated protein react with a trihalo compound such as trichloroethanol included in the gel matrix [30], and a UV-induced reaction produces a derivative that can be detected by the fluorescence imaging system [31, 32]. The concentration of identified translated proteins was determined by the Criterion software based on evaluation of volume and area of pixels contained in the identified translated protein band relative to the same signals for molecular weight size markers of the same approximate size and number of tryptophan residues (Fig. 3).

2. Wheat germ extract 2240 (WEPRO2240) is used when protein purification is not planned. When His-tag purification of the translated protein is anticipated, wheat germ extract 2240H (WEPRO2240H) should be used. This extract has been specially prepared to reduce the level of endogenous wheat germ proteins that adventitiously bind to the metal-activated chromatography resin.
3. If the translated protein possesses a cellulolytic activity or a cellulose binding function, then polyethersulfone (PES) membranes should be used in purification procedures instead of cellulose-derived materials. For example, proteins harboring CBM domains may bind to cellulose-derived materials and be lost during handling.
4. AFEX-treated switchgrass was prepared as described elsewhere [33]. Samples of 1 mm ground material were used in assays. This material was the generous gift of Dr. Bruce Dale (Great Lakes Bioenergy Research Center, Michigan State University).
5. IL-treated switchgrass was prepared as described elsewhere [34]. This material was the generous gift of Dr. Masood Hadi (US Department of Energy Joint BioEnergy Institute, Emeryville, CA).
6. Cthe_0032 has a structure consisting of a signal peptide, one GH26 and two dockerin 1 domains (Fig. 2a). Microarray studies showed the enzyme was highly expressed [12], and mass spectral studies showed it is highly abundant in the secreted proteome [13]. The translocation signal peptide was identified in Uniprot A3DBE4, as calculated using SignalP. The presence of a signal peptide is not needed for cell-free translation. This information was used to define a truncation at the 5′ end of the gene. Likewise, the end of the GH26 domain was also identified from the Uniprot record. The truncation at the 3′ end of the gene was designed to remove the dockerin domain in linker region between the GH26 and dockerin domains. This trimming of the natural gene encodes a variant Cthe_0032 of 327 residues.
7. This protocol is based on the two-step PCR Flexi cloning method developed by the University of Wisconsin Center for Eukaryotic Structural Genomics to amplify genes [17]. Figure 2b shows how primer sequences are aligned with the Cthe_0032 gene. In this example, the original start codon of the Cthe_0032 gene is included. This is not necessary, however, as the vector provides the start codon needed for cell-free translation prior to the His6 tag (Fig. 1). The TEV protease recognition sequence is ENLYFQ/S, with proteolysis occurring between amino acids Q and S [35].
8. The plasmid pEU-HSBC is available by request from the NIH Protein Structure Initiative Materials Repository (http://psi-mr.asu.edu) and is used for translation of proteins containing an

N-terminal MGHHHHHHAIA tag. The wheat germ cell-free translation plasmid pEU-HSBC-CBM3a was created from pEU-HSBC and is used in the translation of proteins containing the abovementioned N-terminal tag and a C-terminal tag consisting of a 40-amino acid linker and a CBM3a domain.

9. X-ray diffraction analysis should be used to confirm that the lyophilized material obtained from this procedure is predominantly amorphous cellulose. A discussion of the method and analysis of X-ray powder diffraction of cellulose is found in [36].
10. Enzyme assays for *C. thermocellum* enzymes have been performed in the pH range of 6.0–8.0; all assays described here were carried out at pH 8.0.
11. The synergy factor for reactions containing two enzymes is the increase in reactivity observed from the combination relative to the result expected by arithmetic combination of the reactivity observed as separate enzymes. Figure 4c provides an example where an endocellulase and a xylanase were combined. This gives rise to increased reaction, i.e., synergy.
12. Our cumulative experience with wheat germ cell-free translation is that over 80 % of protein targets from a wide range of prokaryotic and eukaryotic sources are expressed at levels that can be detected using Coomassie-stained denaturing gel electrophoresis. This overall percentage of successful expression has been observed in other research labs [37]. Enzymes and proteins in the size range of $M_r \approx 34$–140 kDa have been produced using these methods.
13. Some posttranslational modifications are possible in wheat germ extract, such as the removal of N-terminal methionine, Ser/Thr/Tyr phosphorylation, and slow formation of disulfide bonds. In general, glycosylation is not observed because neither the glycosyltransferases nor the nucleotide sugar substrates are present.

Acknowledgments

This work was funded by the DOE Great Lakes Bioenergy Research Center (DOE BER Office of Science DE-FC02-07ER64494). The authors thank researchers in the University of Wisconsin Center for Eukaryotic Structural Genomics for advice and timely access to protocols and equipment (NIGMS U54 GM074901, U54 GM094584, U01 GM094622; J.L. Markley, G.N Phillips, B.G. Fox). The authors thank Dr. Paul Weimer (University of Wisconsin–Madison, Dept. of Bacteriology) for his scientific insights and the generous gifts of phosphoric acid-swollen cellulose and *C. thermocellum* genomic DNA, Dr. Bruce Dale for the generous gift of

AFEX-pretreated switchgrass, Dr. Masood Hadi for the generous gift of ionic liquid-pretreated switchgrass, and Dr. George Phillips and Dr. Chris Bianchetti (Great Lakes Bioenergy Research Center) for the X-ray diffraction analysis of some cellulose substrates.

References

1. Dale BE (2011) Cellulosic biofuels and the road to energy security. Environ Sci Technol 45:9823
2. Hall CAS, Klitgaard KA (2012) Energy and the wealth of nations: understanding the biophysical economy. Springer, Seacaucus, NJ 07096-2485. ISBN 978-1-4419-9397-7
3. Li C, Cheng G, Balan V et al (2011) Influence of physico-chemical changes on enzymatic digestibility of ionic liquid and AFEX pretreated corn stover. Bioresour Technol 102: 6928–6936
4. Merino ST, Cherry J (2007) Progress and challenges in enzyme development for biomass utilization. Adv Biochem Eng Biotechnol 108: 95–120
5. Lynd LR, Weimer PJ, van Zyl WH et al (2002) Microbial cellulose utilization: fundamentals and biotechnology. Microbiol Mol Biol Rev 66:506–577
6. Cantarel BL, Coutinho PM, Rancurel C, Bernard T, Lombard V, Henrissat B (2009) The Carbohydrate-Active EnZymes database (CAZy): an expert resource for glycogenomics. Nucleic Acids Res 37:D233–D238
7. Vuong TV, Wilson DB (2010) Glycoside hydrolases: catalytic base/nucleophile diversity. Biotechnol Bioeng 107:195–205
8. Gao D, Uppugundla N, Chundawat SP et al (2011) Hemicellulases and auxiliary enzymes for improved conversion of lignocellulosic biomass to monosaccharides. Biotechnol Biofuels 4:5
9. Demain AL, Newcomb M, Wu JH (2005) Cellulase, clostridia, and ethanol. Microbiol Mol Biol Rev 69:124–154
10. Gilbert HJ (2007) Cellulosomes: microbial nanomachines that display plasticity in quaternary structure. Mol Microbiol 63:1568–1576
11. Bayer EA, Lamed R, White BA et al (2008) From cellulosomes to cellulosomics. Chem Rec 8:364–377
12. Riederer A, Takasuka TE, Makino S et al (2011) Global gene expression patterns in *Clostridium thermocellum* as determined by microarray analysis of chemostat cultures on cellulose or cellobiose. Appl Environ Microbiol 77:1243–1253
13. Raman B, Pan C, Hurst GB et al (2009) Impact of pretreated switchgrass and biomass carbohydrates on *Clostridium thermocellum* ATCC 27405 cellulosome composition: a quantitative proteomic analysis. PLoS One 4:e5271
14. Schwarz WH (2001) The cellulosome and cellulose degradation by anaerobic bacteria. Appl Microbiol Biotechnol 56:634–649
15. Mingardon F, Chanal A, Lopez-Contreras AM et al (2007) Incorporation of fungal cellulases in bacterial minicellulosomes yields viable, synergistically acting cellulolytic complexes. Appl Environ Microbiol 73:3822–3832
16. Morais S, Barak Y, Caspi J et al (2010) Cellulase-xylanase synergy in designer cellulosomes for enhanced degradation of a complex cellulosic substrate. MBio 1
17. Blommel PG, Martin PA, Wrobel RL et al (2006) High efficiency single step production of expression plasmids from cDNA clones using the Flexi Vector cloning system. Protein Expr Purif 47:562–570
18. Sawasaki T, Ogasawara T, Morishita R et al (2002) A cell-free protein synthesis system for high-throughput proteomics. Proc Natl Acad Sci U S A 99:14652–14657
19. Gallie DR, Sleat DE, Watts JW et al (1987) The 5′-leader sequence of tobacco mosaic virus RNA enhances the expression of foreign gene transcripts in vitro and in vivo. Nucleic Acids Res 15:3257–3273
20. Gay P, Le Coq D, Steinmetz M et al (1985) Positive selection procedure for entrapment of insertion sequence elements in gram-negative bacteria. J Bacteriol 164:918–921
21. Blommel PG, Fox BG (2007) A combined approach to improving large-scale production of tobacco etch virus protease. Protein Expr Purif 55:53–68
22. Beebe ET, Makino S, Nozawa A et al (2011) Robotic large-scale application of wheat cell-free translation to structural studies including membrane proteins. N Biotechnol 28:239–249
23. Vantilbeurgh H, Claeyssens M, Debruyne CK (1982) The use of 4-methylumbelliferyl and other chromophoric glycosides in the study of cellulolytic enzymes. FEBS Lett 149:152–156

24. Potier M, Mameli L, Belisle M et al (1979) Fluorometric assay of neuraminidase with a sodium(4-methylumbelliferyl-alpha-D-N-acetylneuraminate) substrate. Anal Biochem 94:287–296
25. Chundawat SP, Balan V, Dale BE (2008) High-throughput microplate technique for enzymatic hydrolysis of lignocellulosic biomass. Biotechnol Bioeng 99:1281–1294
26. Sharrock KR (1988) Cellulase assay methods: a review. J Biochem Biophys Methods 17:81–105
27. Miller GL (1959) Use of dinitrosalicylic acid reagent for determination of reducing sugar. Anal Chem 31:426–428
28. Shoseyov O, Shani Z, Levy I (2006) Carbohydrate binding modules: biochemical properties and novel applications. Microbiol Mol Biol Rev 70:283–295
29. Sawasaki T, Hasegawa Y, Tsuchimochi M et al (2002) A bilayer cell-free protein synthesis system for high-throughput screening of gene products. FEBS Lett 514:102–105
30. Edwards RA, Jickling G, Turner RJ (2002) The light-induced reactions of tryptophan with halocompounds. Photochem Photobiol 75: 362–368
31. Kazmin D, Edwards RA, Turner RJ et al (2002) Visualization of proteins in acrylamide gels using ultraviolet illumination. Anal Biochem 301:91–96
32. Ladner CL, Yang J, Turner RJ et al (2004) Visible fluorescent detection of proteins in polyacrylamide gels without staining. Anal Biochem 326:13–20
33. Balan V, Bals B, Chundawat SP et al (2009) Lignocellulosic biomass pretreatment using AFEX. Methods Mol Biol 581:61–77
34. Li C, Knierim B, Manisseri C, Arora R et al (2010) Comparison of dilute acid and ionic liquid pretreatment of switchgrass: biomass recalcitrance, delignification and enzymatic saccharification. Bioresour Technol 101: 4900–4906
35. Kapust RB, Tozser J, Copeland TD et al (2002) The P1′ specificity of tobacco etch virus protease. Biochem Biophys Res Commun 294: 949–955
36. Park S, Baker JO, Himmel ME et al (2010) Cellulose crystallinity index: measurement techniques and their impact on interpreting cellulase performance. Biotechnol Biofuels 3:10
37. Goshima N, Kawamura Y, Fukumoto A et al (2008) Human protein factory for converting the transcriptome into an in vitro-expressed proteome. Nat Methods 5:1011–1017
38. Weimer PJ, Lopez-Guisa JM, French AD (1990) Effect of cellulose fine structure on kinetics of its digestion by mixed ruminal microorganisms in vitro. Appl Environ Microbiol 56:2421–2429
39. Finn RD, Mistry J, Tate J et al (2010) The Pfam protein families database. Nucleic Acids Res 38:D211–D222

Chapter 6

Cloning-Independent Expression and Screening of Enzymes Using Cell-Free Protein Synthesis Systems

Yong-Chan Kwon, Jae-Kwang Song, and Dong-Myung Kim

Abstract

We present a strategy for expression and screening of microbial enzymes without involving cloning procedures. Libraries of putative ω-transaminases (ω-TA) and mutated *Candida antarctica* lipase B (CalB) are PCR-amplified from bacterial colonies and directly expressed in an *Escherichia coli*-based cell-free protein synthesis system. The open nature of cell-free protein synthesis system also allows streamlined analysis of the enzymatic activity of the expressed enzymes, which greatly shortens the time required for enzyme screening.

We expect that the proposed strategy will provide a universal platform for bridging the information gap between nucleotide sequence and protein function, in order to accelerate the discovery of novel enzymes. The proposed strategy can also serve as a viable option for the rapid and precise tuning of enzyme molecules, not only for analytical purposes, but also for industrial applications. This is accomplished via large-scale production using microbial cells transformed with variant genes selected from the cell-free expression screening.

Key words Enzyme engineering, Combinatorial mutagenesis, Cell-free protein synthesis, Hot spots

1 Introduction

Recent advances in genome sequencing technology have accumulated enormous amounts of sequence information [1]. Although protein function encoded in nucleotide sequences can be annotated using computational alignment tools, in many cases, significant similarity to proteins with known function is hard to establish [2, 3]. To understand the biological function of these unknown proteins, as well as to validate the computer-annotated results, efficient methods that enable rapid translation of genetic information into protein function are highly demanded. The availability of high-throughput methods for protein generation is also essential for accelerating the discovery and evolution of biocatalysts [4–9] used in industry. While gene cloning and cultivation of transformed cells has long been used as a standard method for production of recombinant proteins, the vast amount of sequence information from various genome sequencing projects now demands a

Kirill Alexandrov and Wayne A. Johnston (eds.), *Cell-Free Protein Synthesis: Methods and Protocols*, Methods in Molecular Biology, vol. 1118, DOI 10.1007/978-1-62703-782-2_6, © Springer Science+Business Media, LLC 2014

throughput of protein expression that exceeds that of the present in vivo expression techniques.

Although the expression and screening of genetic libraries are generally carried out in microbial cells [10, 11], the use of living cells sets intrinsic limitations in numerous aspects. For example, screening protocols should be designed within the boundaries of physiological conditions because cell-based screening is dependent upon cell viability, including that the target enzymes and substrates should not interfere with normal metabolism of the cells. In addition, even when the enzymes are functionally expressed without inhibiting cell growth, the substrates for enzyme screening should be chosen from a narrow range of candidates that are transportable across the cell membrane. Such limitations associated with cell-based expression screening can be effectively overcome by expressing enzyme libraries using the cell-free protein synthesis system.

Compared to cell-based gene expression, cell-free protein synthesis offers substantial advantages in the speed and flexibility for the simultaneous expression of multiple proteins [12–17]. As a part of our efforts to extend the application of cell-free protein synthesis into the field of enzyme technology, we report in this chapter an integrated methodology for fast expression screening of enzymes using ω-transaminases (ω-TA) and *Candida antarctica* lipase B (CalB) as model enzymes.

ω-TA genes from microbial colonies are amplified by PCR and directly expressed in a cell-free protein synthesis system. Expressed enzymes are screened for their activity toward different amine donors by colorimetric measurement of the changes in the concentration of pyruvate, which was used as a common amine acceptor. For the screen of optimized CalB mutants, the targeted sites of CalB are randomized and the resulting library expressed and screened in a cell-free system. After being amplified by colony PCR, the variant genes are directly expressed in a cell-free protein synthesis system prepared in microtiter plates.

As a result, analysis of the substrate specificity of the enzymes encoded in 11 ω-TA genes toward 16 amine-donating compounds and expression and analysis of 1,000 variant genes of CalB can be completed within a matter of hours, and we are able to identify a number of ω-TA enzyme–substrate matches and to select CalB variants that complied with screening criteria in different dimensions including substrate specificity and thermal stability.

2 Materials

2.1 *E. coli* Cell Extract Preparation

1. BL21-Star™ (DE3) strain (Invitrogen).
2. 2×YT medium.
3. Fermenter.
4. French Press Cell or comparable cell-disrupting devices.

2.2 Preparation of PCR-Amplified Expression Templates

1. The primers for colony PCR, second-round PCR using the MEGA primers flanking the T7 promoter, ribosome binding site and T7 terminator, point mutation, and randomized primers (*see* Table 1).
2. DNA polymerase.

Table 1
Oligonucleotide primers used in this study

Gene name	Orientation	Sequence (5′–3′)
ω-TA *At*	Forward Reverse	AAGAAGGAGATATACATATGACCATCCTGCCCAATTCC TAATGATGATGATGATGATGGCCCCTTGCAGCCAGC
ω-TA *Rs*	Forward Reverse	AAGAAGGAGATATACATATGCAACCATCACATGGCCTTA TAATGATGATGATGATGATGGACAAGCCCTTCGGCCC
ω-TAp *Ml0107*	Forward Reverse	AAGAAGGAGATATACATATGTCGAAGACCCAGACCGC TAATGATGATGATGATGATGCCCCCGGCCTTCGACC
ω-TAp *Ml1207*	Forward Reverse	AAGAAGGAGATATACATATGACCTATCAGAATTATTCGCTGA TAATGATGATGATGATGATGCCGCTTGCGCAGATCGG
ω-TAp *Ml1632*	Forward Reverse	AAGAAGGAGATATACATATGTCCAACCGGCTGAAAGTC TAATGATGATGATGATGATGGTCGATCGACCGCAGCAC
ω-TAp *Ml5987*	Forward Reverse	AAGAAGGAGATATACATATGAGGCACCAGATCCTGTC TAATGATGATGATGATGATGGAGTTCCGCAACGGCCG
ω-TAp *Ml5990*	Forward Reverse	AAGAAGGAGATATACATATGAGCCACGTATTTCATCGTTC TAATGATGATGATGATGATGGCCAAGCGCCGCATCAATG
ω-TAp *Ml6101*	Forward Reverse	AAGAAGGAGATATACATATGCGGTTGAGCAATCTGGG TAATGATGATGATGATGATGTATCGCACCCTCGGCCTT
ω-TAp *Ml6963*	Forward Reverse	AAGAAGGAGATATACATATGGTTGAACCTGCTCTCTCG TAATGATGATGATGATGATGTGCCGGCTTCCAAAGTCCC
ω-TAp *Ml7037*	Forward Reverse	AAGAAGGAGATATACATATGCTCGCCAATTCCCTGATC TAATGATGATGATGATGATGTGCCAGAGCCTGGCGGA
ω-TAp *Ml7127*	Forward Reverse	AAGAAGGAGATATACATATGCTCAACCAGTCCAACGAAC TAATGATGATGATGATGATGGAGATTGGCAAACACACTCTTC
MEGA[a]	Forward	TCGATCCCGCGAAATTAATACGACTCACTATAGGGAGA CCACAACGGTTTCCCTCTAGAAATAATTTTGTTTAAC TTTAAGAAGGAGATATACATATG
MEGA	Reverse	CAAAAAACCCCTCAAGACCCGTTTAGAGGCCCCAA GGGGTTATATGTCGACCTCGAGTTATTAATGATGA TGATGATGATG
T7 15UP	Forward	TCGATCCCGCGAAATTAATACGACTCACTATAGG
GTB	Reverse	CAAAAAACCCCTCAAGACCCGTTTA
210-1st	Reverse	CCCGCAAACAGCCTGTGCCTG**GAT**ATTTTTGCCGTT AAACAGATA
210-2nd	Forward	CAGGCACAGGCTGTTTGCGGCC

(continued)

Table 1 (continued)

Gene name	Orientation	Sequence (5′–3′)
219-1st	Reverse	CCGGCATGGTCGATGACGAA**CTG**TGGCCCGCAAACA GCCTGTG
219-2nd	Forward	TTCGTCATCGACCATGCCGG
221-1st	Reverse	GGCCCGCAAACAGCCTGTGCCTG
221-2nd	Forward	CAGGCACAGGCTGTTTGCGGGCCATTATTC**GAC**ATCG ACCATGCCGGT
278-1st	Reverse	ATCGCGGCAGCGGCCGGCGC**CGG**CAAAGCCGCAG CCGCTACCT
278-2nd	Forward	GCGCCGGCCGCTGCCGCGAT
281-1st	Reverse	ACCTGCCACAATCGCGGCAGC**TTC**CGGCGCTAAC AAAGCCGCAGC
281-2nd	Forward	GCTGCCGCGATTGTGGCAGGTCCT
278/281-1st	Reverse	CAAAGCCGCAGCCGCTAC
278/281-2nd	Forward	GTAGCGGCTGCGGCTTTG**CCG**GCGCCG**GAA**GCTGCCG CGATTGTGGCAGGT
221/278/281NNN-1st	Reverse	GGCCCGCAAACAGCCTGTGCCTG
221/278/281NNN-2nd	Forward	CAGGCACAGGCTGTTTGCGGGCCATTATTC**NNN**ATCGA CCATGCCGGT
221/278/281NNN-2nd	Reverse	CAAAGCCGCAGCCGCTAC
221/278/281NNN-3rd	Forward	AAGGTAGCGGCTGCGGCTTTG**NNN**GCGCCG**NNN**GCTG CCGCGATTGTGGCAGGT

Underline: polyhistidine tag (5′–3′)
Each number of primer was used to mutation at amino acids of numbering position in the CalB gene
Bold fonts indicate mutation sites on CalB
[a]MEGA primers were used for adding 5′ and 3′-UTR regions

3. 96-well plate.
4. Colony picker.
5. Liquid handling robot system.
6. PCR instrument.

2.3 Cell-Free Protein Synthesis

1. Reaction mixture for cell-free protein synthesis reaction: 57 mM Hepes-KOH (pH 8.2), 1.2 mM ATP, 0.85 mM each of CTP, GTP, and UTP, 2 mM DTT, 0.17 mg/mL *E. coli* total tRNA mixture (from strain MRE600), 0.64 mM cAMP, 90 mM potassium glutamate, 80 mM ammonium acetate, 12 mM magnesium acetate, 34 μg/mL L-5-formyl-5, 6, 7,

8-tetrahydrofolic acid (folinic acid), 2 mM each of 20 amino acids, 2 % polyethylene glycol (PEG) 8000, 67 mM creatine phosphate, 9.6 μg/mL creatine kinase, 0.01 μM L-[U-^{14}C] leucine (11.3 GBq/mmol), and 27 % (v/v) of S12 extract. Store on ice.

2. 0.1 M NaOH solution.
3. 5 % trichloroacetic acid.
4. 100 % ethanol.
5. Filter discs.
6. Infrared lamp.
7. Liquid scintillation counter.

2.4 Colorimetric Analysis of ω-TA Activity

1. 16 amine donors: S01, α-methylbenzylamine; S02, 1-phenylpropylamine; S03, benzylamine; S04, 3-phenyl-1-propylamine; S05, 4-phenylbutylamine; S06, 1-aminoindan; S07, ethylamine; S08, propylamine; S09, butylamine; S10, amylamine; S11, isopropylamine; S12, *sec*-butylamine; S13, β-alanine; S14, 3-amino-n-butyric acid; S15, phenylalanine; S16, 3-amino-3-phenylporpionic acid. Store at 4 °C.
2. 100 mM sodium pyruvate as amine acceptor and 500 μM pyridoxal-5′-phosphate are dissolved in water. Store at −20 °C.
3. 50 mM Tris–HCl buffer (pH 7.2).
4. 0.05 g 2,4-dinitrophenylhydrazine is dissolved in 2 N HCl and 4 N NaOH is dissolved in water. Store at room temperature.
5. Plate reader.

2.5 Determination of Hydrolytic Activity of the CalB Variants

1. 50 mM *para*-nitropalmitate and 50 mM tributyrin.
2. Sample dilution buffer: 14 mM magnesium acetate, 60 mM potassium acetate, and 10 mM Tris–acetate (pH 8.2).
3. *p*NPP assay buffer: 0.1 mM *p*NPP, 4 % (v/v) ethanol, 50 mM Tris–HCl buffer (pH 8.0).
4. Sonifier (Branson).
5. Tributyrin assay buffer: 0.8 % (v/v) emulsified tributyrin and 0.1 mM of bromothymol blue (BTB) in 5 mM Tris–HCl (pH 8.0).

2.6 Analysis of Wax Ester Synthesis Activity of the CalB Variants

A gas chromatograph with the following characteristics is required:

1. Column, capillary SPB-5 (20 m × 0.32 mm × 1.0 μm) (Supelco).
2. Detector, flame ionization.
3. Carrier gas, hydrogen at a flow rate of 1.3 mL/min.
4. Temperature program, 100–260 °C at 15 °C/min to 315 °C at 5 °C/min (hold 15 min).
5. Detector temperature, 340 °C.

3 Methods

3.1 *E. coli* Cell Extract Preparation

1. 5 mL of LB medium is inoculated with *E. coli* BL21 Star™ (DE3) cell and incubate for 8 h at 37 °C.
2. The culture is transferred to 40 mL of 2×YT medium and grown overnight.
3. The culture is transferred to a fermenter prepared with 4 L of 2×YT medium and grown at 37 °C with vigorous agitation and aeration.
4. Isopropyl thiogalactopyranoside (IPTG, 1 mM) is added to the cell culture media when the cell density reaches 0.5–0.6.
5. Cells are harvested at a mid-log phase by centrifugation at 7,000 RCF for 20 min (*see* **Note 1**).
6. Harvested cells are washed three times in 20 mL of buffer A (10 mM tris–acetate buffer (pH 8.2), 14 mM magnesium acetate, 60 mM potassium glutamate, 1 mM dithiothreitol (DTT), and 0.05 % (v/v) 2-mercaptoethanol (2-ME)) per gram of wet cell paste (*see* **Note 2**).
7. Washed cells are resuspended in buffer B (buffer A without 2-ME) per 10 g of wet cell paste and disrupted in a French Press Cell at a constant pressure of 20,000 psi (*see* **Note 3**).
8. The lysate is centrifuged at 30,000 RCF at 4 °C for 30 min.
9. The supernatant is carefully recovered not to include lipid in the top layer and centrifuged again.
10. The final supernatant is incubated at 37 °C for 30 min.
11. Small aliquots of the supernatant are snap-frozen in liquid nitrogen and stored at −80 °C.

3.2 Quantification of Cell-Free Synthesized Proteins

1. After a cell-free protein synthesis reaction, 5 μL of reaction sample is added to 100 μL of 0.1 N NaOH solutions.
2. After 1 h, the entire solution is spotted on a filter disc and dried under an infrared lamp.
3. Dried filter disc is immersed in 5 % trichloroacetic acid solution for 45 min. This step should be carried out in an ice bath.
4. Disc is washed with 100 % ethanol and dried under an infrared lamp.
5. The radioactivity of the proteins precipitated on a filter disc is determined using a liquid scintillation counter.

3.3 Colony PCR and Cell-Free Expression of ω-TAs

1. Bacterial colonies of *Agrobacterium tumefaciens* str. C58, *Rhodobacter sphaeroides* ATCC 17025, and *Mesorhizobium loti* MAFF303099 were prepared following standard procedures.
2. Single colonies are picked and put into PCR tubes that contain PCR reaction mixture.

3. PCR reaction is conducted under the following conditions. First-round PCR: 95 °C 30 s, 55 °C 30 s, 72 °C 45 s, 25 cycles. Primers for the first-round PCR are listed in Table 1. After the first-round PCR, PCR products are purified by gel extraction method. Second-round PCR is conducted using the MEGA primers listed in Table 1: 95 °C 30 s, 58 °C 30 s, 72 °C 60 s, 30 cycles.
4. 5–10 μL of PCR products are added to the cell-free reaction mixture in 96-well plates.
5. The plates are sealed with a plastic film to prevent evaporation.
6. Cell-free protein synthesis reaction is conducted by incubating the plate at 37 °C for 3 h (*see* **Note 4**).

3.4 Cell-Free Expression of CalB Variants

1. The library of variant genes carrying randomized codons at the predetermined locations was constructed by overlap-extension PCR of the plasmid pK7#33CalB [18] using the degenerate primers shown in Table 1. PCR products were digested with NdeI/SalI, cloned into the pK7 plasmid and transformed into *E. coli* JM109 competent cells.
2. PCR reaction is conducted under the following conditions. First-round PCR: 95 °C 30 s, 55 °C 30 s, 72 °C 60 s, 25 cycles. Primers for the first-round PCR are listed in Table 1. After the first-round PCR, PCR products are purified by gel extraction method. Second-round PCR is conducted using the MEGA primers listed in Table 1: 95 °C 30 s, 58 °C 30 s, 72 °C 60 s, 30 cycles.
3. 5–10 μL of PCR products are added to the cell-free reaction mixture in 96-well plates.
4. The plates are sealed with a plastic film to prevent evaporation.
5. Cell-free protein synthesis reaction is conducted by incubating the plate at 37 °C for 3 h.

3.5 Screening of ω-TAs for Different Amine Donors

1. The amine transfer reaction is initiated by adding 50 μL of assay mixture (50 mM Tris–HCl buffer, pH 7.2, 10 mM sodium pyruvate, 10 mM each of the amine donors, and 20 μM pyridoxal-5′-phosphate) to the completed cell-free synthesis reactions (50 μL) in a 96-well plate.
2. After incubation at 37 °C for 3 h, the assay mixture is diluted with an equal volume of distilled water.
3. 100 μL of the diluted solution is transferred to a fresh plate containing 50 μL of 0.5 mM 2,4-dinitrophenylhydrazine (DNP). Yellow precipitate of pyruvate-dinitrophenylhydrazone (PA-DNPH) derivative is formed instantly.

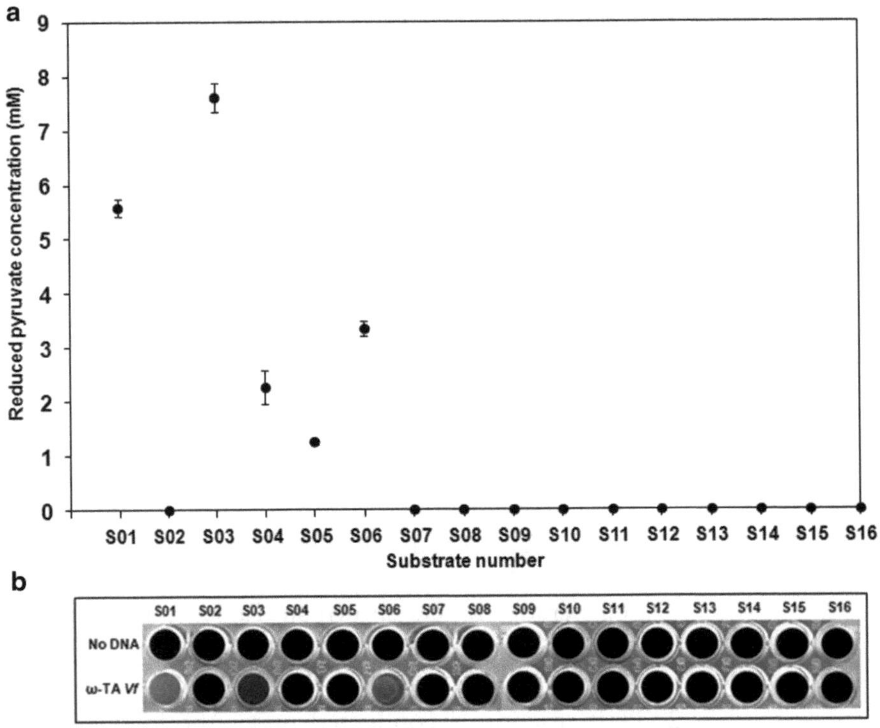

Fig. 1 Enzymatic activity of cell-free synthesized ω-TA Vf. (**a**) Reactivity of 16 amine donors toward *Vibrio fluvialis* ω-TA. Reduced amounts of pyruvate concentration after the amine transfer reactions are plotted. The *insets* show HPLC traces of acetophenone and benzaldehyde after termination of transamination reaction by use of *Vibrio fluvialis* ω-TA against α-methylbenzylamine and benzylamine. *mAU* milliabsorbance units. (**b**) Photo image of the assay plate after the addition of DNP and NaOH. ω-TA *Vf*, ω-TA from *Vibrio fluvialis*

4. The absorbance at 450 nm is measured to determine the amount of residual pyruvate in each well (Fig. 1; *see* **Note 5**).
5. Developed color is further intensified for visual comparison of reactions by adding 100 μL of 4 N NaOH solutions to each well [19, 20].

3.6 Screening for Improved Hydrolytic Activity of Cell-Free Synthesized CalB Using pNPP

1. The reaction samples are diluted 200-fold in a dilution buffer (14 mM magnesium acetate, 60 mM potassium acetate, and 10 mM Tris–acetate (pH 8.2)).
2. 10 μL of the diluted solution is added to 240 μL of assay buffer (0.1 mM *p*NPP, 4 % (v/v) ethanol, and 50 mM Tris–HCl buffer, pH 8.0). The assay buffer is placed to 37 °C incubator for pre-warming.
3. The time-dependent change of absorbance is monitored at 405 nm (Fig. 2a).

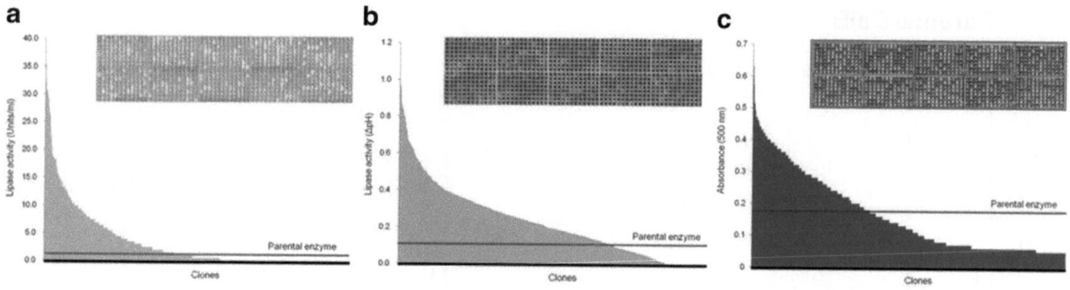

Fig. 2 Colorimetric screening of cell-free synthesized variant CalB library. (**a**) Hydrolytic activity toward pNPP. The variant enzymes prepared from the PCR products of 1,000 variant genes were screened for pNPP hydrolysis; a wide distribution of hydrolytic activity was observed. In particular, five variant clones (clone IDs 230, 463, 577, 867, and 922) exhibited more than 30-fold higher activity than the parental enzyme. (**b**) Hydrolytic activity toward tributyrin. Similar to the case of pNPP hydrolysis assay, the variant enzymes exhibited over a 900-fold range of activity with a few clones showing more than 12-fold enhancement compared to the parental enzyme. (**c**) Wax ester synthesis. After 1 h incubation, the parental CalB enzyme showed the absorbance value of 0.203 ± 0.018. Although approximately two thirds of the CalB variants gave turbidity change less than the parental enzyme, more than twofold increase in the absorbance value was observed from 50 CalB variants. The *arrows* indicate the activity of parental enzyme

3.7 Screening for Improved Hydrolytic Activity of Cell-Free Synthesized CalB Using Tributyrin

1. Tributyrin is emulsified in 5 mM Tris–HCl (pH 8.0) using a sonifier.
2. A total of 10 μL of the diluted reaction sample is added to 240 μL of assay buffer containing 0.8 % (v/v) emulsified tributyrin and 0.1 mM of bromothymol blue (BTB) in 5 mM Tris–HCl (pH 8.0).
3. Progress of the hydrolysis of tributyrin is monitored by colorimetric measurement of pH change (OD_{595}) during the incubation of the assay mixture at 37 °C.

3.8 CalB Variants for Wax Ester Synthesis

1. 200 μL of the tenfold diluted solution of the cell-free synthesized CalB is transferred to individual wells of 96-well plates.
2. Each of the diluted CalB variant solution is overlaid with 100 μL mixture of the same molar amounts of oleic acid and dodecanol.
3. Two-phase reaction for synthesis of dodecyl oleate is conducted at 25 °C with constant shaking at 600 rpm.
4. The turbidity of each well is measured at 500 nm (Fig. 2c).
5. The turbidity data may also be verified with gas chromatography (Fig. 3).
6. The lower phase of two-phase reaction is removed with a pipette and evaporated to dryness.
7. The residue is then dissolved in chloroform and analyzed as described in Subheading 2.

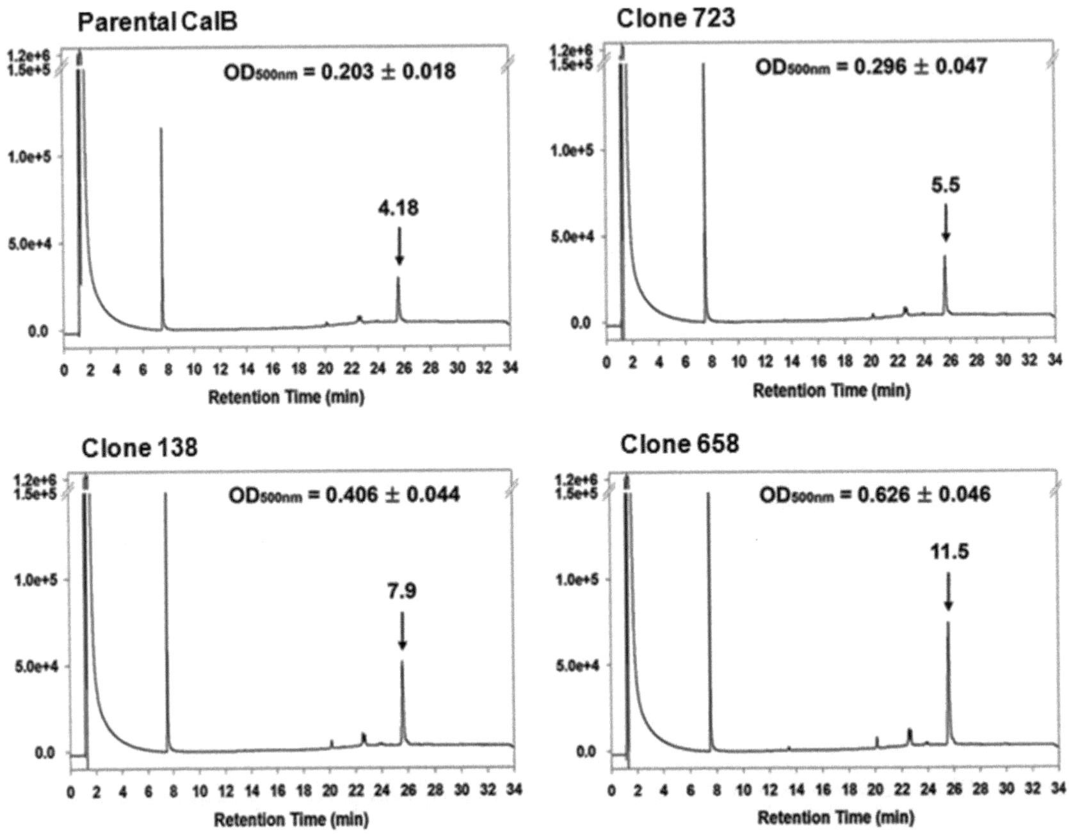

Fig. 3 Correlation of spectrophotometric and gas chromatographic data for wax synthesis. When the aqueous phase of three reactions of them (clone IDs 138, 658, and 723) was withdrawn and analyzed by gas chromatography, the relative peak area of the dodecyl oleate showed good correlation with the observed turbidity change. The areas of the wax peaks were plotted against the turbidity results (OD_{500})

4 Notes

1. In our experience, extracts prepared from the cells harvested at 3.5–5.0 OD_{600} shows best translational activity.
2. The cell pellet can be snap-frozen in liquid nitrogen and stored in a −80 °C freezer at this stage, although streamlined process is recommended.
3. Single passage is recommended. Repeated passages tend to decrease the translational activity of the cell extract.
4. For the expression of aggregation-prone proteins, it is often desirable to prepare molecular chaperone-enriched extracts by overexpressing molecular chaperones in the *E. coli* cells prior to extract preparation. For instance, soluble expression of the ω-transaminases was substantially enhanced by using an extract enriched with GroEL/ES and GrpE (Fig. 4).

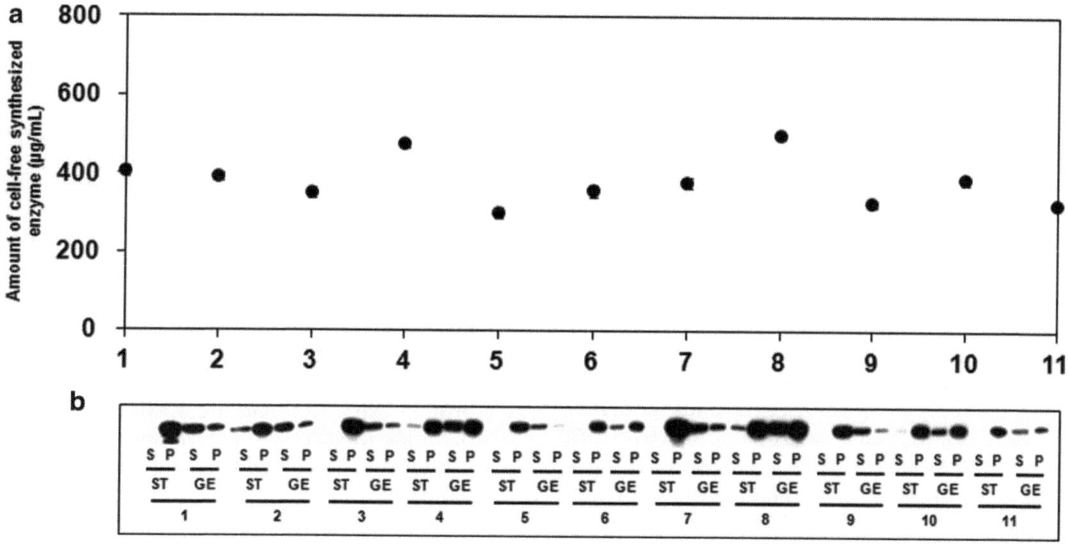

Fig. 4 Expression level and solubility of cell-free synthesized ω-TAs. (**a**) Expression level of individual ω-TAs determined by the TCA-insoluble radioactivity. Amount of cell-free synthesized enzyme is the mean value from the each column the microtiter plates. It was estimated that 300.7 (±13.48) to 500.6 (±9.41) μg/mL of the encoded enzymes were produced after incubation. (**b**) Molecular chaperone-assisted enhancement of the solubility of cell-free synthesized ω-TAs. The relative amounts of the soluble enzymes were markedly improved by using the GroEL/ES-enriched S12 extract. The reaction mixture was centrifuged at 15,000 × *g* for 10 min, and the soluble and pellet fractions were analyzed by Western blot on 12 % tricine–SDS–PAGE gel. *S* soluble fraction, *P* insoluble fraction, *ST* standard S12 extract, *GE* GroEL/ES-enriched S12 extract. Numbers: 1, ωTA *At*; 2, ωTA *Rs*; 3, ωTAp *Ml*0107; 4, ωTAp *Ml*1207; 5, ωTAp *Ml*1632; 6, ωTAp *Ml*5987; 7, ωTAp *Ml*5990; 8, ωTAp *Ml*6101; 9, ωTAp *Ml*6963; 10, ωTAp *Ml*7037; 11, ωTAp *Ml*7127

Table 2
Activity comparison of *Vibrio fluvialis* ω-TA by HPLC and colorimetric method

Amine donor	Acetophenone or benzaldehyde converted from the amine donor (mM)	Reduced amount of pyruvate[a] (mM)
α-Methylbenzylamine	5.88 ± 0.32[b]	5.60 ± 0.46
Benzylamine	8.30 ± 0.55[c]	7.60 ± 0.68

[a]Reduced amount of pyruvate after the amine transfer reactions by the colorimetric assay
[b]Amount of acetophenone converted from α-methylbenzylamine by the HPLC assay
[c]Amount of benzaldehyde converted from benzylamine by the HPLC assay

5. Enzymatic activity of a ω-transaminase can also be determined by HPLC analysis of acetophenone or benzaldehyde converted from the amine donor. In our experiments using *Vibrio fluvialis*, the colorimetric assay and HPLC assay showed good correlations (Table 2).

References

1. Mardis ER (2008) The impact of next-generation sequencing technology on genetics. Trends Genet 24:133–141
2. Gorbalenya AE, Koonin EV, Lai MM (1991) Putative papain-related thiol proteases of positive-strand RNA viruses. Identification of rubi- and aphthovirus proteases and delineation of a novel conserved domain associated with proteases of rubi-, alpha- and coronaviruses. FEBS Lett 288:201–205
3. Schmidt AJ, Ryjenkov DA, Gomelsky M (2005) The ubiquitous protein domain EAL is a cyclic diguanylate-specific phosphodiesterase: enzymatically active and inactive EAL domains. J Bacteriol 187:4774–4781
4. Entzeroth M (2003) Emerging trends in high-throughput screening. Curr Opin Pharmacol 3:522–529
5. Goddard JP, Reymond JL (2004) Recent advances in enzyme assays. Trends Biotechnol 22:363–370
6. Hertzberg RP, Pope AJ (2000) High-throughput screening: new technology for the 21st century. Curr Opin Chem Biol 4:445–451
7. Pereira DA, Williams JA (2007) Origin and evolution of high throughput screening. Br J Pharmacol 152:53–61
8. Turner NJ (2003) Directed evolution of enzymes for applied biocatalysis. Trends Biotechnol 21:474–478
9. Wahler D, Reymond JL (2001) High-throughput screening for biocatalysts. Curr Opin Biotechnol 12:535–544
10. Hibbert EG, Dalby PA (2005) Directed evolution strategies for improved enzymatic performance. Microb Cell Fact 4:29
11. Ottosson J, Rotticci-Mulder JC, Rotticci D et al (2001) Rational design of enantioselective enzymes requires considerations of entropy. Protein Sci 10:1769–1774
12. Jewett MC, Calhoun KA, Voloshin A et al (2008) An integrated cell-free metabolic platform for protein production and synthetic biology. Mol Syst Biol 4:220
13. Katzen F, Chang G, Kudlicki W (2005) The past, present and future of cell-free protein synthesis. Trends Biotechnol 23: 150–156
14. Murthy TV, Wu W, Qiu QQ et al (2004) Bacterial cell-free system for high-throughput protein expression and a comparative analysis of *Escherichia coli* cell-free and whole cell expression systems. Protein Expr Purif 36: 217–225
15. Sawasaki T, Ogasawara T, Morishita R et al (2002) A cell-free protein synthesis system for high-throughput proteomics. Proc Natl Acad Sci U S A 99:14652–14657
16. Shimizu Y, Kuruma Y, Ying BW et al (2006) Cell-free translation systems for protein engineering. FEBS J 273:4133–4140
17. Spirin AS (2004) High-throughput cell-free systems for synthesis of functionally active proteins. Trends Biotechnol 22:538–545
18. Park CG, Kim TW, Oh IS et al (2009) Expression of functional Candida antarctica lipase B in a cell-free protein synthesis system derived from Escherichia coli. Biotechnol Prog 25:589–593
19. Crowley LV (1967) The Reitman-Frankel colorimetric transaminase procedure in suspected myocardial infarction. Clin Chem 13: 482–487
20. Reitman S, Frankel S (1957) A colorimetric method for the determination of serum glutamic oxaloacetic and glutamic pyruvic transaminases. Am J Clin Pathol 28:56–63

Chapter 7

High-Level Cell-Free Production of Membrane Proteins with Nanodiscs

Christian Roos, Lei Kai, Stefan Haberstock, Davide Proverbio, Umesh Ghoshdastider, Yi Ma, Slawomir Filipek, Xiaoning Wang, Volker Dötsch, and Frank Bernhard

Abstract

This chapter addresses two major bottlenecks in cell-free membrane protein production. Firstly, we describe the optimization of expression templates for obtaining membrane proteins in preparative scales. We present details for a newly established tag variation screen providing high success rates in improving expression efficiencies while having only minimal impacts on the target protein structure. Secondly, we present protocols for the efficient co-translational insertion of membrane proteins into defined lipid bilayers. We describe the production of nanodiscs and their implementation into cell-free expression reactions for the co-translational reconstitution of membrane proteins. In addition we give guidelines for the loading of nanodiscs with different lipids in order to systematically analyze effects of lipids on the translocation, functional folding, and stability of cell-free expressed membrane proteins.

Key words Nanodiscs, Tag variation screen, Membrane proteins, Expression optimization, Lipid screen, Reconstitution

1 Introduction

Cell-free (CF) expression has become a routine technique for the production of membrane proteins of both prokaryotic or eukaryotic origin and belonging to a large variety of different families. Transporters, channels, G-protein-coupled receptors, porins, and many others have been synthesized in various CF expression systems in functionally folded state (for review *see* ref. 1). Preparative scale production of MPs is usually achieved with the continuous exchange cell-free (CECF) configuration employing two different reaction compartments [2, 3]. The reaction mixture is separated by a semi-permeable membrane from a feeding mixture compartment holding a fixed volume reservoir of all low molecular weight precursors such as amino acids, nucleotides, or energy regeneration compounds.

Kirill Alexandrov and Wayne A. Johnston (eds.), *Cell-Free Protein Synthesis: Methods and Protocols*, Methods in Molecular Biology, vol. 1118, DOI 10.1007/978-1-62703-782-2_7,

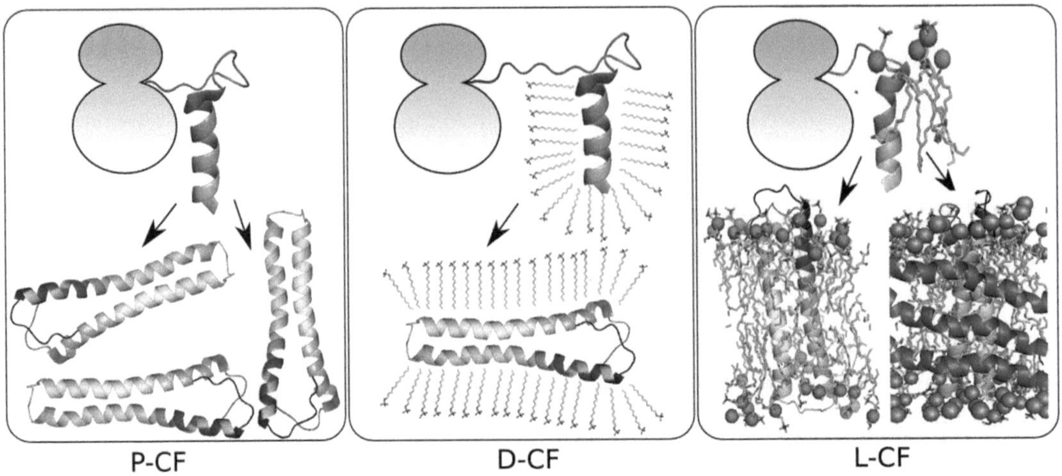

Fig. 1 CF expression modes. Membrane proteins can be synthesized as initial precipitates without any added hydrophobic compounds (P-CF), as proteomicelles in presence of detergents (D-CF) or as membrane protein/membrane complexes after addition of lipids, e.g., as liposomes or nanodiscs (L-CF) [1]

The relatively high molecular weight cut-off (MWCO) of 10–14 kDa of the separating membrane ensures efficient exchange between the two compartments, and thus the feeding mixture reservoir can replenish precursors for a certain time period, while inhibitory byproducts are removed from the reaction mixture. We describe protocols and setup of CECF reactions based on *E. coli* S30 extracts.

CF systems generally provide open accessible reactions allowing the generation of artificial hydrophobic environments for the solubilization and stabilization of the synthesized membrane proteins. Membrane proteins can therefore be produced in several different CF expression modes (*see* Fig. 1). In the precipitate forming (P-CF) mode, no hydrophobic environments are supplied and the membrane proteins precipitate instantly after translation. However, those P-CF precipitates often consist of reversible type I aggregates, and the membrane proteins can be solubilized with detergents into functionally folded conformations [4, 5]. In the detergent-based (D-CF) mode, a variety of detergents or detergent mixtures are supplied directly into the CF reaction, and the synthesized membrane proteins can be solubilized into proteomicelles during translation. In the lipid-based (L-CF) mode, the membrane proteins can insert co-translationally into membrane bilayers provided as vesicles, as liposomes of defined composition, or as nanodiscs [6–8].

In this chapter, we focus on two commonly encountered problems when approaching the CF expression of membrane proteins. Frequently no or insufficient expression of the target protein can be obtained despite control proteins being expressed at high levels. According to our experience, such efficiency problems can in most cases be solved by optimization with either Mg^{2+} ion screens or the newly described tag variation screen. A more complex problem is to obtain membrane protein samples of sufficient quality for functional and structural studies. This could be addressed by the systematic evaluation of membrane protein samples produced in the P-CF and D-CF modes implementing detergent screens [9]. However, some membrane proteins may only fold correctly in presence of lipids or their activity can only be analyzed in membranes. The emerging nanodisc technology offers a completely new avenue of opportunities for the analyses of membrane proteins in their native-like environment [10–13]. Small membrane discs with a diameter of between 10 and 13 nm can be formed by the membrane scaffold protein (MSP) which shields the hydrophobic character of the fatty acid alkyl chain towards the solvent [14]. As a result, membrane protein/nanodisc/ND complexes are soluble particles and can be applied to all techniques of protein analysis including structural methods like X-ray crystallography, NMR, and single particle analysis via electron microscopy. The nanodiscs can be filled with many different lipids, and lipid-dependent effects on membrane protein structure and function can thus be studied. We describe the production of nanodiscs with different lipid types, and we present the guidelines for the implementation of nanodiscs into CF reactions for the efficient co-translational solubilization of membrane proteins.

2 Materials

All stock solutions should be prepared with ultrapure water and stored at −20 °C if not otherwise stated.

2.1 Common Materials

1. Fermenter for 5–10 l of culture volume.
2. French press or other high-pressure cell-disruption equipment.
3. Photometer.
4. Standard centrifuges and set of rotors.
5. Dialysis tubes, 12–14 kDa MWCO.
6. Ultrasonic water bath.
7. Thermo shaker for incubation.
8. Chromatographic system (e.g., Äkta purifier, GE Healthcare).
9. Q-Sepharose column (GE Healthcare).
10. Strep-Tactin Sepharose (IBA).
11. Columns for size exclusion chromatography (SEC): Superdex 200 3.2/30 column.
12. Centriprep filter devices, 10 kDa MWCO (Millipore).

Table 1
Characteristics of lipids for nanodisc preparation

Lipid[a]	Head group	Fatty acid	Charge	MW (kDa)	Origin/source
Synthetic					
DMPC	Choline	14:0	Zwitterionic	677.5	Eucarya
POPC	Choline	16:0–18:1	Zwitterionic	759.6	Eucarya
DOPC	Choline	18:1	Zwitterionic	785.6	Eucarya
DMPG	Glycerol	14:0	Negative	688.4	Eubacteria
DOPG	Glycerol	18:1	Negative	796.5	Eubacteria
DOPE	Ethanolamine	18:1	Zwitterionic	743.6	Eubacteria/ Eucarya
Natural					
$PC_{Soybean}$	Choline	Mixed	Zwitterionic	±776	Soybean
PL_{coli}	Mixed	Mixed	Mixed	±798	*E. coli*
TL_{coli}	Mixed	Mixed	Mixed	±835	*E. coli*
PL_{Brain}	Mixed	Mixed	Mixed	±800	Bovine
TL_{Brain}	Mixed	Mixed	Mixed	±830	Bovine

[a]Abbreviations: *DMPC* 1,2-dimyristoyl-*sn*-glycero-3-phosphocholine, *POPC* 1-palmitoyl-2-oleoyl-*sn*-glycero-3-phosphocholine, *DOPC* 1,2-dioleoyl-*sn*-glycero-3-phosphocholine, *DMPG* 1,2-dimyristoyl-*sn*-glycero-3-phospho-(1′-rac-glycerol), *DOPG* 1,2-dioleoyl-*sn*-glycero-3-phospho-(1′-*rac*-glycerol), *DOPE* 1,2-dioleoyl-*sn*-glycero-3-phospho-ethanolamine, *PC* L-α-phosphatidylcholine, *PL* polar lipid extract, *TL* total lipid extract

13. Plasmid and PCR product purification kits.
14. Dark microplate (96F Nunclon Delta Black Microwell SI, Nunc).
15. Lipids (*see* Table 1).
16. 50× Complete protease inhibitor cocktail (Roche Diagnostics): 1 tablet/ml of MilliQ water.
17. Labsonic homogenizer.
18. Empty gravity flow column.
19. 1.8 ml Nunc cryo-tubes (ThermoScientific).
20. L-[S35]methionine (^{35}S-Met) (Hartmann Analytics).
21. Scintillation vials.
22. Rotiszint®eco scintillation cocktail (Roth).
23. Scintillation counter.

2.2 Materials for CECF Expression Reaction

1. All stock solutions required for CECF reactions are listed in Table 2. Suppliers are: Pyruvate kinase (Roche Diagnostics), RiboLock RNAse inhibitor (Fermentas), and total *E. coli* tRNA (Roche Diagnostics); all other chemicals are from Sigma-Aldrich if not otherwise stated.

   Acetyl phosphate (AcP) is adjusted to pH 7.0 with KOH; Phospho(enol) pyruvic acid (PEP) is adjusted to pH 7.0 with KOH; NTP mixture is adjusted to pH 7.0 with NaOH; 2.4 M Hepes/20 mM EDTA (24×) is adjusted to pH 8.0 with KOH.

Table 2
CECF reaction preparation for 1 ml reaction mixture and 16 ml feeding mixture

Compound	Stock	Final concentration	Volume[a]
Master mixture			
6 Amino acid mix (RCWMDE)	16.7 mM	1 mM	1,020 μl
Amino acid mix (*see* **Note 1**)	25 mM	0.5 mM	340 μl
Acetyl phosphate (Li^+, K^+)	1 M	20 mM	340 μl
Phospho(enol)pyruvic acid (K^+)	1 M	20 mM	340 μl
75× NTP mix	90 mM ATP	1.2 mM	226.7 μl
	60 mM G/C/UTP	0.8 mM	
1,4 Dithiothreitol	500 mM	2 mM	68 μl
Folinic acid (Ca^{2+})	10 mg/ml	0.1 mg/ml	170 μl
Complete protease inhibitor	50×	1×	340 μl
Hepes/EDTA buffer	24×	1×	623.3 μl
$Mg(OAc)_2$	1 M	16 (11.1) mM[b]	274 μl
KOAc	4 M	270 (110) mM[b]	382.5 μl
PEG 8000	40 %	2 %	850 μl
NaN_3	10 %	0.05 %	85 μl
			Total: 5,059.5 μl
Reaction mixture			
Master mixture			297.6 μl
Pyruvate kinase	10 mg/ml	0.04 mg/ml	4 μl
tRNA (*E. coli*)	40 mg/ml	0.5 mg/ml	12.5 μl
T7RNAP	1.4 mg/ml	0.05 mg/ml	35.7 μl
RiboLock	40 U/μl	0.3 U/μl	7.5 μl
DNA template	0.2–0.5 mg/ml	0.015–0.03 mg/ml	60 μl
E. coli S30 extract	1×	0.35×	350 μl
MilliQ water			232.7 μl
			Total: 1 ml
Feeding mixture			
Master mixture			4,762.1 μl
S30-C buffer	1×	0.35×	5,600 μl
Amino acid mix	25 mM	0.5 mM	320 μl
MilliQ water			5,317.9 μl
			Total: 16 ml

[a]Calculated for 1 ml reaction mixture and 16 ml feeding mixture = 17 ml master mixture
[b]Subject to optimization. Volumes are calculated for final total concentrations of Mg^{2+} of 16 mM and K^+ of 270 mM as additional amounts of 4.9 mM Mg^{2+} and 160 mM K^+ are contributed by other compounds. Numbers in parenthesis give the concentrations of the extra added ions

2. Reaction container: Custom-made analytical scale Mini-CECF reactors and preparative scale Maxi-CECF reactors may be used ([15]; *see* **Note 2**). Alternatively, commercially available D-tube™ dialyzer mini, 12–14 kDa MWCO (Merck Biosciences); Slide-A-Lyzer, 10 kDa MWCO (Pierce); or dialysis tubes, 12–14 kDa MWCO, are as efficient.
3. GFP assay buffer: 20 mM Tris, 150 mM NaCl, pH 7.4.

2.3 Materials for S30 Extract and T7RNAP Preparation

1. 40× S30-A/B buffer: 400 mM Tris–acetate pH 8.2, 560 mM $Mg(OAc)_2$, 2.4 M KCl.

   Supplement 1× S30-A buffer with 6 mM ß-mercaptoethanol.

   Supplement 1× S30-B buffer with 1 mM DTT and 1 mM phenylmethanesulfonylfluoride (PMSF).
2. 40× S30-C buffer: 400 mM Tris–acetate pH 8.2, 560 mM $Mg(OAc)_2$, 2.4 M KOAc.

   Supplement 1× S30-C buffer with 0.5 mM DTT.
3. 2× YTPG medium: 22 mM KH_2PO_4, 40 mM K_2HPO_4, 100 mM glucose, tryptone 16 g/l, yeast extract 10 g/l, NaCl 5 g/l.
4. LB medium: Peptone 10 g/l, yeast extract 5 g/l, NaCl 5 g/l.
5. Buffer-T7RNAP-A: 30 mM Tris–HCl pH 8.0, 50 mM NaCl, 10 mM EDTA, 10 mM ß-mercaptoethanol, 5 % glycerol.
6. Buffer-T7RNAP-B: 30 mM Tris–HCl pH 8.0, 50 mM NaCl, 1 mM EDTA, 10 mM ß-mercaptoethanol, 5 % glycerol.
7. Buffer-T7RNAP-C: 30 mM Tris–HCl pH 8.0, 1 M NaCl, 1 mM EDTA, 10 mM ß-mercaptoethanol, 5 % glycerol.
8. Buffer-T7RNAP-D: 10 mM K_2HPO_4/KH_2PO_4 pH 8.0, 10 mM NaCl, 0.5 mM EDTA, 1 mM DTT, 5 % glycerol.
9. 20 % streptomycin sulfate.
10. *E. coli* strain for extract preparation, e.g., A19 (*E. coli* Genetic Stock Center, New Haven, CT) or BL21 (Merck Biosciences).
11. BL21 (DE3) Star x pAR1219 for T7RNAP preparation [15].

2.4 Materials for Nanodisc Preparation

1. pET28b vector with MSP coding sequences modified after [14]. The linker in between the N-terminal poly$(His)_6$-tag and the TEV protease recognition site has been removed (*see* **Note 3**).
2. 1 M isopropyl-β-D-thiogalactopyranoside (IPTG) in H_2O.
3. 30 mM kanamycine.
4. 500 mM PMSF in 100 % ethanol.
5. Syringe filter, sterile, 0.45 μm (Roth).
6. Buffer MSP-A: 40 mM Tris–HCl, pH 8.0, 300 mM NaCl, 1 % TritonX-100 (v/v).
7. Buffer MSP-B: 40 mM Tris–HCl, pH 8.9, 300 mM NaCl, 50 mM cholic acid.
8. Buffer MSP-C: 40 mM Tris–HCl, pH 8.0, 300 mM NaCl.
9. Buffer MSP-D: 40 mM Tris–HCl, pH 8.0, 300 mM NaCl, 50 mM imidazole.
10. Buffer MSP-E: 40 mM Tris–HCl, pH 8.0, 300 mM NaCl, 300 mM imidazole.

11. Buffer MSP-F: 40 mM Tris–HCl, pH 8.0, 300 mM NaCl, 10 % glycerol (v/v).
12. Buffer ND-A: 10 mM Tris–HCl, pH 8.0, 100 mM NaCl.
13. Strep-buffer: 20 mM Tris–HCl, pH 7.5, 100 mM NaCl.
14. Strep-elution-buffer: 20 mM Tris–HCl, pH 7.5, 100 mM NaCl, 5 mM D-desthiobiotin.
15. D-Desthiobiotin (IBA).
16. Sodium cholate (Roth).

2.5 Materials for Tag Variation Screen and Template Preparation

1. PCR purification kit.
2. Plasmid DNA purification kit.
3. Vent polymerase (New England Biolabs).
4. Agarose.
5. Trichloroacetic acid (TCA).
6. Agarose gel extraction kit.

3 Methods

3.1 Preparation of S30 Extract

In order to obtain highly efficient cellular extracts, cultivation of the cells in a fermenter with good aeration is recommended. A growth curve of the selected strain should first be determined in the individual facility, and efficient chilling of the broth media after fermentation should be ensured (*see* **Note 4**).

3.1.1 Cell Fermentation

1. Inoculate 10 l of sterilized YTPG medium in a fermenter with 100 ml of a fresh *E. coli* overnight culture, e.g., strain A19.
2. Incubate the cells at 37 °C with intensive aeration and stirring.
3. Monitor the cell growth by continuously measuring the OD_{600}.
4. Start to chill the cell broth before the cells reach mid-log phase (*see* **Note 4**) (approximately OD_{600} 3–5).
5. Quickly cool the fermenter broth to 14–10 °C.
6. Harvest the cells by centrifugation at 7,000 × *g* for 15 min at 4 °C.
7. Keep the cell pellets at 4 °C for all following steps. Alternatively, freeze the cell paste in thin layers wrapped in aluminum foil at −80 °C for later processing.

3.1.2 Cell Extraction

1. Gently resuspend the cell pellet in approximately 300 ml precooled S30-A buffer and centrifuge at 7,000 × *g* for 10 min at 4 °C. Discard supernatant and repeat this washing step twice. Extend the final centrifugation step to 30 min.

2. Discard the supernatant and resuspend the cell pellet in 110 % (v/w) precooled S30-B buffer.
3. Disrupt the cells by a high-pressure cell disrupter, e.g., French press at 1,000 psi (*see* **Note 5**).
4. Centrifuge the lysate at 30,000 × *g* for 30 min at 4 °C. Transfer the supernatant in a fresh tube and repeat the centrifugation step.
5. Harvest the supernatant and adjust stepwise to a final concentration of 400 mM NaCl. Gently mix and incubate at 42 °C for 45 min in a water bath (*see* **Note 6**).
6. Dialyze the turbid extract overnight against 100-fold excess of precooled S30-C buffer using a dialysis membrane with a 12–14 kDa MWCO and with two changes of dialysis buffer.
7. Centrifuge the extract at 30,000 × *g* for 30 min at 4 °C. Harvest the supernatant and dispense suitable aliquots into plastic tubes.
8. Shock-freeze the aliquots in liquid nitrogen and store at −80 °C. Aliquots are stable at −80 °C for at least 1 year but should not be repeatedly refrozen.
9. The final total protein concentration of the S30 extract should be in between 20 and 40 mg/ml. Each new batch of extract should be adjusted to its optimal concentration of Mg^{2+} (12–25 mM) and K^{+} (250–350 mM) ions. Perform Mg^{2+} concentration screens in 2 mM steps and K^{+} concentration screens in 20 mM steps with GFP expression as monitor.

3.2 T7RNAP Preparation

1. T7RNAP is produced by conventional overexpression in *E. coli* cells in Erlenmeyer flasks [16]. On average, approximately 20,000–40,000 units can be isolated out of 1 l culture. Inoculate 1 l of LB medium 1:100 with a fresh overnight culture of strain BL21 (DE3) Star x pAR1219. Let the cells grow on a shaker at 37 °C until OD_{600} of 0.6–0.8 and induce T7RNAP production by addition of 1 mM IPTG. Incubate the cells for further 5 h and harvest by centrifugation at 8,000 × *g* for 15 min at 4 °C. The cell pellet can be stored at −80 °C until further usage.
2. Resuspend the cell pellet in 30 ml of T7RNAP-A buffer and disrupt the cells by one passage through a French press cell at 1,000 psi or by sonication. Remove cell debris by centrifugation at 20,000 × *g* for 30 min at 4 °C. All subsequent purification steps should be performed at 4 °C.
3. Adjust supernatant to a final concentration of 4 % streptomycin sulfate by stepwise addition of a 20 % stock solution. Mix gently, incubate on ice for 5 min, and centrifuge at 20,000 × *g* for 30 min at 4 °C.

4. Load the supernatant on a 40 ml Q-sepharose column equilibrated with T7RNAP-B buffer and wash the column extensively with T7RNAP-B buffer.
5. Elute the T7RNAP with a gradient from 50 to 500 mM NaCl using T7RNAP-C buffer for 10 column volumes at a flow rate of 3–4 ml/min. Collect the fractions and analyze aliquots by SDS-PAGE (*see* **Note 7**).
6. Pool T7RNAP-containing fractions and dialyze against T7RNAP-D buffer overnight. Adjust to a final concentration of 10 % glycerol and concentrate the T7RNAP fraction to a total protein concentration of 3–4 mg/ml by ultrafiltration (*see* **Note 8**). Adjust to a final concentration of 50 % glycerol and store aliquots at −80 °C.
7. For each new batch of T7RNAP, perform concentration optimization screens in the CECF and batch configurations with GFP as reporter.

3.3 Template Production and Yield Optimization by Tag Variation Screen

CF systems using *E. coli* extract require either circular plasmids or linear DNA as template for expression. No or insufficient expression of proteins is usually caused by an inefficient translation/transcription process. Rational template design (*see* **Note 9**) is therefore important for successful protein synthesis.

High quality and purity of the plasmid or linear DNA is crucial for efficient CF expression. As the vectors do not replicate during the reaction, specific selection markers are not of importance. The final template concentration for optimal expression efficiency should be determined for each new target with an initial concentration screen in the range in between 0.1 and 10 ng/μl of reaction mixture (*see* Fig. 2).

3.3.1 DNA Template Preparation (Plasmid)

1. Inoculate 100–500 ml of LB medium supplemented with the specific antibiotic of choice with the *E. coli* strain containing the desired plasmid and incubate overnight at 37 °C on a shaker.
2. Purify the plasmid using commercial available kits like "Midi" or "Maxi" kits. "Mini" preparations are not suitable due to the low quality of the purified DNA.
3. Dry the DNA overnight and dissolve it into MilliQ water. Optimal concentration of stock DNA solution is in between 0.2 and 1.5 mg/ml. DNA stocks can be stored at −20 °C. In case of precipitation due to freezing/thawing cycles, the DNA concentration must be checked again.

3.3.2 DNA Template Preparation (Linear DNA)

If the target gene is already present in a suitable vector under control of T7 elements, fragments containing the T7 regulatory sequence and the target gene can be amplified by standard PCR and directly used in the CF reaction. If the target gene is not under

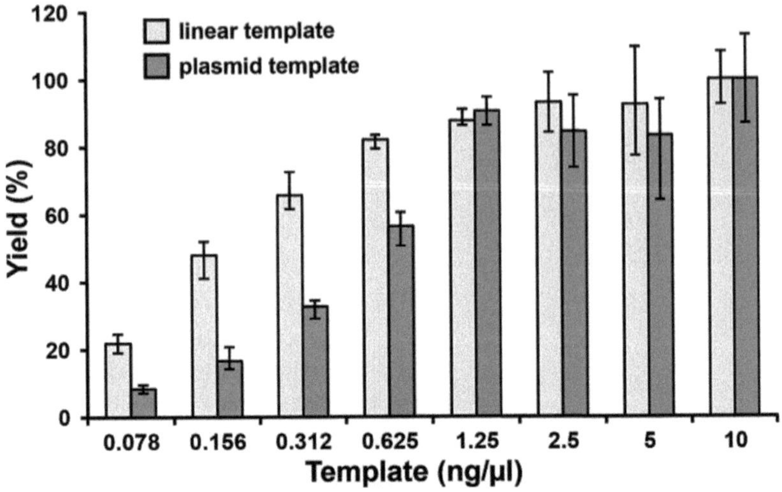

Fig. 2 Optimization of DNA template concentration. The G-protein-coupled receptor CCR3 was expressed in analytical scale P-CF reactions in presence of ^{35}S-Met from linear PCR and circular plasmid DNA templates and quantified by scintillation counting. The expression yield is shown in percent with 10 ng/µl DNA template concentration set as 100 %. Plateau phase expression started for linear template at 0.625 ng/µl and plasmid template at 1.25 ng/µl

T7 control or if expression/purification tags or expression monitors have to be added, a multistep PCR strategy such as for the tag variation screen (*see* Subheading 3.3.3) can be applied [17].

1. For the PCR two primers are needed: a forward primer annealing at the T7 promoter and a reverse primer annealing at the T7 terminator.
2. The reaction can be performed in 50 µl volume with 20 ng of plasmid DNA as template, 0.4 µM each of the forward and reverse primers, 200 µM of dNTPs, 0.5 units of Vent DNA polymerase (NEB), and 1× ThermoPol reaction buffer. The PCR program starts with a 2 min denaturation step at 94 °C followed by 32 cycles of denaturation for 15 s at 94 °C, primer annealing for 30 s at 55 °C, and primer extension of 1 min for each kb of target gene at 72 °C. A last step of final elongation is performed for 10 min at 72 °C.
3. PCR products should be purified using standard PCR purification kits.

3.3.3 Tag Variation Screen for Improved Expression

Small expression tags comprising 6–12 codons are fused to the translational start site of the target coding sequence by an overlap PCR approach (*see* Fig. 3). The expression tags are optimized in suppressing secondary structure formation which may prevent the initiation of translation [18]. As secondary structures also depend

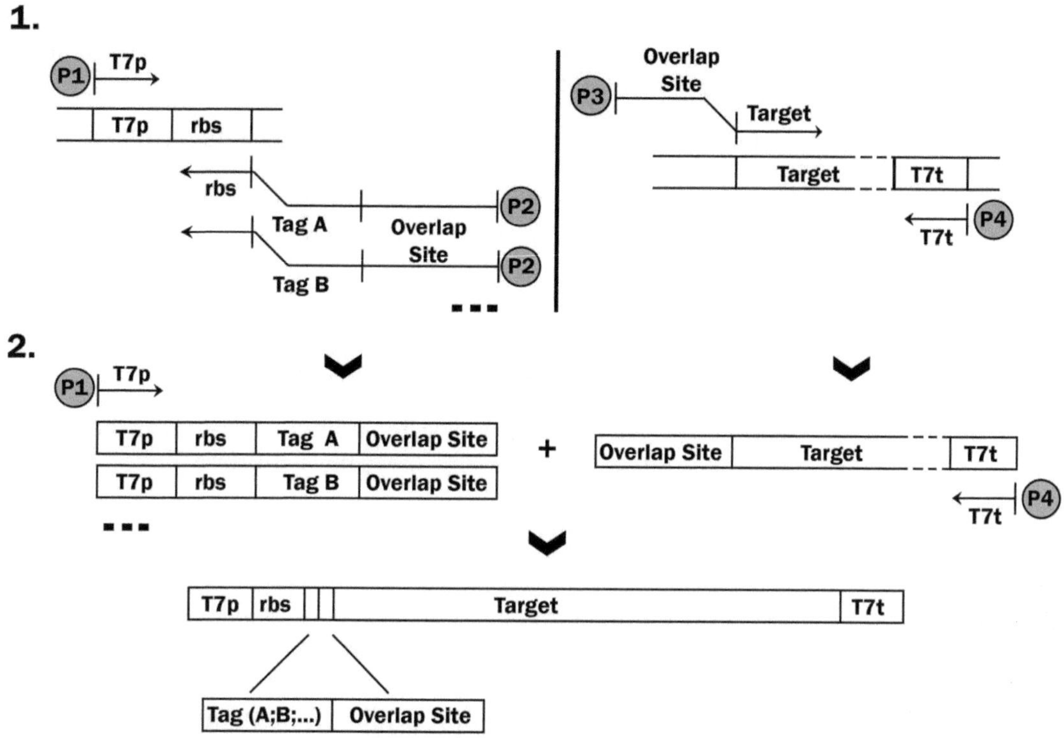

Fig. 3 Strategy for the tag variation screen. Two subsequent PCR steps are illustrated. 1: The tag fragment(s) with primers P1/P2 and the target fragment(s) with primers P3/P4 are generated with appropriate overlap sites. 2: Corresponding tag fragments and target fragments are fused by an overlap PCR reaction using primers P1/P4. The products are suitable as CF expression templates for the tag variation screen

Table 3
Expression tags for tag variation screen

Name	Nucleotide sequence	Amino acid sequence	AT (%)
AT	AAATAT TATAAATATTAT	KYYKYY	100
SER	AAATCATCATCATCATCA	KSSSSS	72
H	AAACCATACGATGGTCCA	KPYDGP	55
G	AAAAGTAAAGGAGAA GAA	KSKGEE	72

on the individual target sequence, it is recommended to empirically screen a mini library of optimized tag fragments (*see* Table 3). The entire tag variation screen can be performed in 1 day, and if the small expression tags have no detectable impact on structure and function, they can usually remain on the target protein. However, the length of the finally selected expression tag may further be systematically reduced by changing the primer P2 in order to allow maximum expression yield with minimum sequence modification of the membrane protein ([17], *see* **Note 10**).

Four different primers are needed for the first step of the PCR reaction in order to generate a set of tag fragments and a target fragment (*see* Fig. 3).

Primer 1 (P1): Universal forward primer annealing upstream of the T7 promoter.

Primer 2 (P2): Reverse tag-specific primer unique for each tag. P2 anneals at the ribosome binding site and contains a 5-prime linker with the specific tag sequence (*see* Fig. 3). For the PCR overlap region, protease cleavage sites, e.g., for TEV or PreScission protease, could be used, allowing the option to remove the tags after protein expression.

Primer 3 (P3): Target gene-specific primer annealing at the start of the coding sequence and carrying a 5-prime linker with the PCR overlap region.

Primer 4 (P4): Universal reverse primer annealing downstream of the T7 terminator.

1. A tag fragment library is produced by PCR using a vector containing the T7 promoter as template and the primers P1 and P2 (*see* Fig. 3).
2. A linear target fragment is produced by PCR with the primers P3 and P4 and with a vector carrying the coding sequence of the target protein under T7 control.
3. All synthesized PCR fragments are purified using either standard PCR purification kits or agarose gel extraction kits.
4. In the overlap PCR step, the linear target fragment is mixed with the individual fragments from the tag library in equal molar ratios using 50 ng of the target fragment and corresponding amounts of tag fragments. The overlap PCR is performed with the primers P1 and P4. The resulting PCR fragments are purified and analyzed for purity and integrity on a 1 % agarose gel. The purified PCR fragments can be directly added into the CF reaction as templates. Expression tags may be removed after CF expression and purification of the target protein by taking advantage of the protease recognition site used in the overlap region. Alternatively, identified beneficial expression tags may be further shortened (*see* **Note 10**).

3.4 General Setup of CF Expression Reactions

Analytical scale CECF reactions are suitable for any kind of screens such as Mg^{2+} optimization screens, tag variation screens, or nanodisc lipid screens and can be performed in Mini-CECF reactors or in suitable D-tube™ dialyzers with reaction mixture volumes of 50–100 μl. The identified optimal reaction parameters can be scaled up in a 1:1 ratio to preparative scale CECF reactions with reaction mixture volumes of several ml by using Maxi-CECF reactors.

Master mixes of common compounds of reaction and feeding mixture should be prepared in order to minimize variations. Nevertheless, reactions should always be performed in duplicate. We recommend reaction mixture:feeding mixture ratios in between 1:15 and 1:20.

3.4.1 Analytical Scale CECF Reactions

1. Calculate the individual compound volumes according to the desired number of reactions and design an appropriate pipetting scheme for the experiment using standard software packages such as MS Excel, LibreOffice Calc, or Gnumeric.
2. Prepare standard 24-well microplates and appropriate numbers of Mini-CECF reactors and pieces of dialysis membrane with a MWCO of 12–14 kDa (*see* **Note 11**).
3. Prepare a common master mix and split required volumes into reaction mixture and feeding mixture (*see* Table 2).
4. Reconstitute reaction and feeding mixture with the appropriate volumes of individual components and complete by addition of MilliQ water (*see* Table 2). Mix feeding mixture by vortexing and the reaction mixture by gentle shaking or pipetting.
5. Fill reaction and feeding mixture aliquots into reaction containers. A: Mini-CECF reactors [15] with reaction mixture volumes of 30–100 μl can be used in combination with standard 24-well microplates holding feeding mixture volumes of up to 1.5 ml in their cavities. A piece of fresh dialysis membrane is fixed to the Mini-CECF reactors with a Teflon ring (*see* **Note 12**). Avoid air bubbles or residual water in the container and check for leakage. B: Commercial D-tube™ dialyzers for reaction mixture (RM) volumes of 100 μl can be used in combination with suitable tubes (e.g., 1.8 ml Nunc cryotubes) holding the appropriate volume of FM.
6. CECF reactions are incubated over night at 30 °C with gentle shaking or rolling in order to ensure efficient substance exchange between reaction and feeding mixture.

3.4.2 Preparative Scale CECF Reactions

To allow expression in a larger volume, e.g., for structural investigation via NMR, the described analytical scale CECF setup (*see* Subheading 3.4.1) can be scaled up to a desired final volume of several milliliters in a 1:1 ratio. As reaction containers for preparative scale CECF reactions, commercial Slide-A-Lyzer devices with a MWCO of 10 kDa and holding up to 3 ml reaction mixture volumes are suitable. As feeding mixture containers, either Maxi-CECF- reactors [15] or any small plastic boxes or beakers having suitable sizes for the required feeding mixture volume may be used. As alternative reaction setup, D-tube™ dialyzer or simply appropriate pieces of dialysis tubes sealed at both ends by knots may be used as reaction mixture container in combination with, e.g., 15–50 ml Falcon tubes as feeding mixture containers.

3.4.3 Quantification of Target Production by ^{35}S-Met Incorporation

If exact determination of expression efficiency is necessary, then labeling with ^{35}S methionine should be performed (*see* **Note 13**). For the CECF expression of radioactively labeled proteins, disposable D-tube™ dialyzer mini, MWCO 12–14 kDa, is recommended as reaction mixture container and 1.8 ml Nunc cryo-tubes as suitable feeding mixture containers.

1. The standard reaction mixture volume in analytical scale reactions of 50 μl (*see* Subheading 3.4.1) has to be increased with 17.5 μl *E. coli* S30 extract and 32.5 μl feeding mixture up to a final volume of 100 μl in order to ensure sufficient compound exchange between reaction and feeding mixture in the D-tube™ dialyzers mini. A feeding mixture volume of 750 μl can be kept if 1.8 ml Nunc cryo-tubes are used.
2. Feeding and reaction mixture are first prepared without labeled isotope and filled into the cryo-tubes and D-tube™ dialyzers mini, respectively. The D-tube™ dialyzers mini are stored on ice until addition of ^{35}S-Met.
3. Reaction and feeding mixture are supplemented with 50 nM ^{35}S-Met resulting in a final ratio of labeled to unlabeled Met of 1:40,000 (*see* **Note 14**).
4. The D-tube™ dialyzers mini are closed with the screw caps and placed into the corresponding feeding mixture containing cryo-tubes which are then capped as well. The CECF reaction is then incubated in a suitable rack in an upright position and incubated at 30 °C with gentle shaking for 16–20 h.
5. The reaction mixture is transferred into a 1.5 ml Eppendorf tube. The D-tube™ dialyzer mini is washed twice with 100 μl S30-C buffer, and the washing solution is combined with the reaction mixture in order to collect the expressed membrane protein quantitatively.
6. For soluble membrane protein expression, e.g., with supplied nanodiscs, the diluted reaction mixture is precipitated by mixing with 600 μl of 15 % ice-cold TCA to reach a final concentration of 10 % TCA and then centrifuged for 10 min at 22,000 × *g*, 4 °C. The pellet is then washed twice with 900 μl 10 % TCA and finally with 900 μl 95 % ethanol.
7. For P-CF reactions, the pellet is suspended in 600 μl H_2O and washed three times with 900 μl H_2O (*see* **Note 15**).
8. The supernatants are carefully removed, and the tubes containing the protein pellets are transferred into 20 ml scintillation vials filled with 5 ml scintillation cocktail. The vials are gently inverted and incubated for 1–2 h at room temperature.

9. The scintillation is counted for 1 min in a scintillation counter.
10. The obtained counts per minute (CPM) are then converted into milligram protein per milliliter RM (*see* **Note 16**).

3.5 Preparation of Nanodiscs as Supplement for CF Reactions

Nanodiscs are highly water soluble and stable. Different diameters of nanodiscs are possible depending on the selected type of MSP [14, 19]. Most frequently used are MSP1 (9.7 nm with POPC) fitting membrane proteins containing at least up to 21 transmembrane segments [20] and MSP1E3 (12.1 nm with POPC). Essential for the efficient solubilization of CF-expressed membrane proteins are highly concentrated and homogenous nanodisc stock solutions [21]. The nanodiscs can be filled with a large variety of different lipids, and the selected type of lipid can have crucial effects on the solubilization efficiency of individual membrane proteins. We describe the basic protocol for MSP1 production containing an N-terminal His_6-tag cleavable by the TEV protease. The protocol can also be applied to the preparation of the larger MSP1E1, MSP1E2, and MSP1E3D1 constructs (*see* **Note 17**).

3.5.1 MSP1 Expression

1. The MSP1 expression plasmid is freshly transformed into BL21 (DE3) Star cells (*see* **Note 18**). Inoculate 4 × 600 ml sterile LB medium supplemented with filter sterilized (0.2 μm) glucose (0.5 %; w/v) each in 2 l Erlenmeyer flasks with 50 ml of fresh LB overnight cultures (37 °C, 180 rpm).
2. Grow cells at 37 °C and induce expression with 1 mM IPTG at OD_{600} = 1. Incubate cells for 1 h at 37 °C and then reduce temperature to 28 °C for an additional 4 h. Harvest cells by centrifugation (10 min, 6,000 × *g*). The pellet can be stored at −20 °C at least up to 1 month.

3.5.2 MSP1 Purification

1. Thaw pellet of 2.4 l fermentation on ice and resuspend in 50 ml buffer MSP-C supplemented with one tablet Complete protease inhibitor cocktail and 1 mM freshly prepared PMSF. Add Triton X-100 from 10 % (v/v) stock to a final concentration of 1 % (v/v).
2. Disrupt cells by ultrasonification (3 × 60 s and 3 × 45 s) with equal cooling times in between each cycle (*see* **Note 19**). Centrifuge at 30,000 × *g* for 20 min. The supernatant is filtered through 0.45 μm prior to further purification.
3. Equilibrate immobilized metal affinity chromatography (IMAC) column (10 ml bed volume, Sepharose 6 FF, GE Healthcare) with 5 column volumes (CV) buffer MSP-A. Load the filtered supernatant of cell disruption onto the column and wash the column with 5 CV buffer MSP-A, 5 CV buffer MSP-B, 5 CV buffer MSP-C, and 5 CV buffer MSP-D.
4. Elute MSP1 with buffer MSP-E in 1 ml fractions. Pool MSP1-containing fractions and immediately add 10 % glycerol (v/v)

Table 4
Conditions for optimal nanodisc assembly

Lipid	MSP1::lipid	MSP1E3::lipid	Sodium cholate (mM)[a]
Synthetic			
DMPC	1:80	1:115	100
POPC	1:55	1:85	200
DOPC	1:30	1:80	300
DMPG	1:70	1:110	300
DOPG	1:30	1:90	200
DOPE	1:30	1:80	300
Natural			
$PC_{Soybean}$	1:40	1:60	300
PL_{coli}	1:40	1:60	200
TL_{coli}	1:40	1:50	500
PL_{Brain}	1:45	1:70	500
TL_{Brain}	1:35	1:60	300

[a]For solubilization of lipid

to prevent precipitation. Dialyze pooled fractions over night at 4 °C against 5 l of buffer MSP-F with one buffer exchange after 2 h. MSP1 concentration is determined using A_{280} with the molar extinction coefficient $\varepsilon = 24{,}750$ l/mol/cm and MW = 25.3 kDa. MSP1 is aliquotted and frozen at −80 °C for long-time storage (*see* **Note 20**).

3.5.3 Nanodisc Assembly

The assembly of empty nanodiscs requires extensive optimization if homogenous samples are required. As the MSP associates with a fixed number of lipids, the MSP to lipid ratio needs to be properly adjusted for each MSP/lipid combination (*see* Table 4).

1. Lipid stocks are prepared by suspending lipids in water to a final concentration of 50 mM supplemented with sodium cholate for complete solubilization. For each individual lipid, a particular final concentration of sodium cholate may be required in order to obtain a clear suspension (*see* Table 4) (*see* **Note 21**).
2. For analytical assembly, combine MSP with lipids and detergent in a final volume of 100 μl (*see* **Note 22**). MSP (25 μM), lipids (e.g., 2 mM with MSP1 and DMPC), and detergent (e.g., DPC 0.1 % w/v) are mixed and incubated for 1 h at room temperature. Induce nanodisc assembly by dialysis (*see* **Note 23**) against a large volume (e.g., 5 l) of buffer ND-A at room temperature for 12 h (10 kDa MWCO) by using Slide-A-Lyzers. Refresh buffer and dialyze for further 24 h at 4 °C. Centrifuge supernatant (20 min, 22,000 × *g*) and apply to SEC analysis.

3. Load supernatant of the assembly mixture to a suitable SEC column (e.g., Superdex 200 3.2/30). Equilibrate the column with buffer ND-A. At optimal MSP to lipid ratios, the SEC elution profiles should show evenly shaped elution peaks indicating homogenous nanodisc populations. Typical MSP to lipid ratios are given in Table 4.
4. For preparative scale nanodisc assembly, volumes can be scaled up at least to 25 ml. Volumes >10 ml might require extended dialysis time and more frequent buffer changes.

3.5.4 Nanodisc Concentration, Storage, and Stability

1. Equilibrate Centriprep filter devices (10 kDa MWCO) with buffer ND-A and concentrate at 4 °C to the minimal volume of 200 μl as recommended by the manufacturer. After concentration centrifuge the nanodisc stock for 10 min at 22,000 × *g*. The final protein concentration is determined by the Bradford assay (*see* **Note 24**).
2. For short-time storage (<12 h), the nanodisc stock can be kept on ice. Long-time storage (up to weeks) can be achieved by flash freezing in liquid nitrogen and storing at −80 °C. Avoid extensive numbers of freeze-thaw cycles as this will destroy the nanodiscs. Thaw nanodisc stock solutions on ice.
3. Stability depends on the type of generated nanodisc. While MSP1-DMPC nanodiscs can be stored at 4 °C for weeks, other types will start to precipitate after a few days. Stability generally decreases with increasing numbers of double bonds in the lipid chains (e.g., DMPC nanodiscs are more stable than DOPC nanodiscs) and with increasing MSP size.

3.6 Co-translational Formation of Membrane Protein/Nanodisc Complexes

A protocol for the co-translational association/insertion of CF-expressed membrane proteins with nanodiscs is described. Solubilization with detergents as in the D-CF mode or the initial precipitation of the membrane proteins as in the P-CF mode can thus be avoided. Nanodiscs are well tolerated by CF systems and usually do not have inhibitory effects on the expression efficiency (*see* **Note 25**). C-terminal fusion constructs of membrane proteins to sGFP allow a fast initial monitoring of the co-translational solubilization of membrane proteins with the nanodiscs, although the folding of sGFP may not always be completely correlated with the folding of the N-terminal membrane protein fusion.

3.6.1 Quantifying sGFP

1. After expression, centrifuge the reaction mixture (10 min, 22,000 × *g*) and use supernatant for fluorescence assay.
2. Add 3 μl of the supernatant in 300 μl of GFP-assay buffer. Measure fluorescence intensity of each reaction by at least triplicates and convert into protein concentration by a calibration curve with purified GFP.

3.6.2 CF Production of Membrane Protein/Nanodisc Complexes

1. The synthesized membrane proteins and the nanodiscs must be present at least in stoichiometric amounts in order to improve the sample homogeneity. In addition, the type of lipid in the nanodiscs can strongly affect the efficiency of complex formation with individual membrane proteins. Initial nanodiscs concentration screens should therefore be performed for each new membrane protein/nanodisc combination in order to determine conditions for best solubilization. The membrane protein is quantified in the soluble and non-soluble fraction after reactions with increasing amounts of supplied nanodiscs. For quantification, either membrane protein-sGFP fusions, detection of immuno-tags after western blotting, or ^{35}S-Met incorporation can be used.
2. Prepare duplicate CF reactions for each nanodisc concentration. A recommended initial screen would comprise final nanodisc concentrations of 0, 20, 40, 60, 80, 100, and 120 μM in the reaction mixture (*see* **Note 26**).
3. After reaction, centrifuge the reaction mixture at 22,000 × *g* for 10 min to separate soluble and insoluble fractions and quantify the expressed membrane protein in supernatant and in the pellet fraction.
4. If sufficient solubilization of the membrane protein cannot be achieved, the effects of different nanodisc types and/or different lipids should be analyzed (*see* **Note 27**).

3.6.3 Purification of Membrane Protein/Nanodisc Complexes

1. For functional analysis in most cases purified membrane protein/nanodisc complexes are required. For optimal sample homogeneity, the nanodiscs are usually added in access to the CF reaction, and empty nanodiscs might need to be removed. Providing unique affinity tags at the membrane protein, e.g., a Strep-II-tag, is therefore recommended (*see* **Note 28**).
2. Centrifuge reaction mixture after expression at 22,000 × *g* for 10 min. Apply fivefold diluted supernatant in Strep-buffer to Strep-Tactin resin (IBA) equilibrated in Strep-buffer. Perform binding in batch setup with 500 μl Strep-Tactin resin (IBA) for each 1 ml supernatant of reaction mixture (*see* **Note 29**). Incubate for 12 h at 4 °C with gentle shaking.
3. Load binding solution on an empty gravity flow column. Wash resin with 5 CV Strep-buffer. Elute protein with Strep-elution-buffer. Apply 1 × 0.5 CV and 4 × 1 CV of Strep-elution-buffer and collect main fractions.

4 Notes

1. Weigh in all compounds and dissolve the powders in water. The stock remains as a turbid suspension and should only be aliquotted immediately after vortexing.

2. Commercial or self-made reaction container as specified may be used with similar efficiencies.
3. N-terminal sequence of MSP contains a poly(His)$_6$-tag, the TEV cleavage site followed by the MSP coding sequence: MGSSHHHHHHENLYFQG-MSP
4. Consider that cells continue to grow during the chilling process, entering the stationary phase of growth should be avoided. Start to harvest the cell culture when a temperature of 18 °C has been achieved.
5. Sonication is not recommended for cell disruption as it may disintegrate translation complexes.
6. This step causes significant precipitation of proteins in the supernatant which is beneficial for the quality of the extract. The high salt concentration causes dissociation of ribosomes from mRNA templates which will subsequently become degraded.
7. T7RNAP should be visible as major band, but the fractions will still contain significant amounts of impurities.
8. T7RNAP becomes unstable at higher concentrations; stop ultrafiltration as soon as the solution becomes turbid.
9. High-level protein production in the *E. coli* CF system is usually directed by standard T7 regulatory sequences. The T7 promoter as well as T7 terminator must be present in the DNA template. Suitable vectors containing the T7 promoter sequence include the pET (Merck Biosciences) and pIVEX (Roche Diagnostic) series. In addition, the following parameters regarding template design should be considered: (a) Purification tags. The presence of small tags at the C-terminal of the target protein is a valuable tool for the detection of expression and for full-length protein purification. Poly(His)$_{10\text{-tags}}$ or Strep-II-tags may be used for protein detection by immunoblotting or for efficient membrane protein purification. (b) Expression monitoring. Fusion with reporter proteins such as GFP allows the fast monitoring of expression and may accelerate the optimization of the reaction conditions for improved protein target production. (c) Codon usage. In rare cases, clusters of non-frequent codons or the formation of unfavorable secondary structures within the coding sequence may cause amino acid mis-incorporation or premature terminations of the protein product. Such problems should be addressed by expression of synthetic genes. (d) Expression tags. In most cases, inefficient initiation of translation is the reason for low expression efficiency. This problem is addressed by the tag variation screen (*see* Subheading 3.3.3) by fusing small tags to the N-terminus of the target protein. A small set of templates containing different optimized tags (*see* Table 3) is synthesized by a two-step PCR approach, and the generated linear PCR products are directly screened in CF reactions [17].

10. Systematic reduction of the initial expression tag may result in even further enhanced expression yields [17]. Even the addition of only one codon could significantly improve the expression yields. However, positive effects are hard to predict, and thus the described systematic screening starting with a small set of sufficiently long tags is recommended.
11. Dialysis membranes for Mini-CECF reactors are cut into 2.5 cm × 2.5 cm pieces and stored in water until required.
12. The membrane pieces are placed on top of the Teflon rings and the Mini-CECF reactors are pushed through the bottom of the ring, thus fixing tightly the dialysis membrane. The assembled Mini-CECF reactors are equilibrated for few minutes in the feeding mixture before adding the reaction mixture.
13. To analyze, e.g., different N-terminal expression tags with regard to expression yield, the same final concentration of DNA should be used for all tested tag-target combinations.
14. ^{35}S-Met is diluted 1:6 in H_2O prior to addition for better handling, and 22.5 μl are added to the feeding mixture and 3 μl to the reaction mixture, respectively. The CF system tolerates labeled isotopes with stabilizer and with dye.
15. 50–100 μl supernatant is retained on top of the protein pellet to prevent loss of pellet by supernatant removal.
16. Calculation of protein concentration: 1 CPM = 4.5^{-13} Ci; 1 Ci = 1 μM ^{35}S-Met; Mol ^{35}S-Met are converted to Mol MP by calculating the Mol S^{35}-Met divided by the number of methionines in the target protein: Calculated Mol ^{35}S-Met by CPM/number of Met in protein = Mol expressed protein. The calculated Mol MP are then multiplied with the dilution factor of ^{35}S-Met (40,000) and with the specific MP mass in order to determine the expression yield in g/RM volume.
17. Typical averaged yields of purified MSP1 and MSP1E3D1 are 20 mg/l and 15 mg/l bacterial culture.
18. Use BL21 (DE3) Star (Invitrogen) only as expression yields in, e.g., BL21 (DE3) are reduced.
19. Chilling is necessary in order to prevent MSP degradation.
20. MSP stocks can be stored at −80 °C for several months without any obvious decrease of quality.
21. Lipid solubilization may be supported by incubating at 40 °C or in an ultrasonic water bath.
22. For nanodisc assembly, detergent is required for initial lipid solubilization. Besides cholate, other detergents which can be removed by dialysis can be used above their critical micellar concentration (e.g., DPC at 0.1 % (w/v) final concentration).
23. Use dialysis only as hydrophobic polymers such as Bio-Beads SM-2 adsorb lipids as well as MSP. As the MSP to lipid ratio is

crucial for homogenous nanodisc assembly, this would result into high variations and inhomogenous nanodisc samples.

24. From the determined mg/ml MSP according to a standard Bradford assay, calculate the mol/l and then divide by 2, as two MSP copies assemble to one nanodisc. Nanodiscs can be concentrated into the millimolar range.
25. sGFP expression can be used as monitor of negative effects on the CF system. Nanodiscs loaded with different lipids can usually be supplied in high concentrations without significant loss of sGFP expression efficiency.
26. To give final concentrations of up to 120 μM of nanodiscs within the reaction mixture, ND stock solutions of at least 525 μM (26.57 mg/ml MSP1) are required.
27. Changing the charge of lipid head groups or increasing the diameter of nanodiscs are most promising alternatives for improving the formation of soluble membrane protein/nanodisc complexes.
28. Alternatively, the N-terminal His_6-tag of the described MSP constructs could be removed by TEV cleavage before nanodisc assembly. However, processing is often incomplete and significant amounts of MSP may remain non-cleaved.
29. Capacity of 1 ml Strep-Tactin resin is 100 nmol recombinant protein, corresponding to 3 mg for a 30 kDa protein.

Acknowledgements

This work was supported by the Collaborative Research Center (SFB) 807 of the German Research Foundation (DFG). We further thank the European Drug Initiative on Channels and Transporters (EDICT), contract number HEALTH-F4-2007-201924; the European initiative on Structural Biology of Membrane Proteins (SBMP), contract number PITN-GA-2008-211800; and the NIH (grant number U54 GM094608) for funding. We further thank Oliver Ernst for helpful advice and Vladimir Shirokov for critical discussions.

References

1. Junge F, Haberstock S, Roos C et al (2010) Advances in cell-free protein synthesis for the functional and structural analysis of membrane proteins. N Biotechnol 28:262–271
2. Spirin AS, Baranov VI, Ryabova LA et al (1988) A continuous cell-free translation system capable of producing polypeptides in high yield. Science 242:1162–1164
3. Kigawa T, Yokoyama S (1991) A continuous cell-free protein synthesis system for coupled transcription–translation. J Biochem 110: 166–168
4. Junge F, Luh LM, Proverbio D et al (2010) Modulation of G-protein coupled receptor sample quality by modified cell-free expression protocols: a case study of the human endothelin A receptor. J Struct Biol 172:94–106
5. Ma Y, Muench D, Schneider T et al (2011) Preparative scale cell-free production and quality optimization of MraY homologues in different

expression modes. J Biol Chem 286: 38844–38853
6. Yang JP, Cirico T, Katzen F et al (2011) Cell-free synthesis of a functional G protein-coupled receptor complexed with nanometer scale bilayer discs. BMC Biotechnol 11:57
7. Lyukmanova EN, Shenkarev ZO, Khabibullina NF et al (2012) Lipid–protein nanodiscs for cell-free production of integral membrane proteins in a soluble and folded state: comparison with detergent micelles, bicelles and liposomes. Biochem Biophys Acta 1818:349–358
8. Katzen F (2008) Cell-free protein expression of membrane proteins using nanolipoprotein particles. Biotechniques 45:190
9. Kai L, Roos C, Haberstock S et al (2012) Systems for the cell-free synthesis of proteins. Methods Mol Biol 800:201–225
10. Glück JM, Koenig BW, Willbold D (2011) Nanodiscs allow the use of integral membrane proteins as analytes in surface plasmon resonance studies. Anal Biochem 408:46–52
11. Borch J, Roepstorff P, Møller-Jensen J (2011) Nanodisc-based co-immunoprecipitation for mass spectrometric identification of membrane-interacting proteins. Mol Cell Proteomics 10:O110.006775
12. Kobashigawa Y, Harada K, Yoshida N et al (2011) Phosphoinositide-incorporated lipid-protein nanodiscs: a tool for studying protein–lipid interactions. Anal Biochem 410:77–83
13. Raschle T, Hiller S, Yu TY et al (2009) Structural and functional characterization of the integral membrane protein VDAC-1 in lipid bilayer nanodiscs. J Am Chem Soc 131:17777–17779
14. Denisov I, Grinkova Y, Lazarides A et al (2004) Directed self-assembly of monodisperse phospholipid bilayer nanodiscs with controlled size. J Am Chem Soc 24:3477–3487
15. Schneider B, Junge F, Shirokov VA et al (2010) Membrane protein expression in cell-free systems. Methods Mol Biol 601:165–186
16. Li Y, Wang E, Wang Y (1999) A modified procedure for fast purification of T7 RNA polymerase. Protein Expr Purif 16:355–358
17. Haberstock S, Roos C, Hoevels Y et al (2012) A systematic approach to increase the efficiency of membrane protein production in cell-free expression systems. Protein Expr Purif 82:308–316
18. De Smit MH, van Duin J (1990) Secondary structure of the ribosome binding site determines translational efficiency: a quantitative analysis. Proc Natl Acad Sci USA 87: 7668–7672
19. Grinkova Y, Denisov I, Sligar G (2010) Engineering extended membrane scaffold proteins for self-assembly of soluble nanoscale lipid bilayers. Protein Eng Des Sel 23: 843–848
20. Zocher M, Roos C, Wegmann S et al (2011) Single-molecule force spectroscopy from nanodiscs: an assay to quantify folding, stability, and interactions of native membrane proteins. ACS Nano 6:961–971
21. Roos C, Zocher M, Müller D et al (2012) Characterization of co-translationally formed nanodisc complexes with small multidrug transporters, proteorhodopsin and with *E. coli* MraY translocase. Biochim Biophys Acta 1818(12):3098–3106

Chapter 8

Cell-Free Protein-Based Enzyme Discovery and Protein–Ligand Interaction Study

Sabrina Guillemer, Cécile Persillon, Jean-Michel Masson, and Gilles Ravot

Abstract

Cell-free expression-based screening is sometimes more suitable than cell-based assays for enzyme discovery. The advantage of cell-free systems for expression of toxic, poorly expressed, or insoluble proteins has already been well documented. Cell-free methods can advantageously replace cell-based ones when screening has to be performed on cell lysates prepared from harvested cells, for instance, when dealing with protein–ligand interactions particularly when the latter is hydrophobic. From our experience, cell-free extracts efficient in both transcription and translation can be prepared from potentially any microorganism. Here we present a general method for preparation of cell-free extracts from prokaryotic and eukaryotic cells, selection of the best systems, and optimized conditions for specific protein expression. The method allows to select proteins for their ability to bind a selected target, to identify the inhibitors of such binding, or to identify novel enzymatic activities.

Key words Cell-free, Protein expression, Recombinant protein, Binding assay, Enzyme discovery, Screening of expression conditions

1 Introduction

Cell-free expression is a quick way to obtain sufficient amount of functional protein for various characterizations. It is especially convenient for screening applications. Prokaryotic and eukaryotic in vitro protein synthesis systems are widely used for rapid production and analysis of protein mutants and enzyme discovery by screening gene-shuffling or genomic libraries [1, 2]. Cell-free expression also provides a convenient way to incorporate unnatural amino acids in proteins for structural studies. Moreover, incorporation of radioactive amino acids will yield specific labeling of the recombinant protein leading to a very sensitive detection method with minimal impact on protein conformation.

To succeed in enzyme discovery through cell-free expression, it is necessary to achieve extract preparation while preserving the

Kirill Alexandrov and Wayne A. Johnston (eds.), *Cell-Free Protein Synthesis: Methods and Protocols*, Methods in Molecular Biology, vol. 1118, DOI 10.1007/978-1-62703-782-2_8, © Springer Science+Business Media, LLC 2014

activity of biosynthetic machinery. However, there are no standard conditions for universal cell-free expression of all proteins. To mimic intracellular machinery and obtain efficient in vitro transcription/translation systems, various components such as energy sources, salts, buffer, and reducing agent have to be optimized. Cell-free systems allow the creation of various environments and are very flexible regarding specific protein folding requirements. However, concurrent optimization of all components is not easy and requires a good understanding of the system and efficient screening methodology. The method for the preparation of a cell-free extract from *E. coli* is based on Zubay et al. [3] that was subsequently improved by Spirin, Kim, and Swartz [4, 5] and allows to obtain highly active cellular machinery in *E. coli* S30 extracts with efficient supply of energy sources. Other organisms have been used to develop systems for proper folding of eukaryotic proteins such as rabbit reticulocyte lysates, wheat germs, or insect cells extracts. Potentially any organism can be used for preparation of extracts efficient in in vitro transcription/translation as recently exemplified by the systems based on human cells [6, 7].

In order to express functional enzymes designed for bio-based processes in the chemical, environmental and biofuel industry, Protéus developed suitable cell-based and cell-free systems. This foremost involves choosing appropriate organisms for the preparation of extracts. For prokaryotic gene expression, we usually obtain good results with *E. coli*, *Bacillus*, or *Pseudomonas* cells. This small panel of species covers a wide array of gene expression conditions. For eukaryotic extracts *Trichoderma* turned out to be the best choice. In addition to cell extract preparation and reaction medium optimization for expression of functional enzymes, activity detection methods are crucial for screening success. This prompted us to develop our own subset of substrates that yield minimal background in in vitro assays.

2 Materials

It is generally recommended to use RNAse-free chemicals (*see* **Note 1**). For prokaryotic cell disruption, we suggest using the Cell Disruptor from Constant Systems Limited (UK) (*see* **Note 2**). For eukaryotic cell disruption, use mortar and pestle chilled in liquid nitrogen (*see* **Note 3**).

2.1 Prokaryotic Culture Medium

1. Buffered rich growth medium: 10 g yeast extract, 5.6 g KH_2PO_4, and 28.9 g K_2HPO_4 per liter (autoclaved).
2. 25 % (w/v) glucose and 100 mg/mL thiamine solutions filter sterilized (150 μL thiamine solution and 40 mL glucose solution per liter of autoclaved medium. Add just before use).

2.2 Prokaryotic S30 Solutions

Prepare the following stock solutions using DEPC MilliQ water:

1. 2 M Tris acetate pH 8.2, filter sterilized, store at +4 °C.
2. 2 M magnesium acetate, filter sterilized, store at +4 °C.
3. 1 M potassium chloride, filter sterilized, store at +4 °C.
4. 6 M potassium acetate, filter sterilized, store at +4 °C.
5. 1 M DTT, filter sterilized, store at −20 °C.
6. 0.1 M PMSF in ethanol, store at −20 °C.
7. 20 amino acid solution (5 mM each), store at −20 °C.
8. 0.5 M ATP store at −20 °C.
9. 2 M Acetyl phosphate store at −20 °C.

   Prepare the following buffered solutions using DEPC MilliQ water:

10. *Solution A*: Wash buffer.

    10 mM Tris acetate pH 8.2, 14 mM magnesium acetate, 60 mM potassium chloride, 6 mM β-mercapto-ethanol.

11. *Solution B*: Disruption buffer.

    10 mM Tris acetate pH 8.2, 14 mM magnesium acetate, 60 mM potassium chloride, 1 mM DTT, 0.1 mM PMSF (add last two components just before use).

12. *Solution C*: "Run off" complement (prepare just before use).

    750 mM Tris acetate pH 8.2, 21 mM magnesium acetate, 0.5 mM amino acid mix, 6 mM ATP, 0.5 M acetyl phosphate, 7.5 mM DTT.

13. *Solution D*: Dialysis buffer.

    10 mM Tris acetate pH 8.2, 14 mM magnesium acetate, 60 mM potassium acetate, 1 mM DTT (add just before use).

Dialysis Slide-A lyzer cassette or dialysis tubing 10,000 MWCO (Pierce).

2.3 Prokaryotic Energy Regeneration System/Salts Mixes

Prepare the following stock solutions using DEPC MilliQ water:

1. 2 M Hepes KOH pH 7.5, filter sterilized, store at +4 °C.
2. 0.5 M EDTA pH 8.2, filter sterilized, store at −20 °C.
3. 4 M potassium glutamate, filter sterilized, store at −20 °C.
4. 2 M ammonium acetate, filter sterilized, store at +4 °C.
5. 1 M magnesium acetate, filter sterilized, store at +4 °C.
6. 2.5 M potassium acetate, filter sterilized, store at +4 °C.
7. 60 % (w/v) PEG-8000, autoclaved, store at RT.
8. 1 M DTT, filter sterilized, store at −20 °C.
9. 10 mg/mL folinic acid, store at −20 °C.

10. 0.5 M spermidine, filter sterilized, store at –20 °C.
11. 16 mg/mL MRE600 tRNA (Roche Applied Bioscience), store at –20 °C.
12. 0.2 M cAMP, store at –20 °C.
13. NTP mix pH 7.5 (125 mM each), store at –20 °C.
14. 19 amino acid solution without methionine (5 mM each), store at –20 °C.
15. 5 mM methionine, store at –20 °C.
16. Methionine, L-[^{35}S]-, (EasyTag) 10.2 NEG709A packaged in a stabilized aqueous solution with blue dye from PerkinElmer.
17. 1 M Acetyl phosphate, store at –20 °C.

2.4 Eukaryotic Culture Medium

1. Germination medium: 30 g maltose, 10 g casein hydrolysate, 5 g yeast extract, 1 g KH_2PO_4, 0.5 g $MgSO_4{\cdot}7H_2O$, 30 mg $ZnCl_2$, 26 mg $CaCl_2{\cdot}2H_2O$, 6 mg $MnSO_4{\cdot}H_2O$, 0.3 g $FeSO_4{\cdot}7H_2O$, 3 g Tween 80 per liter (autoclave).
2. Buffered minimal medium: 15 g KH_2PO_4, 20 g D-glucose, 0.5 g yeast extract, and 0.5 g soya peptone per liter (autoclave).
3. 36 g $MgSO_4{\cdot}7H_2O$ and 300 g $(NH_4)_2SO_4$ per liter (60× solution, filter sterilized).
4. 96 g $CaCl_2{\cdot}2H_2O$ per liter (120× solution, filter sterilized).
5. 5 g $FeSO_4{\cdot}7H_2O$, 1.6 g $MnSO_4{\cdot}H_2O$, 1.4 g $ZnSO_4{\cdot}7H_2O$, 3.7 g $CoCl_2{\cdot}6H_2O$ per liter (1,000× solution, filter sterilized).

2.5 Eukaryotic S30 Solutions

Prepare the following stock solutions using DEPC MilliQ water:

1. 2 M Hepes KOH pH 7.5, filter sterilized, store at +4 °C.
2. 2 M magnesium acetate, filter sterilized, store at +4 °C.
3. 6 M potassium acetate, filter sterilized, store at +4 °C.
4. 0.5 M EDTA pH 8.2, filter sterilized, store at –20 °C.
5. 1 M DTT, filter sterilized, store at –20 °C.
6. 0.1 M PMSF in ethanol, store at –20 °C.
7. Protease inhibitor cocktail for use with fungal and yeast extracts (Sigma P8215).
8. 60 % (w/v) PEG-8000, autoclaved, store at RT.
9. 50 % autoclaved glycerol.

Prepare the following buffered solutions using DEPC MilliQ water:

10. *Solution A*: Wash buffer
    60 mM Hepes KOH pH 7.5, 3 mM magnesium acetate, 80 mM potassium acetate, 0.2 mM EDTA, 2 mM DTT (add just before use).

```
       T7 promotor
   ------------------->
AAATTAATACGACTCACTATAGGGGAATTGTGAGCGGATAACAATTCCCCTCTAGAAATAAT
                                       Met
TTTGTTTAACTTTAAGAAGGAGATATACATATG>your gene
                ------     ------
                 RBS        NdeI
```

Fig. 1 3′ Upstream region for prokaryotic transcription/translation

11. *Solution B*: Extraction buffer
    60 mM Hepes KOH pH 7.5, 3 mM magnesium acetate, 80 mM potassium acetate, 0.2 mM EDTA, 2 mM DTT, 0.1 mM PMSF and 5 % (v/v) protease inhibitors cocktail (the last three components must be added just before use).

Sephadex G-25 column (GE Healthcare).

2.6 Eukaryotic Energy/Salts Mixes

Prepare the following stock solutions using DEPC MilliQ water:

1. 2 M Hepes KOH pH 7, filter sterilized, store at +4 °C.
2. 1 M magnesium acetate, filter sterilized, store at +4 °C.
3. 2.5 M potassium acetate, filter sterilized, store at +4 °C.
4. 1 M DTT, filter sterilized, store at −20 °C.
5. 0.2 M spermine, filter sterilized, store at −20 °C.
6. 10 mg/mL wheat germ tRNA, store at −20 °C (Sigma R7876).
7. NTP mix pH 7.5 (125 mM each), store at −20 °C.
8. 19 amino acid solution without methionine (5 mM each), store at −20 °C.
9. 5 mM methionine, store at −20 °C.
10. Methionine, L-[^{35}S]-, (EasyTag) 10.2 NEG709A packaged in a stabilized aqueous solution with blue dye from PerkinElmer.
11. 1.5 M creatine phosphate, store at −20 °C.
12. 5 mg/mL creatine kinase, aliquoted, store at −20 °C.

2.7 Configuration of the DNA Template

Supercoiling has an effect on yield (*see* **Note 4**). For prokaryotic cell-free systems, the DNA template must contain the gene of interest placed downstream of a T7 promoter and a ribosome binding site (RBS). We recommend cloning your gene at the *NdeI* site of a pET-like vector or downstream the following sequence which is provided in the pET vectors (*see* Fig. 1).

After the gene of interest, a T7 terminator sequence is strongly recommended in particular if a plasmid is used as DNA template. Here again, we suggest to use the T7 terminator sequence which is present in the pET-like vectors.

T7 terminator sequence:
CTAGCATAACCCCTTGGGGCCTCTAAACGGGTCTT
GAGGGGTTTTTTG

For eukaryotic cell-free systems, the DNA template 5′ and 3′ flanking region are very important for efficient translation. In addition to a T7 promoter, we place the gene of interest downstream of a stable 5′ viral enhancer of tobacco mosaic virus and upstream 3′UTR (untranslated region), as described [8–10].

2.8 Purification of the DNA Template

For DNA template purification, we recommend the use of Qiagen kits: Qiagen plasmid midi-kit for supercoiled plasmid preparations, Qiaquick PCR purification kits for PCR, or linearized plasmid cleanup.

2.9 Cell-Free Reaction

1. DNA template.
2. Energy salt mixes (*see* Subheadings 2.3 and 2.6).
3. Ribonuclease inhibitors (recommended RNasin®, Promega).
4. T7 RNA polymerase (from Invitrogen or any other commercial sources).

2.10 Auto-radiography

1. ^{35}S-labeled methionine (EasyTag™ from PerkinElmer).
2. Prechilled acetone at −20 °C.
3. Laemmli sample buffer.
4. Prestained Protein Ladder (from any commercial source).
5. Precast SDS-PAGE Hepes gels 4–20 %.
6. Cellophane sheets and frame for drying gels.

2.11 Binding Assay

1. Cell-free reactions.
2. EIA microtiter plate coated with appropriate ligand.
3. Phosphate buffered saline (PBS).
4. Bovine serum albumin (BSA).
5. Scintillation counter.

3 Methods

3.1 Guiding Principles

S30 extract preparation is based on a method described by Zubay and coworkers [3]. S30 extracts efficient for cell-free transcription/translation can be prepared from various organisms. The most frequently used are *E. coli* cells, rabbit reticulocytes, or wheat germ [11–13], but we have prepared extracts from other sources such as hyperthermophilic bacteria. For the preparation of "routine" extracts, we choose organisms that we know have a good level of expression in cell-based systems for our applications (mostly expression of prokaryotic, yeast, or fungal enzymes). We choose microorganisms that allow high cell density culture that allow preparation of concentrated extracts. In addition to *E. coli* extracts, we developed *Bacillus* and *Pseudomonas* extracts in order to create different backgrounds for protein folding. For eukaryotic gene expression, we routinely use *Trichoderma* extracts (*see* **Note 5**).

Cells have to be harvested in the middle of exponential phase when the cellular machinery has the best performances and when proteases are expressed at low levels. The extraction has to be performed quickly after harvesting, keeping all reagents cold unless otherwise mentioned, to retain high specific activity.

Cell disruption has to keep intact all the elements of cell machinery essential for transcription and translation. Prokaryotic cell disruption can be done in different ways such as cell homogenizer or French press. In contrast, eukaryotic cell disruption is much more difficult and has to be performed in liquid nitrogen with mortar and pestle or other device that reproduces this action (*see* **Note 2**).

Steps of the S30 preparation:

1. Cells culture and harvesting.
2. Cells washing.
3. Cells disruption.
4. 30,000 × *g* centrifugation: Collect clear supernatant (S30).
5. Getting rid of engaged mRNA: "Run off" (*see* **Note 6**).
6. Elimination of small inhibitors through dialysis or size exclusion chromatography.
7. Deep freezing and storage.

Methods described hereafter are for the preparation of *E. coli*, *Bacillus subtilis* and *Pseudomonas fluorescens* S30 extract for prokaryotic expression, and *Trichoderma* S30 extract for eukaryotic expression.

3.2 Preparation of DEPC-Treated MilliQ Water

Diethyl pyrocarbonate is used to inactivate RNAse in solution. Prepare stock solution of DEPC-treated MilliQ water as follows. Add 1 mL DEPC per liter of MilliQ water and agitate overnight at +4 °C. The day after, the solution has to be autoclaved 45 min at 121 °C to inactivate DEPC and aliquoted into sterile containers. For long-term storage, aliquots can be stored at −20 °C.

3.3 Preparation of S30 Prokaryotic Extracts

Use sterilized medium and stock S30 solutions (*see* Subheading 2.1).

1. Harvest cells grown to mid-log phase by centrifuging 10 min at 10,000 × *g*, +4 °C.
2. Weigh the wet cell pellet and wash the pellet at least two times with 10 mL of wash buffer (*Solution A*) per gram of wet cells. For each wash, mix by pipetting and centrifuge 10 min at 10,000 × *g*, +4 °C.

   Washed cells can be stored at −80 °C for several months.

3. Thoroughly resuspend the cells pellet in disruption buffer, *Solution B* (1 mL per gram)
4. Disrupt the cell suspension at ~15,000 psi in the Constant Cell Disruptor.
5. Centrifuge 30 min at 30,000 × *g*, +4 °C and carefully collect the clear supernatant without pipetting up the interphase layer.
6. Add 70 μL of "run off" complement (*Solution C)* per mL of clear lysate and incubate 40 min at 37 °C under gentle agitation.
7. Chill on ice
8. Transfer the lysate in appropriate dialysis device and dialyze in 50 volumes of *Solution D* (two times for at least 2 h, changing the dialysis buffer in between)
9. Aliquot the dialysate into working fractions and deep freeze in liquid nitrogen (*see* **Note 7**).

3.4 Preparation of Eukaryotic (Trichoderma) S30 Extracts

Use sterilized medium and stock S30 solutions (*see* Subheading 2.2).

To obtain efficient extracts, you should use young mycelia and avoid pellets. For this, make spores germinate (*see* **Note 8**), and then perform successive inoculations of culture media increasing the volume step by step.

1. Harvest mycelium with help of gauze bandage and funnel. Press the gauze to eliminate most culture media. Wash the mycelia at least twice with wash buffer, *Solution A*
2. Make a thin flat cake with the washed cells, deep freeze in liquid nitrogen, and store at −80 °C until use.
3. Disrupt the cells in liquid nitrogen with mortar and pestle. Grind the cells to a fine powder (15–20 min) (*see* **Note 2**).
4. Transfer the powder in a sterile container and add 1 mL of extraction buffer *Solution B* per gram of ground cells.
5. Homogenize and incubate on ice for 20 min
6. Centrifuge for 45 min at 30,000 × *g*, +4 °C and carefully collect the clear supernatant without aspirating the upper and interphase layers.

To eliminate small molecular weight inhibitors, the extract is purified by size exclusion chromatography. Use appropriate column packed with Sephadex G-25 Medium.

7. Equilibrate the column with wash buffer *Solution A*.
8. Load the clear lysate (maximum 10 % of the column volume).
9. Push with *Solution A*; collect and pool the fractions with the highest OD_{260} (*see* **Note 9**).
10. Aliquot in working fractions and deep freeze in liquid nitrogen (*see* **Note 7**).

Optional nuclease treatment (*see* **Note 10**).

3.5 Preparation of DNA Template

Use the method of your choice or a commercially available kit to purify your plasmid or linear DNA template for protein expression. Resuspend purified DNA such that the final concentration is 0.5–1 mg/mL (150–300 nM). Take care to avoid contamination with RNases.

We recommend the use of Qiagen kits (*see* Subheading 2 and **Note 4**).

3.6 Reaction Mixes

Reaction mixes may have to be adjusted regarding final concentration of different reagents for each batch of extract. For this, we use classical reporter genes such as GFP for prokaryotic systems and luciferase for eukaryotic systems. The most important parameters to adjust are energy source, magnesium, and extract concentration. You can vary energy source for best performance of your cell-free system. In our hands *E. coli* and *Bacillus* extracts perform well with acetyl phosphate [14], whereas *Pseudomonas* extracts need either creatine phosphate with creatine kinase (respectively around 30 mM of creatine phosphate and 0.2 mg/mL of creatine kinase) or phosphoenolpyruvate with pyruvate kinase. Extract amount used in the reaction depends on the concentration reached during the preparation (*see* **Note 11**). We routinely use around 25 % of *E. coli* extract per reaction, 40 % for *Bacillus* extract and *Pseudomonas* extract, and 50 % for *Trichoderma* extract.

Master mixes are prepared in advance in working aliquots and can be stored at −20 °C for months.

1. Prokaryotic extracts master mix = PK-MMTT 3.2 X (*see* Table 1)
2. Eukaryotic extracts master mix 5× = EK-MMTT 5 X (*see* Table 2)

3.7 Transcription/Translation Reactions

We provide standard conditions of reaction mixes used for prokaryotic and eukaryotic cell-free systems in our laboratory.

Prokaryotic reaction mixes: *see* Table 3

1. Add the previous reaction mix to the DNA template (10–20 nM final concentration).
2. Add the extract and homogenize by gently aspirating and discharging it with a pipette.
3. Incubate for 1–3 h at 30 °C (*see* **Note 12**).

Eukaryotic reaction mixes: *see* Table 4

1. Add the previous reaction mix to the DNA template (10–50 nM).
2. Add the extract and homogenize by gentle sucking up and pushing back with your pipette.
3. Incubate for 1–3 h at 20–24 °C.

Table 1
Prokaryotic extracts master mix = PK-MMTT 3.2 X

Stock solutions	Mix concentrations
Hepes KOH (pH 7.5)	480 mM
EDTA (pH 8.2)	0.64 mM
NTPs	16 mM
Folinic acid	76.8 μg/mL
19 Amino acid mix	0.8 mM
PEG-8000	9.6 %
tRNA *E. coli*	540 μg/mL
cAMP	1.92 mM
Potassium glutamate	0.64 M
Ammonium acetate	112 mM
Spermidine	1.6 mM

Table 2
Eukaryotic extracts master mix 5× = EK-MMTT 5 X

Stock solutions	Mix concentrations
Hepes KOH (pH 7)	200 mM
19 Amino acid mix	1.25 mM
tRNA *Wheat Germ*	50 μg/mL
NTPs	12.5 mM
Spermine	2.5 mM
Creatine phosphate	180 mM

3.8 Screening for Cell-Free "Host" and Condition Selection for a Specific Protein

Once lysate has been produced and before initiating a screening campaign, one may have to further optimize the reaction conditions in order to improve the expression yield or protein folding. For this purpose, you can adjust the following parameters: DNA template amount (especially for eukaryotic systems), magnesium concentration, and incubation temperature. Folding environment can also be generated for a particular protein by using glutathione redox buffers or by the addition of helper molecules such as protein disulfide isomerase [15].

Table 3
Prokaryotic reaction mix

Stock solutions	Mix concentrations
PK-MM-TT 3.2×	1×
Magnesium phosphate	20 mM
Acetyl phosphate	40 mM
DTT	10 mM
Rnase inhibitor	0.4 U/μL
T7 RNA Pol	3 U/μL
Methionine or Met ^{35}S	0.25 mM or 0.2 μCi/μL

Table 4
Eukaryotic reaction mix

Stock solutions	Mix concentrations
EK-MM-TT 5×	1×
Magnesium phosphate	5 mM
DTT	4 mM
Creatine kinase	150 μg/mL
Rnase inhibitor	0.4 U/μL
T7 RNA Pol	3 U/μL
Methionine or Met ^{35}S	0.25 mM or 0.2 μCi/μL

To select the best extract and expression conditions for a specific protein, we perform multiplexed optimization expressions. To screen for the most appropriate host and conditions, we use as much as possible functional activity assays. When a functional assay is not available, we usually perform multiplexed radiolabeled expressions analyzed by autoradiography of SDS-PAGE or scintillation counting. We also adapted binding assays with radiolabeled cell-free expression reactions.

3.9 Enzymatic Activity Assay

You can use any appropriate enzymatic activity assay to evaluate the functional expression of target protein.

To standardize and facilitate the screening of enzymes expressed through in vitro transcription/translation, we mainly use a proprietary activity assay named CLIPS-O™. Here fluorogenic substrates consist of a fluorescent phenolic structure that is masked by an enzyme labile group. A spacer separates the fluorogenic reporter structure from the enzyme labile group. The spacer is attached to the phenol through an ether bond which is completely stable towards hydrolysis. Due to the chemical nature of the spacer, the bond between the enzyme labile group and the spacer is virtually inert towards nonenzymatic hydrolysis allowing for a low background ultra sensitive enzyme assay. This substrate design enables the assay of a wide variety of enzymatic activities, including dehydrogenases [16], aldolases [17], and hydrolases such as lipases [18] and esterases [19], proteases, phosphatases and epoxide hydrolases, amidases, sulfatases, nitrilases, acylases, and xylanases [20]. This method allows developing assays for specific catalysts or classes of catalysts. CLIPS-O™ closely mimics the chemical structure and energetic state of natural and industrial substrates. They can be prepared in optically pure form of either enantiomers, which allows assay for enantioselectivity. *See* Fig. 2, for example, CLIPS-O™ substrates.

3.10 Autoradiography of SDS-PAGE Gels

When a functional assay is not available, we perform multiparallel radiolabeled expressions in a very small volume (20 μL) and analyze through autoradiography of SDS-PAGE gels (*see* **Note 13**). After migration, to obtain more accurate results, it is better to dry the gel between two cellophane sheets. In a dark room, place the gel on a Biomax film and expose at least overnight. *See* Fig. 3 for an example of in vitro expression profiling. For eukaryotic expression (*see* Fig. 3e), the extract has not been treated with micrococcal nuclease (*see* **Note 10**).

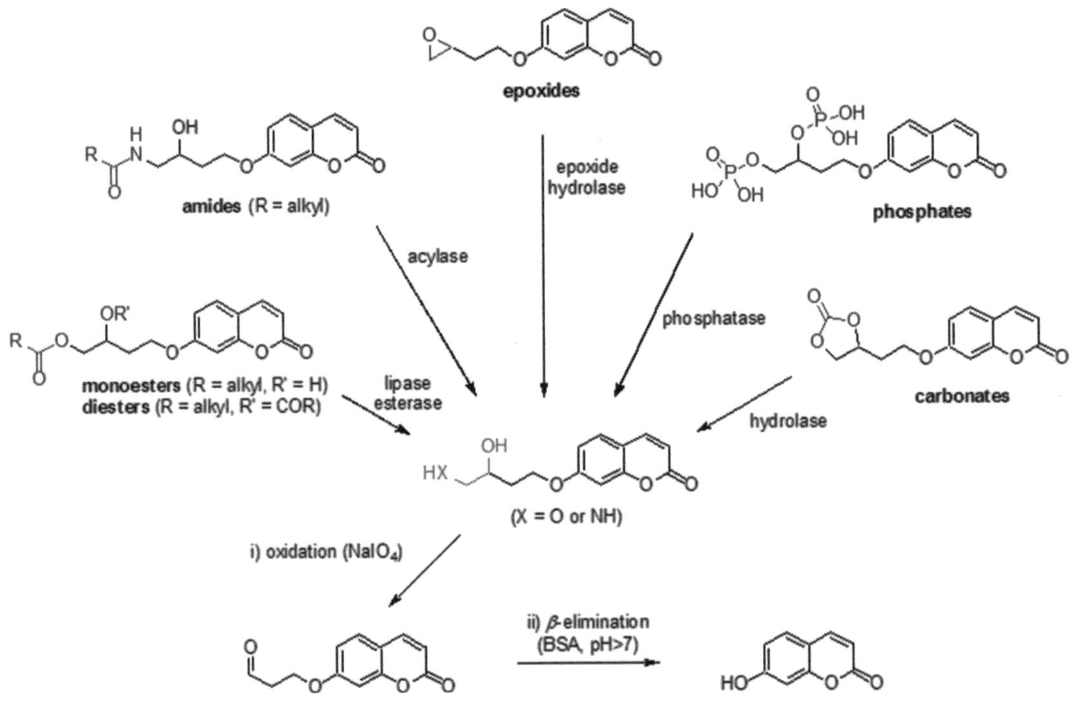

Fig. 2 Example of CLIPS-O™ substrates

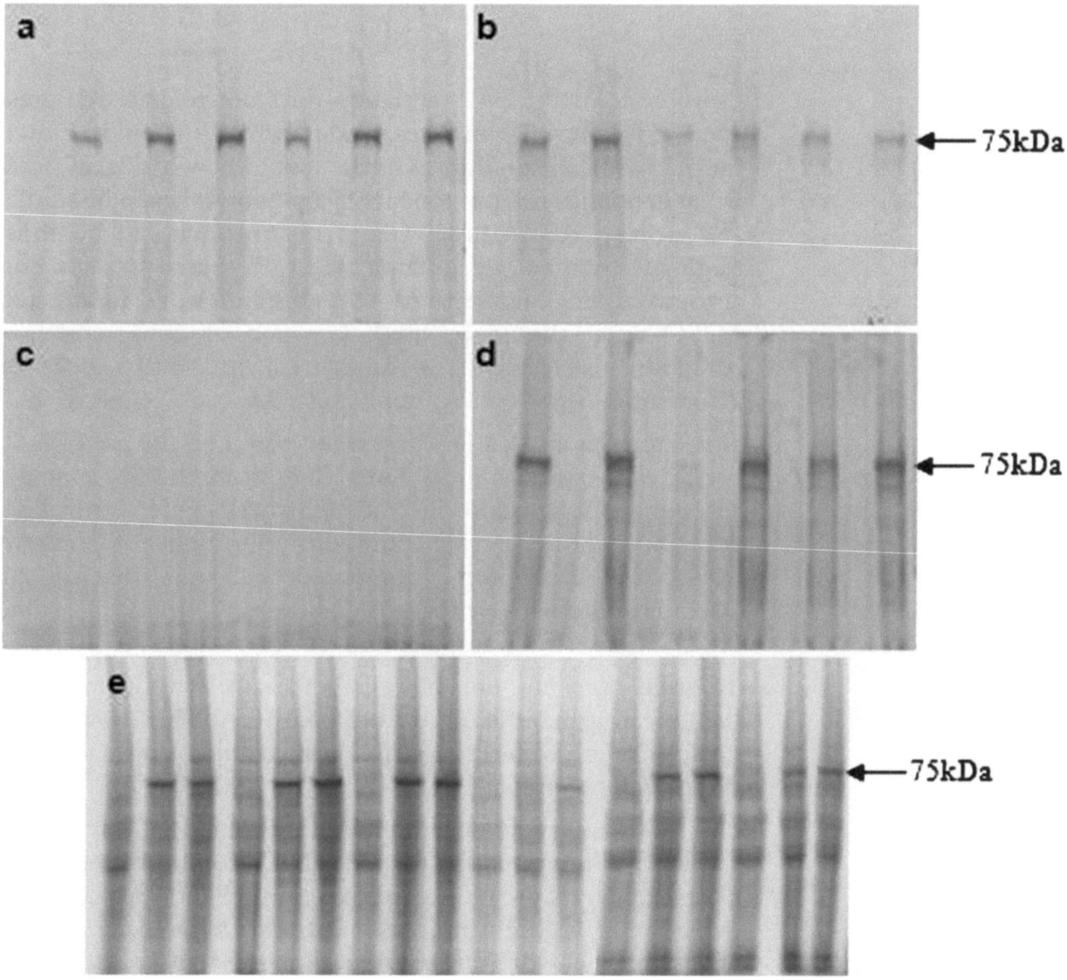

Fig. 3 Example of in vitro expression profiling. Autoradiography of SDS-PAGE gels. Samples correspond to different cell-free expression conditions. For prokaryotic extracts (**a**) *E. coli* stain 1, (**b**) *E. coli* strain 2, (**c**) *Bacillus,* and (**d**) *Pseudomonas* extracts, six different salt mix compositions have been evaluated. Each two consecutive samples correspond respectively to negative control (in vitro transcription translation without DNA) and target protein expression reactions. For the eukaryotic extract (**e**), in addition to salt concentrations, two concentrations of DNA were evaluated. Each three consecutive samples correspond respectively to negative control (in vitro transcription translation without DNA) and target protein expression reactions with 16 nM and 32 nM DNA

3.11 Binding Assay with Scintillation Counting

Coat the ligand and BSA (as control) for nonspecific interaction to the plates through any classical method and store at +4 °C until use. Prepare radiolabeled transcription translation reaction as described before with optimal conditions defined. Prepare at least tenfold excess volume of cold reaction to measure specific protein–ligand interaction. Desalt your cell-free expression reactions on a desalting column equilibrated with PBS buffer containing 1 % BSA (NAP-25 column from GE Healthcare or for smaller volumes Micro Bio-Spin® P6 BIORAD) (*see* **Note 14**).

4 Notes

1. Cell-free transcription/translation does not tolerate the presence of RNAses. RNAse-free grade chemicals should be used for preparing all solutions. Gloves should be worn at all times when handling the components. Disposables should be sterile, DNAse and RNAse free. Glass containers should be thoroughly washed and decontaminated with special solution and autoclaved. For the preparation of all solution, DEPC-treated MilliQ water should be used. For transcription/translation reactions, use standard sterilized 1.5 mL micro-tubes or U-bottom 96-wells microtiter plates such as COSTAR 3799.
2. Constant System Cell Disruptors are based on the use of high pressure are based on the use of high pressure to force a sample through a small fixed orifice at high speed under controlled, contained, and repeatable conditions. Our model of Constant System Cell Disruptor is composed of 2 units that can be assembled alternatively for small volumes, up to 10 mL (One Shot) or larger volumes, from dozens of milliliters to liters.
3. To grind larger amounts of fungal cakes, we recommend using the mortar grinder RM100 from RETSCH. This allows cryogenic grinding with reproducible results due to adjustable pestle and scraper pressure.
4. As DNA template, you can use either desalted PCR template, linear DNA template, or plasmid DNA. From our experience, plasmid DNA is the most efficient template and less subject to degradation. Moreover, plasmid template preparation can affect the yield of transcription. For efficient transcription, DNA has to be supercoiled. For this purpose we recommend QIAGEN midiprep preparation kit. All the reaction mixes given in this publication are given for plasmid DNA template. For PCR templates, you should optimize parameters such as magnesium, DTT, or buffer concentration.
5. For eukaryotic extracts we chose *Trichoderma* instead of wheat germ, due to the set of enzymes we have to express. Moreover it allows us to overcome the difficulties in obtaining reproducible high-quality wheat germ supplies.
6. In addition to functional cellular machinery cell-free lysates, contain endogenous mRNA that can compete with target mRNA during in vitro translation. These mRNAs are either free or engaged in translation complexes. For prokaryotic extracts, they are most often engaged in translation complexes (as transcription and translation are coupled). Nevertheless the "run off" has two advantages: first, it completes the translation of engaged mRNAs and make the ribosomes totally available for target mRNA and, second, it degrades endogenous free mRNA as their lifetime in cellular medium is very short.

7. Always make small aliquots of S30 extract to maintain homogeneity during freezing. For preparation of larger volumes, you can drop the extract in liquid nitrogen in a strainer. You can then easily collect beads of S30 extract and transfer them into tubes for storage.
8. For better preservation, fungal strains are usually stored as spores. Therefore, a first step of growth in a specific medium is necessary to make spores germinate into young mycelia before culture steps.
9. If the extract is not concentrated enough, it may be poorly efficient and fragile. We recommend to concentrate it by a two step dialysis (10,000 MWCO). The first step is against eukaryotic *Solution B* containing around 10 % PEG and 10 % glycerol for 1 h 30 min and a second step against eukaryotic *Solution B* containing around 10 % glycerol for 20 min.
10. "Run off" preincubation is not possible for eukaryotic S30 extracts as their efficiency will drastically decrease. Use micrococcal nuclease treatment if needed to reduce background. The endogenous mRNA can be eliminated by incubation with Ca^{2+}-dependent micrococcal nuclease (S7 Nuclease), which is later inactivated by chelation of the Ca^{2+} by EGTA. Micrococcal nuclease is a relatively nonspecific endo-exonuclease that digests single-stranded and double-stranded nucleic acids, although it is more active on single-stranded substrates. It can thus be hard to control without affecting the efficiency of the extract. Therefore, this treatment is not recommended when screening for functional activity but only for labeled amino acid incorporation if necessary.
11. For prokaryotic extracts OD_{260} is generally from 300 to 500 with an OD_{260}/OD_{280} ratio of 1.7–1.8. For eukaryotic extracts OD_{260} is generally from 50 to 80 for non-concentrated extracts and from 150 to 250 for concentrated ones with an OD_{260}/OD_{280} ratio of 1.6–1.8.
12. When performing *E. coli* S30 based in vitro transcription/translation, incubating at 30 °C is more appropriate than 37 °C, in order to slow down transcription and maintain the coupling of transcription and translation. This insures mRNA preservation and sometimes better folding of translated proteins. Agitation during cell-free reaction often improved the yield of expression.
13. For safe manipulation of the radiolabeled reaction, we eliminate free methionine that has not been incorporated into proteins through acetone precipitation. After incubation, add 80 μL of acetone to the 20 μL reaction. Vortex to homogenize and place at least 30 min at −20 °C. Centrifuge 5 min at 6,000 × *g*, throw out the supernatant with respect of your lab

procedure for radioactive waste, and resuspend the pellet in 20 μL Laemmli 1×. Load 10 μL of each sample on SDS-PAGE precast gels.

14. Desalting the cell-free radiolabeled reactions on the appropriate columns reduces the level of radioactivity in solutions, is safer, and facilitates the treatment of radioactive waste. In this way, free labeled methionine will be entrapped into the desalting gel, whereas labeled proteins will pass through. This technique can also decrease background for scintillation counting.

Acknowledgements

We thank Dr. Lyuba Ryabova for her useful contribution to the development of some of the mentioned protocols. We also thank Stephanie Pallas and Elisabeth Senaux for technical contribution.

References

1. Ravot G, Wahler D, Favre-Bulle O et al (2003) High throughput discovery of alcohol dehydrogenases for industrial biocatalysis. Adv Synth Catal 345:691–694
2. Ryabova LA, Guillemer S, Pallas S et al (2008) A quick in vitro pathway from prokaryotic genomic libraries to enzyme discovery. Biotechniques 45:63–68
3. Zubay G (1973) *In vitro* synthesis of protein in microbial systems. Annu Rev Genet 7: 267–287
4. Kim DM, Swartz JR (2000) Prolonging cell-free protein synthesis by selective reagent additions. Biotechnol Prog 16:385–390
5. Spirin AS, Swartz JR (eds) (2008) Cell free protein synthesis: methods and protocols. Wiley-VCH, Weinheim
6. Mikami S, Kobayashi T, Masutani M et al (2008) A human cell-derived in vitro coupled transcription/translation system optimized for production of recombinant proteins. Protein Expr Purif 62:190–198
7. Kobayashi T, Mikami S, Yokoyama S et al (2007) An improved cell-free system for picornavirus synthesis. J Virol Methods 142: 182–188
8. Gallie DR, Walbot V (1992) Identification of the motifs within the tobacco mosaic virus 5′-leader responsible for enhancing translation. Nucleic Acids Res 20(17):4631–4638
9. Sawasaki T, Ogasawara T, Morishita R et al (2002) A cell-free protein synthesis system for high-throughput proteomics. Proc Natl Acad Sci U S A 99(23):14652–14657
10. Kovtun AA, Shirokikh NE, Gudkov AT et al (2007) The leader sequence of tobacco mosaic virus RNA devoid of Watson–Crick secondary structure possesses a cooperatively melted, compact conformation. Biochem Biophys Res Commun 358(1):368–372
11. Pelham HRB, Jackson RJ (1976) An efficient mRNA dependent translation system from reticulocyte lysates. Eur J Biochem 67: 247–256
12. Madin K, Sawasaki T, Ogasawara T et al (2000) A highly efficient and robust cell-free protein synthesis system prepared from wheat embryos: plants apparently contain a suicide system directed at ribosomes. Proc Natl Acad Sci U S A 97(2):559–564
13. Endo Y, Sawasaki T (2004) High-throughput, genome-scale protein production method based on the wheat germ cell-free expression system. J Struct Funct Genomics 5(1–2): 45–57
14. Ryabova LA, Vinokurov LM, Shekhovtsova EA et al (1995) Acetyl phosphate as an energy source for bacterial cell-free translation systems. Anal Biochem 226:184–186
15. Ryabova LA, Desplancq D, Spirin AS et al (1997) Functional antibody production using cell-free translation: effects of protein disulfide isomerase and chaperones. Nat Biotechnol 15(1):79–84
16. Klein G, Reymond JL (1999) Enantioselective fluorogenic assay of acetate hydrolysis for detecting lipase catalytic antibodies. Helv Chim Acta 82:400–407

17. Carlòn RP, Jourdain N, Reymond JL (2000) Fluorogenic polypropionate fragments for detecting stereoselective aldolases. Chem Eur J 6:4154–4162

18. Grognux J, Wahler D, Nyfeler E et al (2004) Universal chromogenic substrates for lipases and esterases. Tetrahedron Asym 15:2981–2989

19. Ravot G, Buteux D, Favre-Bulle O et al (2004) Screening for thermostable esterases: from deep sea to industry. Eng Life Sci 4:533–538

20. Badalassi F, Wahler D, Klein G et al (2000) A versatile periodate-coupled fluorogenic assay for hydrolytic enzymes. Angew Chem Int Ed 39:4067–4070

Chapter 9

Human Cell Extract-Derived Cell-Free Systems for Virus Synthesis

Tominari Kobayashi, Kodai Machida, and Hiroaki Imataka

Abstract

Cell-free synthesis of an infectious virus is an ideal tool for elucidating the mechanism of viral replication and for development of antiviral drugs. In this chapter, the synthesis of *Encephalomyocarditis virus* (EMCV) from RNA and DNA in a HeLa cell extract-derived in vitro protein expression system is described. When a synthetic EMCV RNA with a hammerhead ribozyme sequence at its 5′-end is incubated with a HeLa cell extract using a dialysis system, EMCV particles are progressively synthesized. For EMCV synthesis from DNA, a plasmid harboring the full-length cDNA of EMCV with the T7 promoter/terminator unit is incubated in the HeLa cell extract supplemented with T7 RNA polymerase.

Key words Cell-free, *Encephalomyocarditis* virus, RNA virus, Virus particles, Virus synthesis human cells, HeLa cells, Coupled transcription/translation

1 Introduction

Cell-free technology has made it possible to produce picornaviruses, such as poliovirus, mengovirus, and *Encephalomyocarditis virus* (EMCV), in mammalian cell extracts using genomic or synthetic RNAs as templates. Since the in vitro virus synthesis system faithfully recapitulates the in vivo biogenesis, this technology has enabled detailed analyses of replication of the picornavirus [1–5]. A recent advance in the cell-free technique enables the synthesis of EMCV even from a DNA template, circumventing handling of the viral RNA [6]. In this chapter, we describe HeLa cell-derived cell-free systems configured for the synthesis of EMCV from RNA and DNA templates.

Kirill Alexandrov and Wayne A. Johnston (eds.), *Cell-Free Protein Synthesis: Methods and Protocols*, Methods in Molecular Biology, vol. 1118, DOI 10.1007/978-1-62703-782-2_9, © Springer Science+Business Media, LLC 2014

2 Materials

2.1 Cell Culture

1. The HeLa S3 cell line (*see* **Note 1**).
2. Minimum essential medium for suspension culture S-MEM, SMEM, or JMEM. Store at 4 °C.
3. Calf serum (*see* **Note 2**). Store in 50 ml aliquots at −20 °C.
4. GlutaMAX. Store at 4 °C.

2.2 Cell Extracts

1. Washing buffer: 35 mM HEPES–KOH (pH 7.5), 140 mM NaCl, 11 mM glucose. Store at 4 °C.
2. Extraction buffer: 20 mM HEPES–KOH (pH 7.5), 45 mM potassium acetate, 45 mM KCl, 1.8 mM magnesium acetate, 1 mM dithiothreitol (DTT). Store at 4 °C and add DTT just before use.
3. High-potassium buffer: 20 mM HEPES–KOH (pH 7.5), 945 mM potassium acetate, 945 mM KCl, 1.8 mM magnesium acetate, 1 mM DTT. Store at 4 °C and add DTT just before use.
4. Mini-bomb cell disruption chamber.

2.3 Plasmid and mRNA

1. Plasmid: pUC18 EMCV Rbz [2] (*see* **Note 3**).
2. A commercially available mRNA synthesis kit.
3. RNA purification column: Chroma Spin + TE-30 column.
4. TE buffer: 10 mM Tris–HCl (pH 7.5), 1 mM EDTA.
5. 10 % SDS, 0.5 M EDTA.
6. Proteinase K.
7. Phenol, chloroform, ethanol.

2.4 Cell-Free System

1. ATP, GTP, UTP, CTP, and creatine phosphate: Each nucleotide and creatine phosphate are dissolved in 240 mM HEPES–KOH, pH 7.5, to the final concentrations of 200 mM each and 1.0 M, respectively. Store at −80 °C.
2. The amino acid mixture consisting of 20 amino acids: 12 mM L-arginine; 6.7 mM L-glutamine, L-isoleucine, L-leucine, L-lysine, L-threonine, and L-valine; 3.3 mM L-alanine, L-ASPARAGINE, L-aspartic acid, L-glutamic acid, L-glycine, L-HISTIDINE, L-phenylalanine, L-proline, L-serine, and L-tyrosine; 1.7 mM L-cystine, and L-methionine; 0.8 mM L-tryptophan. Store at −80 °C.
3. Creatine kinase is dissolved in 20 mM HEPES–KOH, pH 7.5, 50 % glycerol to the final concentration of 30 mg/ml. Store at −30 °C.
4. Calf liver tRNA: Store at −80 °C.
5. K3L and GADD34 proteins [7, 8]. Store at −80 °C.

6. Mixture-1: 6 μl of GADD34 (0.2 mg/ml), 6 μl of K3L (0.4 mg/ml). Store at −80 °C (*see* **Note 4**).
7. Mixture-2: 56 μl of magnesium acetate (63 mM), 354 μl of potassium acetate (157 mM), 82.5 μl of DTT (100 mM), and 158 μl of HEPES–KOH (400 mM), pH 7.5. Store at 4 °C without DTT and add DTT just before use.
8. Mixture-3: 11.3 μl of ATP (200 mM), 1.1 μl of GTP (200 mM), 36 μl of creatine phosphate (1 M), 18 μl of creatine kinase (6 mg/ml), 18 μl of calf liver tRNA (9 mg/ml), 18 μl of the mixture of 20 amino acids (*see* **item 2** in Subheading 2.4), and 77.6 μl of H_2O. Store at −80 °C.
9. Mixture-4: Consists of the same ingredients as in mixture-2 except that the concentration of magnesium acetate is 98 mM. Store at 4 °C without DTT and add DTT just before use.
10. Mixture-5: 11.3 μl of ATP (200 mM), 7.5 μl each of GTP, CTP, and UTP (200 mM each), 36 μl of creatine phosphate (1 M), 18 μl of creatine kinase (6 mg/ml), 18 μl of calf liver tRNA (9 mg/ml), 18 μl of a mixture of 20 amino acids (*see* **item 2** in Subheading 2.4), 18 μl of spermidine (50 mM), 20 μl of T7 RNA polymerase (1.0 mg/ml) (*see* **Note 5**), and 18.2 μl of H_2O. Store at −80 °C.
11. Dialysis membrane (molecular weight cutoff 50,000 daltons, regenerated cellulose).
12. Outer buffer-1: 1,480 μl of H_2O; 12.5 μl of ATP (200 mM); 1.25 μl of GTP (200 mM); 40 μl of creatine phosphate (1 M); 20 μl of the mixture of 20 amino acids; 10 μl of DTT (1 M); 10 μl of EGTA, pH 7.5 (300 mM); 5.4 μl of magnesium acetate (1 M); 0.3 ml of HEPES–KOH (400 mM), pH 7.5, 13.3 μl of KCl (3 M); and 108 μl of potassium acetate (1.5 M). Prepare fresh every time before use.
13. Outer buffer-2: 1,453 μl of H_2O; 12.5 μl of ATP (200 mM); 8.3 μl each of GTP, CTP, and UTP (200 mM each); 40 μl of creatine phosphate (1 M); 20 μl of the mixture of 20 amino acids; 10 μl of DTT (1 M); 10 μl of EGTA, pH 7.5 (300 mM); 1 μl of spermidine (1 M); 7.6 μl of magnesium acetate (1 M); 0.3 ml of HEPES–KOH (400 mM), pH 7.5; 13.3 μl of KCl (3 M); and 108 μl of potassium acetate (1.5 M). Prepare fresh every time before use.

2.5 Viral Titer

1. RNase A solution (10 mg/ml) and TurboDNase. Store at −30 °C.
2. The BHK-21 cell line (*see* **Note 1**).
3. Dulbecco's modified minimum essential medium (DMEM). Store at 4 °C.
4. Fetal calf serum. Store in 50 ml aliquots at −20 °C.
5. Agar.

3 Methods

We describe the cell-free systems for EMCV synthesis using extracts from HeLa cells. There are two choices for the template: RNA and DNA (Fig. 1). For the RNA template system, the full-length EMCV RNA with the ribozyme at the 5′-end is synthesized from the corresponding plasmid and is incubated with the HeLa cell extract supplemented with various reagents and enhancing factors (GADD34 and K3L). For the DNA template system, the same plasmid is directly incubated with the HeLa cell extract further supplemented with T7 RNA polymerase so that the coupled transcription/translation can proceed efficiently. In both cases, the dialysis system that continuously supplies the substrates and energy source for protein synthesis and removes waste products through

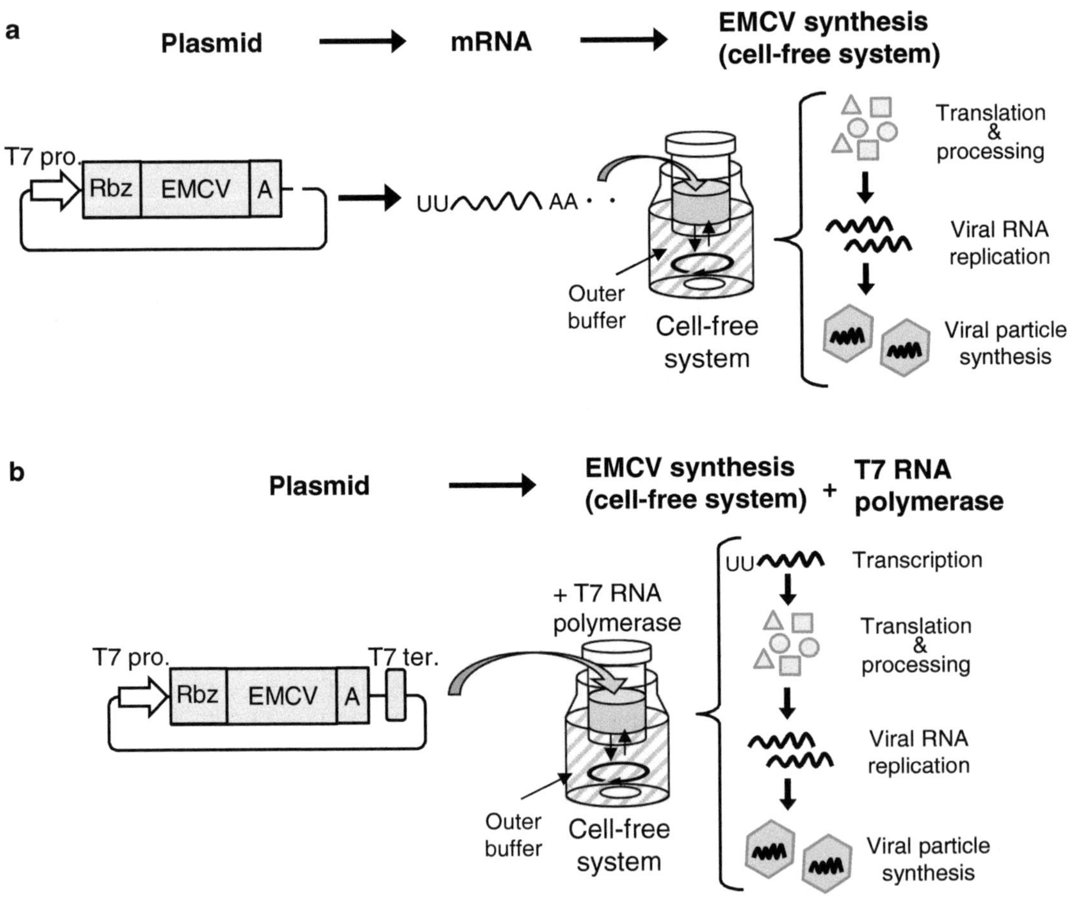

Fig. 1 RNA-dependent and DNA-dependent cell-free systems for EMCV synthesis. (**a**) mRNA-dependent system. EMCV RNA is synthesized in vitro and purified. The purified EMCV RNA is incubated with the HeLa cell extract by the dialysis system. (**b**) DNA-dependent system. The plasmid encoding the EMCV RNA is directly incubated with the HeLa cell extract by the dialysis system. *T7 pro* T7 promoter, *Rbz* ribozyme, *T7 ter* T7 terminator, *A* poly-A tail

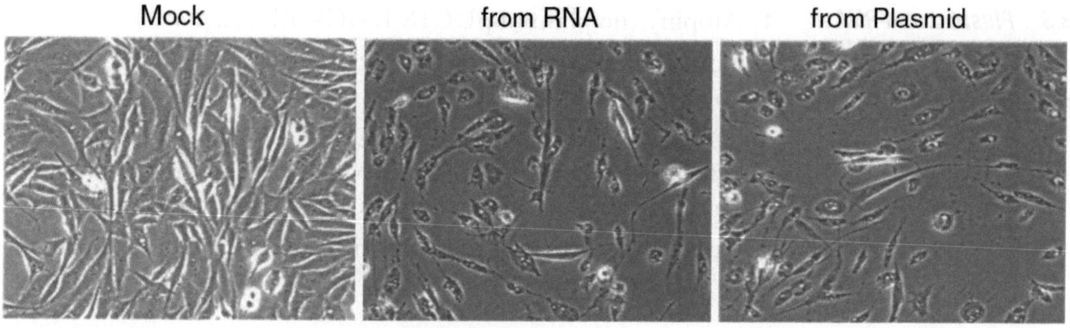

Fig. 2 Infection of BHK-21 cells with EMCV synthesized by the cell-free system. BHK-21 cells grown on a 6-well plate were incubated with medium (600 μl; DMEM plus 10 % fetal bovine serum) containing the RNase-/DNase-treated sample (*see* **step 1** in Subheading 3.5.3) (10 μl) from the RNA- or plasmid-programmed system for 1 h at 37 °C. After culture with additional medium (3.4 ml) for 20 h, cells were observed by microscopy. Mock: untreated BHK-21 cells

a dialysis membrane is preferred to the batch system (Fig. 1). Synthesis of the virus can be easily tested by inoculating cultured BHK-21 cells with the sample (Fig. 2). The titer of the synthesized virus is estimated by plaque-forming assay.

3.1 Cell Culture

1. Culture HeLa S3 cells at 37 °C in S-MEM, SMEM, or JMEM supplemented with 10 % heat-inactivated calf serum and GlutaMAX (2 mM) (*see* **Note 6**) using a spinner flask connected with a cell culture controlling system (*see* **Note 7**) with the control values of temperature (37 °C), pH (7.2), oxygen density (6.7 ppm), and stirring speed (50 rpm).
2. Harvest the cells when the cell density reaches 0.8–1.0×10^6 cells/ml (*see* **Note 8**).

3.2 Preparation of Cell Extracts

1. Wash the cells three times with the washing buffer and once with the extraction buffer (ten times volume of the cell pellet, for each wash).
2. Resuspend the cell pellet in an equal volume of the extraction buffer (to approximately 3.0×10^8 cells/ml).
3. Disrupt cells by nitrogen pressure (1.0 MPa, 30 min saturation time) in the mini-bomb cell disruption chamber on ice.
4. Mix the cell homogenate with the high-potassium buffer (1/29 volume of the cell homogenate).
5. Centrifuge twice at $1{,}200 \times g$ for 5 min at 4 °C and recover the supernatant.
6. Divide the supernatant into aliquots, and freeze them in liquid nitrogen.
7. Store the frozen aliquots at −80 °C (*see* **Note 9**).

3.3 Plasmid and RNA

3.3.1 Plasmid Preparation

1. Amplify the plasmid pUC18 EMCV Rbz [2].
2. Purify the plasmid using a commercially available kit, and dissolve it in 450 μl TE buffer.
3. Add 25 μl SDS (10 %), 10 μl EDTA (0.5 M), and 10 μl proteinase K (20 mg/ml) and incubate for 30–60 min at 37–50 °C to inactivate possibly contaminating RNases.
4. Purify the plasmid by treating with phenol and chloroform followed by ethanol precipitation.

3.3.2 RNA Preparation

1. Digest the plasmid pUC18 EMCV Rbz with Sal I.
2. Synthesize RNA using the digested plasmid as the template with a commercially available RNA synthesis kit.
3. Purify the synthesized RNAs by using Chroma Spin + TE-30 column.

3.4 Cell-Free mRNA-Dependent Translation

3.4.1 Batch Methods

1. Preincubate the HeLa cell extract (7.5 μl) with mixture-1 (1.2 μl) and mixture-2 (6.5 μl) at room temperature for 10 min (Fig. 2) (*see* **Note 4**).
2. Add mixture-3 (1.8 μl) and mRNA (1.0 μl, 36 ng/μl final concentration).
3. Incubate the mixture (18 μl, total volume) at 32 °C for 1.0 h.

3.4.2 Dialysis Methods

1. Inject the incubation mixture (180 μl, total volume) described in the batch method section (*see* **steps 1–3** in Subheading 3.4.1) into a chamber with a dialysis membrane (molecular weight cutoff 50,000, regenerated cellulose).
2. Incubate the chamber at 32 °C for 3–12 h being dialyzed against outer buffer-1 (2 ml).

3.5 Cell-Free Coupled Transcription/Translation

3.5.1 The Batch Method

1. Preincubate the HeLa cell extract (7.5 μl) with mixture-1 (1.2 μl) and mixture-4 (6.5 μl) at room temperature for 10 min (*see* **Note 4**).
2. Add mixture-5 (1.8 μl) and the plasmid (1.0 μl) (15 ng/μl, final concentration).
3. Incubate the mixture (18 μl, total volume) at 32 °C for 1–3 h.

3.5.2 Dialysis Method

1. Inject the incubation mixture (180 μl, total volume) described in the batch method section (*see* **steps 1–3** in Subheading 3.5.1) into a cup with a dialysis membrane (*see* **item 11** in Subheading 2.4).
2. Incubate the chamber at 32 °C for 3–12 h being dialyzed against outer buffer-2 (2 ml) (*see* **Note 10**).

3.5.3 Viral Titer

1. The sample after translation is treated with RNase A (final concentration, 100 μg/ml) and TurboDNase (20 units/ml) for

30 min at room temperature to inactivate the input RNA and plasmid.

2. The nuclease-treated sample is serially diluted (10^{-2}–10^{-6}) with DMEM (>1 ml).
3. Confluent BHK-21 cells on a 60 mm dish are incubated with each diluted sample (1 ml) for 1 h for infection.
4. Following removal of the sample medium, an agar-containing medium [0.5 ml agar (2.5 % in water) plus 2 ml DMEM with 10 % FCS] is overlaid on the cells (*see* **Note 11**).
5. After the agar layer becomes solid (about 30 min at room temperature), 2.5 ml of DMEM with 10 % FCS is overlaid and incubated for 30 h at 37 °C.
6. Count plaques formed on each dish and estimate the number of the virus particles synthesized by the cell-free system (*see* **Note 12**).

4 Notes

1. When purchased, cells should be propagated to make more than 50 frozen cell stocks. Avoid continued culture of cells for longer than 1 month.
2. It is important to use the serum lot which promotes cells to grow with a doubling time of ~1 day at the concentration of 10 %.
3. A ribozyme sequence is encoded between the T7 RNA promoter and the 5′-end sequence of the EMCV genome so that the RNA transcribed by T7 RNA polymerase can possess the authentic 5′-end of the EMCV genome RNA [2].
4. Preincubation with K3L/GADD34 prior to addition of mixture-3 is important, because ATP/creatine phosphate present in mixture-3 enhances phosphorylation of eIF2α [7, 8].
5. T7 RNA polymerase can be expressed in bacteria and purified to homogeneity with a comparable activity to commercially available T7 RNA polymerase [9].
6. Addition of GlutaMAX to the culture medium helps cells grow and leads to an enhanced protein synthesis in the cell-free systems, even when L-glutamine is already included.
7. If an appropriate controlling system for culturing cells is not available, culturing cells on Petri dishes (20–25 ml medium per 150 mm dish) is recommended.
8. It is very important to recover cells when they are in the logarithmic growth phase.
9. The extracts can be kept at −80 °C at least for 6 months without a loss of translation activity.

10. For a preparative purpose, 1.0 ml of the reaction mixture in dialysis tube is dialyzed against 20–40 ml of outer buffer-2.
11. Agar is overlaid to restrict the cell-to-cell infection within neighboring cells.
12. Each plaque is considered to derive from one infectious viral particle.

References

1. Kobayashi T, Machida K, Mikami S et al (2011) Cell-free RNA replication systems based on a human cell extracts-derived in vitro translation system with the encephalomyocarditis virus RNA. J Biochem 150:423–430
2. Kobayashi T, Mikami S, Yokoyama S et al (2007) An improved cell-free system for picornavirus synthesis. J Virol Methods 142:182–188
3. Molla A, Paul AV et al (1991) Cell-free, de novo synthesis of poliovirus. Science 254:1647–1651
4. Svitkin YV, Sonenberg N (2003) Cell-free synthesis of encephalomyocarditis virus. J Virol 77: 6551–6555
5. Svitkin YV, Herdy B, Costa-Mattioli M et al (2005) Eukaryotic translation initiation factor 4E availability controls the switch between cap-dependent and internal ribosomal entry site-mediated translation. Mol Cell Biol 25:10556–10565
6. Kobayashi T, Nakamura Y, Mikami S et al (2012) Synthesis of encephalomyocarditis virus in a cell-free system: from DNA to RNA virus in one tube. Biotechnol Lett 34:67–73
7. Mikami S, Kobayashi T, Yokoyama S et al (2006) A hybridoma-based in vitro translation system that efficiently synthesizes glycoproteins. J Biotechnol 127:65–78
8. Mikami S, Kobayashi T, Machida K et al (2010) N-terminally truncated GADD34 proteins are convenient translation enhancers in a human cell-derived in vitro protein synthesis system. Biotechnol Lett 32:897–902
9. Mikami S, Kobayashi T, Masutani M et al (2008) A human cell-derived in vitro coupled transcription/translation system optimized for production of recombinant proteins. Protein Expr Purif 62:190–198

Chapter 10

Cell-Free Protein Synthesis in Microfluidic 96-Well Plates

Kirsten Jackson, Ruba Khnouf, and Z. Hugh Fan

Abstract

Cell-free protein synthesis (CFPS) enables rapid protein expression for the structural and functional characterization of proteins. Implementation of CFPS in a microfluidic platform has additional benefits such as reduced reaction volumes and simultaneous expression of multiple proteins. Here, we describe a microfluidic device that is composed of 96 continuous-exchange cell-free protein expression units and produces a protein synthesis yield up to 87 times higher than a conventional batch system.

Key words Cell-free protein synthesis, Microplates, Protein expression, Microfluidics, Proteomics, Microarray, Wheat germ extract

1 Introduction

Methods for rapidly expressing proteins have become a popular area of research to characterize the functional and structural properties of proteins. The current methods available for protein expression include chemical, cell-based, and cell-free protein synthesis; however, the chemical and cell-based systems are limited to the synthesis of short and nontoxic polypeptide chains, respectively [1]. Cell-free protein synthesis (CFPS) systems eliminate these limitations by operating without a complete cell. To have sufficient synthesis yield, CFPS systems generally require two components: a reaction solution containing transcriptional and translational machinery (ribosomes, DNA, mRNA, tRNAs, RNA polymerase, etc.) and a feeding solution containing energy and nutrients (amino acids, ATP, GTP, etc.). These systems enable rapid expression, simple operation, low reaction volume, reduced proteolytic degradation, and production of proteins with cytotoxicity [1–3].

CFPS systems have been studied in several forms including conventional batch process, bilayer diffusion, continuous-flow cell-free (CFCF) schemes, and continuous-exchange cell-free (CECF) methods. In the conventional batch system, the duration of protein synthesis is limited by the consumption of nutrients and production

Kirill Alexandrov and Wayne A. Johnston (eds.), *Cell-Free Protein Synthesis: Methods and Protocols*, Methods in Molecular Biology, vol. 1118, DOI 10.1007/978-1-62703-782-2_10, © Springer Science+Business Media, LLC 2014

of by-products (inorganic phosphates) that inhibit protein synthesis. In 1999, Kim and Swartz used a batch system in which all components necessary for cell-free protein synthesis were combined to synthesize human lymphotoxin for up to 2 h [4]. Compared to the conventional batch system, the bilayer diffusion system enables the continuous supply of nutrients and removal of by-products through a bilayer resulting from the feeding solution overlaying the reaction solution [5]. Sawasaki et al. developed such a system in 2002 that synthesized proteins for up to 14 days [5, 6]. Similar to the bilayer diffusion system, the CFCF system enables the continuous supply of nutrients and removal of by-products by incorporating a semipermeable membrane in an active system, capable of supporting protein expression for several hours. In this system, the feeding solution is continuously provided to a reaction solution using a pump. As the mixture proceeds through the system, a semipermeable membrane separates the synthesized protein from the reaction by-products, eliminating their inhibitory effects. Spirin et al. created an active CFCF translation system that had total synthesis duration of 20 h [7, 8]. The CECF system also enables prolonged protein synthesis in comparison to the conventional batch system. Unlike the CFCF system, the CECF system involves the passive exchange of substrates and by-products through a dialysis membrane separating the reaction and feeding solutions. As a result, all of the nutrients and energy molecules for protein synthesis are continuously supplied while the obstructive by-products are removed in a simple system [2]. This format has been extensively studied by several researchers [9–16].

In this chapter, high-throughput protein expression is demonstrated with a miniaturized CECF system that is capable of producing up to 96 proteins simultaneously within hours. The device consists of three layers. The bottom and top layers are composed of an array of 96 wells. The bottom well of each unit functions as a feeding chamber, and it is accessed through three access ports in the top layer. The well in the top layer functions as a reaction chamber. The middle layer is a dialysis membrane that enables continuous exchange of the feeding solution and the reaction by-products. Each protein is produced through a coupled transcription and translation reaction in the 96-well continuous-exchange cell-free system. The device has been used for the production of six proteins with a combination of five enzymatic assays. Compared to a conventional microplate, this miniaturized CECF system was able to produce up to 87 times higher protein synthesis yields [13, 15, 16].

2 Materials

2.1 96-Well CECF Device

1. Top layer: a 4.8 mm thick polypropylene sheet (Plastruct).
2. Bottom layer: a 6.35 mm thick polypropylene sheet (Plastruct).

3. Middle layer: a dialysis membrane with an 8,000 Da molecular cutoff (Spectrum Labs) (*see* **Note 1**).
4. A commercial CNC mill (*see* **Note 2**).
5. Ethanol (*see* **Note 3**).
6. Poly(dimethylsiloxane) (PDMS) (Sylgard 184, Dow Corning).
7. C-clamps.
8. A commercial spin coater (*see* **Note 4**).
9. A commercial furnace capable of 60 °C.

2.2 Protein Expression

1. Nuclease-free water (*see* **Note 5**).
2. Cell-free systems (*see* **Note 6**): RTS 100 wheat germ kit (5 Prime) (*see* **Note** 7) and RTS 500 *E. coli* kit (5 Prime) (*see* **Note 8**). Store at –20 °C.
3. Control vectors: T7 control vector (Promega) and pDNR-LacZ vector (Clontech). Store at 20 °C.
4. A commercial PCR machine.
5. Polymerase chain reaction (PCR) reagents for producing β-lactamase (β-lac) and alkaline phosphatase (AP) vectors: PUC 18 template DNA vector and chromosomal DNA from *Pseudomonas aeruginosa*, primers β-lac (5′-AAAAGCGGCCGC zATGAGTATTCAACATTTCCGTG-3′ and 5′-GGTCGGAT CCTTACCAATGCTTAATCAGTG-3′) and AP (5′-AGGGA GCGGCCGCATGACCCAGGTTATCCCCTCGCCCTC-3′ and 5′-TCGTTCGGATCCGATCAGTCGCGCAGG TTCAG TGCGC-3′), 10× PCR buffer, dimethylsulfoxide (DMSO), deoxynucleotide triphosphates (dNTPs), and Taq polymerase enzyme.
6. Agarose gel electrophoresis apparatus.
7. A commercial gel and PCR cleanup kit (*see* **Note 9**).
8. GEM-T easy vector system (Promega).
9. A commercial water bath.
10. Materials used for centrifugation: a commercial centrifuge, washing buffer, resuspension buffer, cell lysis buffer, and neutralization buffer.
11. Reagents for ligation reactions: 10× buffer, 2× rapid ligation buffer, pGEM vector, restriction enzymes Notl and BamH1, GUS vector fragment, ligase enzyme, and T4 DNA ligase.
12. Reagents for colony growth and verification: DH5α *E. coli* cells, L-broth, a glass tube, agar plates, ampicillin, a commercial miniprep system (*see* **Note 10**), glycerol, sterile sticks, and X-gal. L-Broth was prepared by adding 10 g of tryptone, 5 g of yeast extract, and 10 g of NaCl into pure H_2O to make 1 L of solution, followed by thorough mixing and immediate

autoclave. Store the DH5α *E. coli* cells at −80 °C and thaw on ice for 5 min before use.

13. Standard 384-well black microplate.
14. A commercial orbital shaker.
15. Commercial PCR tapes (*see* **Note 11**). Store in a cool and dry location.

2.3 Protein Detection

1. A commercial plate reader (*see* **Note 12**).
2. Substrates for enzymatic assays: 4-methylumbelliferyl β-D-glucuronide (MUG, Marker Gene Technologies), luciferase assay reagent (LAR, Promega), fluorescein mono-β-D-galactopyranoside (FMG, Marker Gene Technologies), 3-phenylumbelliferone 7-O-phosphate hemipyridinium (PPH, Marker Gene Technologies), and *m*-([(phenylacetyl)glycyl] oxy)benzoic acid (PBA, Calbiochem).

3 Methods

3.1 Device Fabrication

1. For the top layer, machine a 12 by 8 matrix of 4.5 mm diameter reaction chambers surrounded by three 2 mm diameter access ports (*see* **Note 13**) in the 4.8 mm thick polypropylene sheet (*see* **Note 14**).
2. For the bottom layer, machine a 12 by 8 array of 8 mm diameter wells in the 6.35 mm thick polypropylene sheet (*see* **Note 15**).
3. Rinse the top and bottom layers of polypropylene thoroughly with ethanol.
4. Prepare 15 g of PDMS monomer mixture according to the manufacturer's instructions for a 10:1 ratio of the base solution to the curing agent.
5. Spin approximately 5 g of PDMS into a thin film using the spin coater at a speed of 500 RPM for 30 s.
6. Transfer PDMS to the bottom of the top layer by pressing the bottom of the top layer onto the spun PDMS (*see* **Note 16**).
7. Spin another thin film of 5 g of PDMS at 500 RPM for 30 s.
8. Transfer PDMS to the top of the bottom layer by pressing the top of the bottom layer onto the spun PDMS (*see* **Note 17**).
9. Quickly assemble the three layers together and secure them using C-clamps (*see* **Note 18**). No preparation of the dialysis membrane is necessary.
10. Cure them in an oven at 60 °C overnight. Following curing, the completed device may be stored at room temperature (Fig. 1).

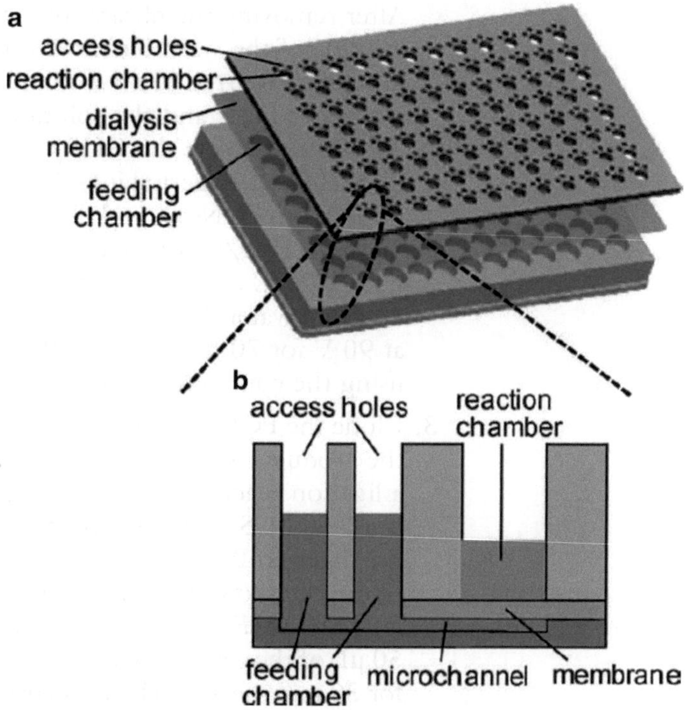

Fig. 1 Schematic of the 96-well device. (**a**) Illustration of the assembled device with the bottom layer consisting of feeding chambers, a dialysis membrane for the middle layer, and a top layer with reaction chambers and access ports. (**b**) A cross-sectional view of one unit (drawing is not to scale). Reproduced with permission from [13]

3.2 Protein Expression

1. Prepare the PCR components for β-lac vector production by mixing 0.5 μL of PUC 18 template DNA vector, 5 μL of 10× PCR buffer, 0.5 μL of the primers 5′-AAAAGCGGCCGCATG AGTATTCAACATTTCCGTG-3′ and 5′-GGTCGGATCCTT ACCAATGCTTAATCAGTG-3′, 2.5 μL of dimethylsulfoxide (DMSO), 1 μL of deoxynucleotide triphosphates (dNTPs), 0.5 μL of Taq polymerase enzyme, and 39.5 μL of water. Incubate the mixture in a PCR machine with a melting temperature of 65 °C.
2. Verify and clean the PCR product by agarose gel electrophoresis as follows. Cut and extract the fragment using the commercial gel and PCR cleanup kit by adding an equal amount of the membrane binding solution to the gel with the desired DNA fragments and incubating the mixture in a 50 °C water bath for 10 min to dissolve the agarose gel. Using the column provided in the kit, the mixture is spun in the column in a centrifuge at 370 × *g*, and the filtrate is discarded. The remaining mixture in the column is washed by adding 700 μL of the wash solution and centrifuging it at 24,000 × *g* for 1 min.

After removing the filtrate, the mixture is again washed using 500 μL of the wash buffer. To dissolve the plasmid, the column is transferred to a 1.5 ml microcentrifuge tube, 50 μL of water is added, and the solution is centrifuged for 1 min at 24,000 ×*g*. Isolate the GUS vector fragment by digesting the RTS 100 wheat germ kit GUS control vector with the restriction enzymes Notl and BamH1. The reaction consists of 20 μL of 1 μg/μL of concentration GUS plasmid, 5 μL of 10× buffer, 2 μL of Notl enzyme, 2 μL of BamH1 enzyme, and 21 μL of water. Run this restriction reaction on an agarose gel at 90 V for 70 min, and cut the vector fragment from the gel using the commercial gel and PCR cleanup system kit.

3. Clone the PCR product into the pIVEX 1.4 plasmid by adding the product to the purified GUS vector fragment, and undergo a ligation reaction by combining 3 μL of 10× ligation buffer, 2 μL of GUS vector fragment, 4 μL of insert sequence from the digested PCR fragment, 1 μL of ligase enzyme, and 20 μL of water, followed by incubation overnight at 4 °C. Transform into DH5α *E. coli* cells by mixing the ligation solution with 50 μL of the cells in a glass tube, incubating the mixture on ice for 30 min, heat shocking the bacteria at 42 °C in a water bath for 2 min to allow uptake of the plasmids, and placing the mixture on ice for 5 min to reduce the temperature. To grow the transformed bacteria, add 1.5 ml of L-broth liquid media to the glass tube and incubate at 32 °C on an orbital shaker for an hour. Then divide the grown cells into two agar plates with an ampicillin concentration of 100 μg/ml. The first plate contains 100 μL of the cell media retrieved directly following the cell growth. For the second plate, the cell media is centrifuged at 9,300 ×*g* for 1 min and the supernatant is removed. Resuspend the resulting pellet in 100 μL of L-broth and plate on the second agar plate for a 1:100 dilution ratio relative to the first plate. Incubate both plates at 37 °C overnight. Select colonies from both agar plates using a sterile stick, grow in a glass tube containing L-broth and 100 μg/ml ampicillin, and incubate overnight on an orbital shaker at 37 °C.
4. Isolate the DNA/plasmid from the DH5α *E. coli* colonies using the commercial miniprep system by centrifuging the miniprep contents at the top speed (25,000 ×*g*) for 1 min and removing the supernatant. Resuspend these cells in 250 μL of resuspension buffer and vortex, and produce protein precipitates by adding 250 μL of cell lysis buffer and then 350 μL of neutralization buffer. Spin the mixture at the top speed for 10 min, and transfer the supernatant into a spin column inserted into a collection tube. Process this supernatant by spinning the tube at the top speed for 1 min, discarding the precipitates, adding 750 μL of washing solution, and spinning

the tube again at the top speed for 1 min. Repeat this process with 250 μL of washing solution, and insert the spin column into a 1.5 ml centrifuge tube after adding 100 μL of water and centrifuging at the top speed for 1 min to dissolve those DNA bound to the spin column's membrane. Verify those colonies that have the correct plasmid by digesting a sample plasmid using the restriction enzymes Notl and BamH1 through a reaction consisting of 20 μL of extracted plasmids, 5 μL of 10× buffer, 2 μL of Notl enzyme, 2 μL of BamH1 enzyme, and 21 μL of water. Incubate this mixture in a water bath at 37 °C for 2 h and run on an agarose gel to indicate restriction sites for both Notl and BamH1. Discard the colony with the incorrect plasmid, and add glycerol to the cells with the correct plasmid.

5. Prepare the PCR components for AP vector production by mixing 2 μL of chromosomal DNA from *P. aeruginosa*, 5 μL of 10× PCR buffer, 0.5 μL of the primers 5′-AGGGAGCGGCCG CATGACCCAGGTTATCCCCTCGCCCTC-3′ and 5′-TCG TTCGGATCCGATCAGTCGCGCAGG TTCAGTGCGC-3′, 2.5 μL of dimethylsulfoxide (DMSO), 1 μL of deoxynucleotide triphosphates (dNTPs), 0.5 μL of Taq polymerase enzyme, and 39.5 μL of water. Incubate the mixture in a PCR machine with a melting temperature of 65 °C.
6. Verify the PCR product by agarose gel electrophoresis. Cut and extract the fragment using the commercial gel and PCR cleanup system kit as done for the β-lac construction (**step 2** above).
7. Clone the PCR product into pGEM vector (*see* **Note 19**) using the GEM-T easy vector system and transform into DH5α *E. coli* cells through a ligation reaction consisting of 5 μL of 2× rapid ligation buffer, 1.5 μL of T4 DNA ligase, 1.5 μL of pGEM vector, and 1.5 μL of PCR fragment. Incubate the resulting mixture overnight at 4 °C. Use this mixture to transform and grow pretreated DH5α *E. coli* cells as outlined for β-lac construction (**step 3** above). The grown cells are then divided into two agar plates applied with an ampicillin concentration of 100 μg/ml and X-gal for blue-white screening. The first plate contains 100 μL of the cell media retrieved directly following cell growth. For the second plate, the cell media is centrifuged at 9,300 ×*g* for 1 min and the supernatant is removed. Resuspend the resulting pellet in 100 μL of L-broth and plate on the second agar plate for a 1:100 dilution factor in comparison to the first plate. Both plates are then incubated at 37 °C overnight. Colonies that remain white are selected and then grown in a glass tube containing L-broth and 100 μg/ml ampicillin and incubated overnight on a shaker at 37 °C.

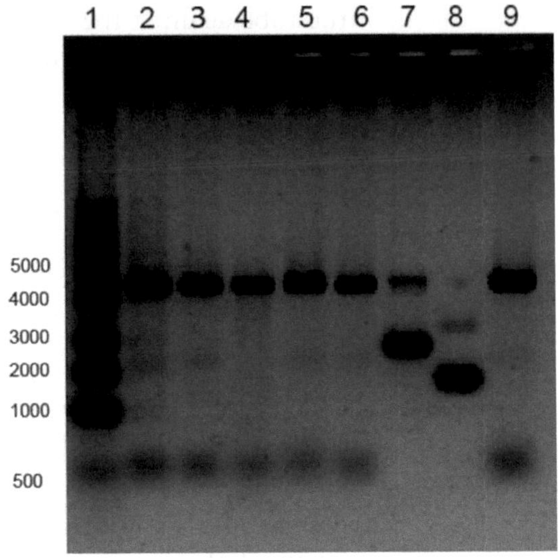

Fig. 2 Agarose gel results for the verification of AP plasmids in DH5α *E. coli* cells. Lane 1 is the ladder, and the other lanes contain colonies with the AP plasmid digested using the restriction enzyme Pst1. Lanes 2–6 and 9 indicate that colonies contain correct plasmid, whereas Lanes 7 and 8 are colonies with the incorrect plasmid. Reproduced with permission from [15]

8. Isolate the plasmids from the selected colonies using the commercial miniprep system as described for β-lac construction (**step 4** above). Verify those colonies that have the correct plasmid by digesting the samples using the restriction enzyme Pstl (*see* **Note 20**) (Fig. 2).
9. Mix the reaction solution for luciferase, β-glucuronidase (GUS), AP, β-lac, and β-galactosidase (lacZ), consisting of 15 μL of wheat germ lysate, 15 μL of reaction mix, 4 μL of amino acids, 1 μL of methionine, and 15 μL of the individual DNA vector. In the negative control, the individual DNA vector is replaced with an equal volume of water. The feeding solution is composed of 900 μL of feeding mix, 80 μL of amino acids, and 20 μL of methionine from the RTS 100 wheat germ kit by following the manufacturer's instructions (*see* **Note 21**).
10. Mix the reaction solution for green fluorescent protein (GFP), consisting of 10 μL of a solution containing 2 μg of the GFP vector and 40 μL of an aliquot made by mixing 0.525 ml of *E. coli* lysate, 0.255 ml of reaction mix, 0.27 ml of amino acid mix without methionine, and 30 μL of methionine. The feeding solution is composed of 2.65 ml of amino acid mix without methionine, 0.3 ml of methionine, and 8.1 ml of feeding mix from the RTS 500 *E. coli* kit (*see* **Note 22**).
11. In the 96-well device, pipette 10 μL of the reaction solution into the reaction chamber and add 200 μL of the feeding

solution to the feeding chamber through one of the access holes (*see* **Note 23**). Seal the device with PCR tape to prevent evaporation.

12. For comparison, place positive and negative controls in a 384-well microplate by combining the same volume of the reaction solution and feeding solution into wells. The reaction solutions for the positive controls are identical to those used in the device; however, the negative control reaction solutions have water in place of the DNA vector. Seal the microplate with PCR tape.
13. Place the device and microplate on an orbital shaker at a speed of 30 RPM for 4 h. Keep GFP samples at room temperature for an additional 2 h to allow complete folding.

3.3 Protein Detection

1. Measure the fluorescence of GFP using a filter with excitation and emission wavelengths of 485 and 535 nm, respectively.
2. Add 30 μL of luciferase assay reagent to the luciferase reaction chambers using one of the plate reader injectors, shake the microplate for 2 s, and measure the resulting luminescence over 10 s.
3. Inject 30 μL of 100 μM MUG into the GUS wells, incubating for 6 min, and measure the resulting fluorescence with excitation and emission wavelengths of 355 and 460 nm, respectively.
4. Inject 30 μL of 10 μM PPH into the AP wells, incubating for 6 min, and measure the resulting fluorescence with excitation and emission wavelengths of 355 and 460 nm, respectively.
5. Inject 30 μL of 200 μM FMG into the LacZ wells, incubating for 10 min, and measure the resulting fluorescence with excitation and emission wavelengths of 485 and 535 nm, respectively.
6. Add 90 μL of 2 mM PBA into the β-lac wells, incubating for 5 min, and measure the resulting absorbance at 314 nm.
7. An example of protein expression in the microfluidic 96-well devices is shown in Fig. 3.

4 Notes

1. Dialysis membranes with various molecular cutoffs and thicknesses were examined to determine the optimum protein synthesis yield. Standard dialysis discs with a molecular cutoff of 6,000–8,000 Da were determined to have the highest protein synthesis yield.
2. The Sherline 5400-CNC mill and Flashcut CNC Version 4.0 control software were used.

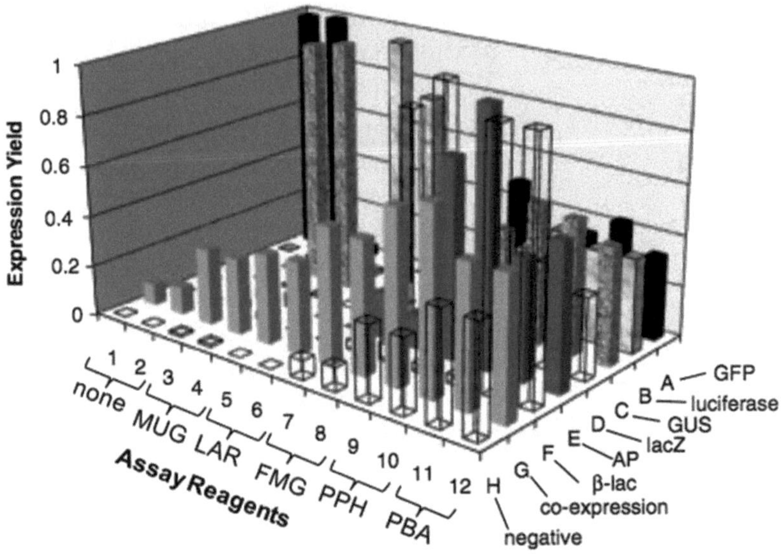

Fig. 3 Simultaneous expression and detection of GFP, luciferase, GUS, lacZ, AP, and β-lac using the assay reagents MUG, LAR, FMG, PPH, and PBA. The protein expression yields were normalized against the highest signal. Reproduced with permission from [16]

3. Waste disposal regulations should be followed with all materials.
4. The Laurell Technologies 150 mm spin coater is recommended.
5. Nuclease-free water should be used to prevent exposure of genetic templates to DNA or RNA nucleases.
6. Protein kits and components should be properly stored as specified by the manufacturer. Once prepared, the samples can be kept at room temperature, unless otherwise specified. The CFPS systems used in the device included wheat germ, *E. coli*, and rabbit reticulocyte extracts. Ideally, the extract type should be compatible with the proteins to be expressed.
7. The vector for GUS is the control vector, included with RTS 100 wheat germ kit.
8. The vector for GFP is the control vector included in the RTS 500 *E. coli* kit.
9. The Wizard SV gel and PCR cleanup kit (Promega) is recommended.
10. The Wizard Plus Miniprep system (Promega) is recommended.
11. A PCR tape is a clear sealing tape capable of covering the wells in a standard 384-well microplate. The PCR tape prevents evaporation, leaves no residue when removed, and is devoid of DNase or RNase.
12. A Mithras LB 940 plate reader (Berthold Technologies) was used.

13. In order for the device to be compatible with commercial fluid dispensers and microplate readers, the reaction wells must be in the same location and size of a standard 384-well microplate.
14. For precision, AutoCAD designs of the top and bottom layers are recommended. This AutoCAD file can be opened and automatically run with the Flashcut CNC Version 4.0. A feed rate of 127 mm/min, spindle speed of 2,400 RPM, and depth of cut of 1 mm/pass are recommended.
15. When assembled, the feeding chambers must line up underneath the reaction chamber and access holes of the top layer, engulfing them all.
16. This technique is referred to as the PDMS microstamping technique and involves transferring PDMS to the desired material by stamping the surface of the material in a thin layer of PDMS. The microstamping technique is used to avoid transferring PDMS onto the access holes and reaction chambers of the top layer. The detail can be found in reference [17].
17. The microstamping technique is used to avoid transferring PDMS onto the feeding chambers.
18. Ensure that the device layers align once clamped together and prior to curing. Clamp all sides of the device to secure the alignment.
19. The pGEM vector self-ligates, making differentiation between the pGEM vector and pGEM vector containing the PCR fragment difficult. X-gal is used to differentiate between these two. Under normal conditions, the pGEM vector is opened in the middle of the ß-galactosidase gene for target DNA vector insertion. However, when pGEM self-ligates, this gene is intact and the protein cleaves X-gal, resulting in the bacterial colonies turning blue in color.
20. The plasmid concentrations for ß-lac and AP can be found by diluting 5 μL of each solution in 95 μL of water and measuring the mixture in a spectrophotometer at an absorbance wavelength of 260 nm. The resulting absorbance measurement is then multiplied by 20 to account for the dilution factor.
21. All of these components are included in the RTS 100 kit. Follow the manufacturer's instructions for reconstituting these components.
22. All of these components are included in the RTS 500 kit. Follow the manufacturer's instructions for reconstituting these components.
23. In order to dispense the feeding solution into the feeding chamber, the dialysis membrane in the access holes must be punctured by a sharp tip. Pipette the feeding solution slowly to avoid the formation of air bubbles in the feeding chamber.

Acknowledgements

This work was supported in part by Defense Advanced Research Projects Agency (DARPA), National Science Foundation (OISE-0968313), and the University of Florida via UF Opportunity Fund. Assistance from Dr. Shouguang Jin and help in device fabrication from Daniel Olivero are greatly appreciated.

References

1. Sawasaki T, Gouda MD, Kawasaki T et al (2005) The wheat germ cell-free expression system: methods for high-throughput materialization of genetic information. In: Zanders E (ed) Chemical genomics: reviews and protocols. Humana Press, New Jersey, pp 131–144
2. Katzen F, Chang G, Kudlicki W (2005) The past, present and future of cell-free protein synthesis. Trends Biotechnol 23:150–156
3. Reckel S, Sobhanifar S, Durst F et al (2010) Strategies for the cell-free expression of membrane proteins. In: Endo Y (ed) Cell-free protein production: methods and protocols. Humana Press, New Jersey, pp 187–212
4. Kim DM, Swartz JR (1999) Prolong cell-free protein synthesis with a novel ATP regeneration system. Biotechnol Bioeng 66:180–188
5. Sawasaki T, Hasegawa Y, Tsuchimochi M et al (2002) A bilayer cell-free protein synthesis system for high-throughput screening of gene products. FEBS Lett 514:102–105
6. Endo Y, Sawasaki T (2003) High-throughput, genome-scale protein production method based on the wheat germ cell-free expression system. Biotechnol Adv 8:695–713
7. Spirin AS, Baranov VI, Ryabova LA et al (1988) A continuous cell-free translation system capable of producing polypeptides in high yield. Science 242:1162–1164
8. Ryabova LA, Morozov IY, Spirin AS (1998) Continuous-flow cell-free translation, transcription-translation, and replication-translation systems. In: Martin R (ed) Protein synthesis: methods and protocols. Humana Press, New Jersey, pp 179–193
9. Kim DM, Choi CY (1996) A semicontinuous prokaryotic coupled transcription/translation system using a dialysis membrane. Biotechnol Prog 12:645–649
10. Mei Q, Fredrickson W, Jin S et al (2005) Toxin detection by a miniaturized in vitro protein expression array. Anal Chem 77: 5494–5500
11. Mei Q, Fredrickson CK, Lian W et al (2006) Ricin detection by biological signal amplification in well-in-a-well device. Anal Chem 78: 7659–7664
12. Mei Q, Fredrickson CK, Simon A et al (2007) Cell-free protein synthesis in microfluidic array devices. Biotechnol Prog 23:1305–1311
13. Khnouf R, Olivero D, Jin S et al (2010) Miniaturized fluid array for high-throughput protein expression. Biotechnol Prog 26: 1590–1596
14. Khnouf R, Olivero D, Jin S et al (2010) Cell-free expression of soluble and membrane proteins in an array device for drug screening. Anal Chem 82:7021–7026
15. Khnouf R (2010) Fluidic platforms for high-throughput protein synthesis and their applications in drug screening. University of Florida, Florida
16. Khnouf R, Chapman BD, Fan ZH (2011) Fabrication optimization of a miniaturized array device for cell-free protein synthesis. Electrophoresis 32:3101–3107
17. Chueh BH, Huh D, Kyrtsos CR et al (2007) Leakage-free bonding of porous membranes into layered microfluidic array systems. Anal Chem 79:3504–3508

Chapter 11

Preparation of Multiple Site-Specific Mutant Proteins for NMR Studies by PCR-Directed Cell-Free Protein Synthesis

Kiyoshi Ozawa and Ruhu Qi

Abstract

Cell-free protein synthesis (CFPS) offers a fast and inexpensive approach to selectively label proteins with isotopes that can then be detected by nuclear magnetic resonance (NMR) spectroscopy directly in the translation mixture. We describe a PCR-based approach for production of protein-coding circularized DNA templates that can be expressed in *Escherichia coli* extract in CFPS dialysis system. This approach typically yields target protein concentrations close to 1 mg/mL, which is sufficient for subsequent analysis by 2D ^{1}H,^{15}N-NMR. Furthermore, this PCR-based technique also enables parallel preparation of mutant proteins in a high-throughput mode, enabling rapid assignments of NMR signals. This chapter describes the general CFPS protocol that we used to rapidly assign residue-specific cross peaks from 2D ^{1}H,^{15}N-NMR spectra obtained from 12 Ile/Ala substituted mutants of the 40 kDa protein complex, αCTS:τ_c16.

Key words Cell-free protein synthesis, High-throughput protein production, Selective isotope labeling, PCR-generated DNA template, NMR spectroscopy

1 Introduction

Cell-free protein synthesis (CFPS) has won great interest in structural biology since Kigawa et al. demonstrated milligram per microliter production of recombinant proteins in coupled cell-free transcription–translation system based on *Escherichia coli* cell extract [1]. These concentrations are sufficient for application of nuclear magnetic resonance (NMR) spectroscopy on modern high-field NMR spectrometers, providing a convenient and rapid way to analyze protein structures [2–4].

In molecular biology, "alanine scanning" is a technique widely used to determine the contribution of a specific residue to the stability or the function of protein [5]. Alanine is used because of its non-bulky, chemically inert, methyl functional group that nevertheless mimics the secondary structure preferences that many of the other amino acids possess. Sometimes bulky amino acids such as valine or leucine are also used in cases where conservation

Kirill Alexandrov and Wayne A. Johnston (eds.), *Cell-Free Protein Synthesis: Methods and Protocols*, Methods in Molecular Biology, vol. 1118, DOI 10.1007/978-1-62703-782-2_11, © Springer Science+Business Media, LLC 2014

of the size of mutated residues is needed. This technique can also be used to determine whether the side chain of a specific residue plays a significant role in biological function. This is usually accomplished by site-directed mutagenesis or randomly by creating a mutant library by PCR. Moreover, computational methods to estimate thermodynamic parameters based on theoretical alanine substitutions have been developed [5].

Here we demonstrate a practical application of this "alanine scanning" technique for rapid cross-peak assignments in NMR spectra, by coupling it with a preparative *E. coli* CFPS system using PCR-generated DNA templates [6, 7].

2 Materials

2.1 Two-Step Site-Directed Mutagenesis and Preparation of PCR-Generated DNA Template

1. Target gene: The encoding gene and the deduced amino acid sequence of the C-terminal segment ([964]His–[1160]Asp) of *E. coli* DNA polymerase III (Pol III) subunit α (hereafter αCTS) are shown below for reference (Table 1). The gene encoding αCTS is located between *Nde*I (CATATG) and *Eco*RI (GAATTC) sites in the T7 expression vector, pETMCSIII [8].

Table 1
The encoding gene and the deduced amino acid sequence of the C-terminal segment ([964]His–[1160]Asp) of *E. coli* DNA polymerase III (Pol III) subunit α (hereafter αCTS) are shown below for reference (12 *Ile* residues are shown in *bold* letters)

										Met	His	Pro	**Ile**	Asn	Gln	Tyr	7
									\CAT	ATG	CAC	CCT	ATC	AAC	CAG	TAT	
Leu	Lys	Glu	**Ile**	Glu	Arg	Tyr	Val	Gly	Gly	Val	Arg	Leu	Lys	Asp	Met	His	24
TTA	AAA	GAG	ATT	GAG	CGT	TAT	GTC	GGA	GGC	GTA	AGG	CTG	AAA	GAC	ATG	CAC	
Pro	Thr	Glu	Arg	Gly	Lys	Val	**Ile**	Thr	Ala	Ala	Gly	Leu	Val	Val	Ala	Ala	41
CCG	ACA	GAA	CGT	GGT	AAA	GTC	ATC	ACG	GCT	GCG	GGG	CTC	GTT	GTT	GCC	GCG	
Arg	Val	Met	Val	Thr	Lys	Arg	Gly	Asn	Arg	**Ile**	Gly	**Ile**	Cys	Thr	Leu	Asp	58
CGG	GTT	ATG	GTC	ACC	AAG	CGC	GGC	AAT	CGT	ATC	GGT	ATC	TGC	ACG	CTG	GAT	
Asp	Arg	Ser	Gly	Arg	Leu	Glu	Val	Met	Leu	Phe	Thr	Asp	Ala	Leu	Asp	Lys	75
GAC	CGT	TCC	GGG	CGG	CTG	GAA	GTG	ATG	TTG	TTT	ACT	GAC	GCC	CTG	GAT	AAA	
Tyr	Gln	Gln	Leu	Leu	Glu	Lys	Asp	Arg	**Ile**	Leu	**Ile**	Val	Ser	Gly	Gln	Val	92
TAC	CAG	CAA	TTG	CTG	GAA	AAA	GAC	CGC	ATA	CTT	ATC	GTC	AGC	GGA	CAG	GTC	
Ser	Phe	Asp	Asp	Phe	Ser	Gly	Gly	Leu	Lys	Met	Thr	Ala	Arg	Glu	Val	Met	109
AGC	TTT	GAT	GAC	TTC	AGC	GGT	GGG	CTT	AAA	ATG	ACC	GCT	CGC	GAA	GTG	ATG	
Asp	**Ile**	Asp	Glu	Ala	Arg	Glu	Lys	Tyr	Ala	Arg	Gly	Leu	Ala	**Ile**	Ser	Leu	126
GAT	ATT	GAC	GAA	GCC	CGG	GAA	AAA	TAT	GCT	CGC	GGG	CTT	GCT	ATC	TCG	CTG	
Thr	Asp	Arg	Gln	**Ile**	Asp	Asp	Gln	Leu	Leu	Asn	Arg	Leu	Arg	Gln	Ser	Leu	143
ACG	GAC	AGG	CAA	ATT	GAT	GAC	CAG	CTT	TTA	AAC	CGA	CTC	CGT	CAG	TCT	CTG	
Glu	Pro	His	Arg	Ser	Gly	Thr	**Ile**	Pro	Val	His	Leu	Tyr	Tyr	Gln	Arg	Ala	160
GAA	CCC	CAC	CGC	TCT	GGG	ACA	ATT	CCA	GTA	CAT	CTC	TAC	TAT	CAG	AGG	GCG	
Asp	Ala	Arg	Ala	Arg	Leu	Arg	Phe	Gly	Ala	Thr	Trp	Arg	Val	Ser	Pro	Ser	177
GAT	GCA	CGC	GCG	CGG	TTG	CGT	TTT	GGC	GCG	ACG	TGG	CGT	GTC	TCT	CCG	AGC	
Asp	Arg	Leu	Leu	Asn	Asp	Leu	Arg	Gly	Leu	**Ile**	Gly	Ser	Glu	Gln	Val	Glu	194
GAT	CGT	TTA	TTA	AAC	GAT	CTC	CGT	GGC	CTC	ATT	GGT	TCG	GAG	CAG	GTG	GAA	
Leu	Glu	Phe	Asp	Stop													198
CTG	GAG	TTT	GAC	TAA	\GAATTC\												

2. Synthetic oligonucleotides for Ile/Ala mutagenesis (Table 2).
3. Thermal cycler (e.g., *i*Cycler/C1000 Thermal Cycler PCR apparatus from Bio-Rad).

Table 2
PCR and Ile/Ala mutagenesis primers used[a]

1 (I4A)	5′-CATATGCACCCTGCCAACCAGTATTTAAAAGAG-3′ (33-mer)
1′ (I4A)	5′-TACTGGTTGGCAGGGTGCATATGTATATC-3′ (29-mer)
2 (I11A)	5′-TTAAAAGAGGCGGAGCGTTATGTCGGAGG-3′ (29-mer)
2′ (I11A)	5′-CATAACGCTCCGCCTCTTTTAAATACTGGTT G-3′ (32-mer)
3 (I32A)	5′-GTAAAGTCGCCACGGCTGCGGGGCTCGTTG-3′ (30-mer)
3′ (I32A)	5′-GCAGCCGTGGCGACTTTACCACGTTCTGTC-3′ (30-mer)
4 (I52A)	5′-CAATCGTGCCGGTATCTGCACGCTGGATG-3′ (29-mer)
4′ (I52A)	5′-GTGCAGATACCGGCACGATTGCCGCGCTTG-3′ (30-mer)
5 (I54A)	5′-ATCGTATCGGTGCCTGCACGCTGGATGACCGTTC-3′ (34-mer)
5′ (I54A)	5′-CAGCGTGCAGGCACCGATACGATTGCCGCGCTTG-3′ (34-mer)
6 (I85A)	5′-AAAGACCGCGCGCTTATCGTCAGCGGACAG-3′ (30-mer)
6′ (I85A)	5′-GACGATAAGCGCGCGGTCTTTTTCCAGCAATTG-3′ (33-mer)
7 (I87A)	5′-ACCGCATACTTGCCGTCAGCGGACAGGTCAG-3′ (31-mer)
7′ (I87A)	5′-TCCGCTGACGGCAAGTATGCGGTCTTTTTC-3′ (30-mer)
8 (I111A)	5′-GTGATGGATGCGGACGAAGCCCGGGAAAAATATG-3′ (34-mer)
8′ (I111A)	5′-GGCTTCGTCCGCATCCATCACTTCGCGAG-3′ (29-mer)
9 (I124A)	5′-CGCGGGCTTGCTGCCTCGCTGACGGACAGGCAAATTG-3′ (37-mer)
9′ (I124A)	5′-CAGCGAGGCAGCAAGCCCGCGAGCATATTTTTC-3′ (33-mer)
10 (I131A)	5′-GACAGGCAAGCGGATGACCAGCTTTTAAAC-3′ (30-mer)
10′ (I131A)	5′-CTGGTCATCCGCTTGCCTGTCCGTCAGCGAG-3′ (31-mer)
11 (I151A)	5′-CTCTGGGACAGCGCCAGTACATCTCTACTATC-3′ (32-mer)
11′ (I151A)	5′-GATGTACTGGCGCTGTCCCAGAGCGGTGGGGTTC-3′ (34-mer)
12 (I188A)	5′-CTCCGTGGCCTCGCGGGTTCGGAGCAGGTGGAAC-3′ (34-mer)
12′ (I188A)	5′-GAACCCGCGAGGCCACGGAGATCGTTTAATAAAC-3′ (34-mer)
P1	5′-PO4-TTAGCTGGTCGATCCCGCGAAATTAATACG-3′ (30-mer)
P2	5′-PO4-CCAGCTAACAAAAAACCCCTCAAGACCCG-3′ (29-mer)
P3	5′-PO4-TCGATCCCGCGAAATTAATACG-3′ (22-mer)
P4	5′-PO4-CAAAAAACCCCTCAAGACCCG-3′ (21-mer)

[a]Codons of mutated amino acids are underlined. Primers P1–P4 corresponding to the primers 1–4, respectively, in our original publication. P1 and P3 include the T7 ϕ10 promoter [6]

4. *Vent* DNA polymerase (accompanied by 10× ThermoPol Reaction Buffer from New England Biolabs) or similar thermostable *high-fidelity* DNA polymerase (e.g., *Phusion* DNA polymerase from New England Biolabs).
5. Agarose gel electrophoresis equipment including agarose, loading dye, and staining solution (e.g., ethidium bromide (EtBr), SYBR™ Green from Bio-Rad).
6. NucleoSpin Extract II kit (Macherey-Nagel, Düren, Germany).
7. *Optional*: Heat block (or PCR apparatus) to enable DNA denaturation and re-annealing.
8. NanoDrop spectrophotometer (Thermo Scientific, USA).

2.2 Preparation of E. coli S30 Extracts Expressing T7 RNA Polymerase

1. Spectra/Por #2 (10 mm flat width corresponding to 6.4 mm diameter) and #4 dialysis tubing, 12–14,000 MWCO (Spectrum Laboratories Inc., Rancho Dominguez, CA, USA).
2. French press.
3. Centrifuges with rotors suitable for 50 mL, 500 mL, and 1 L tubes.
4. Optical densitometer.
5. 10 mL centrifugal ultrafiltration units, 10,000 MWCO, e.g., Millipore Ultra-4.
6. 10 and 50 mL disposable centrifuge tubes.
7. Benchtop orbital shaker.
8. 20 L capacity fermenter.
9. Vortex mixer.
10. Tris(hydroxymethyl)-aminomethane (Tris).
11. Potassium acetate.
12. Magnesium acetate.
13. Acetic acid.
14. Potassium dihydrogen phosphate.
15. Dipotassium hydrogen phosphate.
16. Yeast extract.
17. Phenylmethanesulfonyl fluoride (PMSF).
18. Antifoam 204 (Sigma).
19. Milli-Q (MQ) water.
20. 1 mg/mL thiamine.
21. 2 M glucose.
22. 1 M Tris acetate pH 8.2.
23. 50 % PEG 8000 w/w: 1,000 g PEG 8000 and 1 L of 1× S30 buffer (V_{total} ~ 2 L).

24. LBT: 10 g/L tryptone, 5 g/L yeast extract, 5 g/L sodium chloride, 2.5 mL/L of 1 M sodium hydroxide, 0.5 mg/L thymine.
25. 4× Z-medium: 165 mM potassium dihydrogen phosphate, 664 mM dipotassium hydrogen phosphate, 40 g/L yeast extract [9].
26. 10× S30 buffer: 100 mM Tris acetate, pH 8.2, 160 mM potassium acetate, 140 mM magnesium acetate, pH 8.3 with potassium hydroxide.
27. S30 buffer α: 1× S30 buffer containing 0.5 mM PMSF, 1 mM dithiothreitol, and 7.2 mM β-mercaptoethanol.
28. S30 buffer β: 1× S30 buffer containing 1 mM dithiothreitol.
29. *E. coli* strain expressing the T7 RNA polymerase gene, e.g., BL21 Star™ (DE3) (Invitrogen).
30. Isopropyl-β-D-1-thiogalactoside (IPTG).

2.3 CFPS Using PCR-Generated DNA Templates Results in Production of ^{15}N-Ile-Labeled αCTS Protein and Its 12 Ile/Ala Mutants in the Presence of Non-labeled Purified τ_c16 Protein

1. 10× Protein synthesis buffer: 580 mM HEPES–KOH (pH 7.5), 17 mM dithiothreitol, 12 mM ATP, 8 mM GTP, 8 mM CTP, 8 mM UTP, 6.4 mM cyclic AMP, 0.68 mM folinic acid, 276 mM ammonium acetate, 2.08 M potassium glutamate (20 mL).
2. *Freshly prepared* mixture of 15 mM each of 20 amino acids, with Ile substituted by ^{15}N-Ile (14 mL) [10] (*see* **Note 1**).
3. 1.07 M magnesium acetate (4 mL).
4. 1 M creatine phosphate (Sigma) (14 mL).
5. 10 mg/mL creatine kinase (Roche) (1.4 mL): This enzyme was dissolved in MQ water.
6. 0.1–1.5 mg/mL of PCR-generated template DNA: *See* Subheading 3.1.
7. Purified τ_c16 protein (40 mg) [11].
8. *E. coli* cell-free S30 extract (*see* Subheading 3.2).
9. 17.5 mg/mL of total tRNA mixture (from *E. coli* MRE 600; Roche).
10. L-Isoleucine-^{15}N, 98 % (^{15}N-Ile, FW = 132.17) (ISOTEC—Sigma-Aldrich).
11. ProPur IMAC Mini spin columns (Nalgene Nunc International, USA).
12. All buffers and solutions are stored at −20 °C.

2.4 Preparation of the Samples for NMR Analysis

1. NMR buffer (10 mM sodium phosphate, pH 6.8, 1 mM dithiothreitol, and 100 mM NaCl).
2. 10 mL centrifugal ultrafiltration units, 10,000 MWCO, e.g., Millipore Ultra-4.

3. Deuterated water, D_2O.
4. SDS-PAGE equipment.
5. NMR tube (3 mm in diameter).
6. 800 MHz/600 MHz NMR spectrometer equipped with a cryoprobe.

3 Methods

3.1 Two-Step Site-Directed Mutagenesis and Preparation of PCR-Generated DNA Template

Twelve Ile residues were strategically chosen for mutation to Ala because they are spread over the entire αCTS protein, and their locations would give us some valuable information on binding with the partner subunit τ_c16 within the overall unsolved structure of the αCTS:τ_c16 complex. The overall gene configuration and locations of PCR primers (Table 2) are depicted in Fig. 1. The strategy relies on PCR to generate overlapping fragments encoding the target gene, in which the desired mutation is located in the overlap region. Overlap/extension PCR is then performed on these DNA fragments to create the complete expression template encoding the mutant protein.

3.1.1 PCR to Generate Two Overlapping DNA Fragments Encoding Each Mutation

1. Prepare the *P3 mix* and the *P4 mix* for 13 (12 + 1 extra) samples (Table 3):

   These following procedures should be carried out on ice:
2. Pipette 48 μL each of the *P4 mix* into 12 separate 0.2 mL PCR tubes and mix with 2 μL each of the *forward primers* (10 μM) *1–12*, respectively.
3. Pipette 48 μL each of the *P3 mix* into 12 separate 0.2 mL PCR tubes and mix with 2 μL each of the *backward primers* (10 μM) *1′–12′*, respectively.
4. Run the Hot Start PCR with following 35 cycling of 94 °C, 30 s; 50 °C, 30 s; 72 °C, 40 s with a preheating step at 94 °C, 5 min and a post extra incubation at 72 °C, 10 min.
5. Subject a 5 μL aliquot of 24 PCR reactions to 1.0 % agarose gel electrophoresis (AGE) analysis, and stain the gel with an appropriate staining solution (e.g., EtBr, SYBR™ Green).
6. The AGE profiles from the first-round PCR are shown in Fig. 2a, b.

3.1.2 1.0 % Agarose Gel Electrophoresis and Isolation of the PCR Products from the Gel

1. Isolate the 24 PCR-amplified DNA bands separately from 1.0 % AGE gel and purify with NucleoSpin Extract II kit (following the manufacturer's protocol): This should yield 50 μL each of 24 DNA solutions (*see* **Note 2**).
2. Determine the concentrations of the 24 gel-purified DNA solutions individually by measuring absorbance at 260 nm using NanoDrop spectrophotometer.

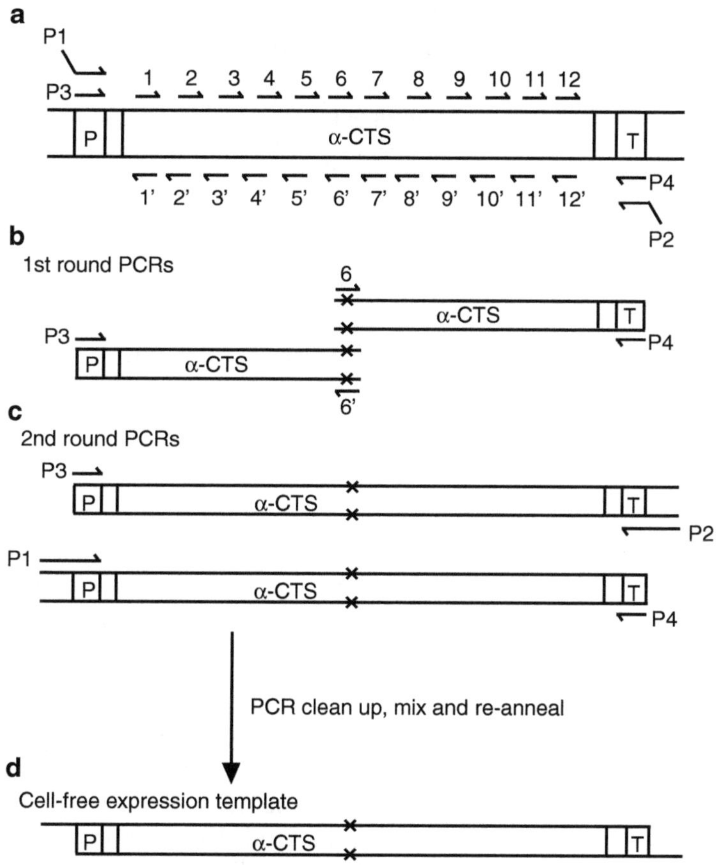

Fig. 1 A site-specific mutagenesis strategy using an overlap/extension PCR approach. P, T7 promoter; T, T7 terminator; α-CTS, αCTS structural gene; the *arrows* numbering from 1 to 12 represent the forward mutating primers, and the *arrows* numbering from 1′ to 12′ represent the backward mutating primers. Primers P1 and P3 are both forward primers and are 5′ phosphorylated. Primers P2 and P4 are both backward primers and are 5′ phosphorylated. Primers P1 and P2 are longer than primers P3 and P4, respectively, by eight nucleotides at their 5′-termini. The phosphorylated 8-nucleotide overhangs between primers P1 and P2 are complementary to each other; this is believed to enable circularization of the template by endogenous DNA ligase present in the *E. coli* cell-free extract [6]. All primers used in this protocol are listed in Table 2. (**a**) Overall gene configuration and rough locations of primers on the T7 expression vector, pETMCSIII-αCTS. (**b**) An example of the first-round overlap/extension PCRs using pairs of primers 6 and P4, and P3 and 6′; *X* indicates the location of the mutation. Two overlap/extension PCR products are gel-purified, and they were mixed together in equal amounts and then subjected to the second-round PCR as the template DNA. (**c**) The second-round PCRs using pairs of primers P3 and P2, and P1 and P4. Mixing and re-annealing of these products produce a DNA template with complementary overhangs that undergoes ligation in the cell-free reaction to stable circular template [6]. (**d**) Cell-free expression template produced by overlap/extension PCR

Table 3
Forward P4 (or reverse P3) mutagenesis PCR master mix

P3 or P4 mix	(×1) (μL)	(×13) (μL)
Primer P3 (*P3 mix*) or P4 (*P4 mix*) (10 μM)	2	26
Template DNA (pETMCSIII-αCTS, 2 ng/μL)	2	26
50× dNTP (12.5 mM)	1	13
10× ThermoPol Reaction Buffer	5	65
Vent DNA polymerase (2 U/μL)	0.5	6.5
MQ water	37.5	487.5
Total	48	624

3. Subject a 5 μL aliquot of the isolated DNA solutions to 1.0 % AGE analysis again to double check that the recovered DNA yields of 24 samples are similar and sufficient (*see* **Note 3**).
4. Mix the isolated DNA pairs from the first-round PCR (e.g., 1 and 1′, 2 and 2′, 3 and 3′, and so on) together in an approximately 1:1 ratio (judging from the band intensities and your eyes).
5. The resulting 12 separate (1:1) DNA mixtures (about 90 μL each) prepared from the first-round PCR are subjected to the second-round PCR as the template DNA.

3.1.3 Overlap/Extension PCR to Generate the Cell-Free Expression Templates for Each Mutant

1. Prepare the *P1+P4 mix* (Table 4) and the *P2+P3 mix* (Table 5) for 13 (12+1 extra) samples.

   Keep solutions on ice.
2. Pipette 144 μL each of the *P1+P4 mix* into 12 separate 0.2 mL PCR tubes and mix with 6 μL each out of the 12 separate (1:1) DNA mixtures (90 μL each) prepared from the first-round PCR, resulting in **A–L**, respectively. Split 150 μL each of the total mixture into 75 μL each in two PCR tubes.
3. Pipette 144 μL each of the *P2+P3 mix* into 12 separate 0.2 mL PCR tubes and mix with 6 μL each out of the 12 separate (1:1) DNA mixtures (90 μL each) prepared from the first-round PCR, resulting in *A′–L′*, respectively. Split 150 μL each of the total mixture into 75 μL each in two PCR tubes.
4. Run the Hot Start PCR of the total 48 PCR tubes with the following 35 cycling of 94 °C, 30 s; 55 °C, 30 s; and 72 °C, 1 min, with a preheating step at 94 °C, 5 min and a post extra incubation at 72 °C, 10 min.
5. Combine the equivalent PCR samples together into 150 μL each in one PCR tube.

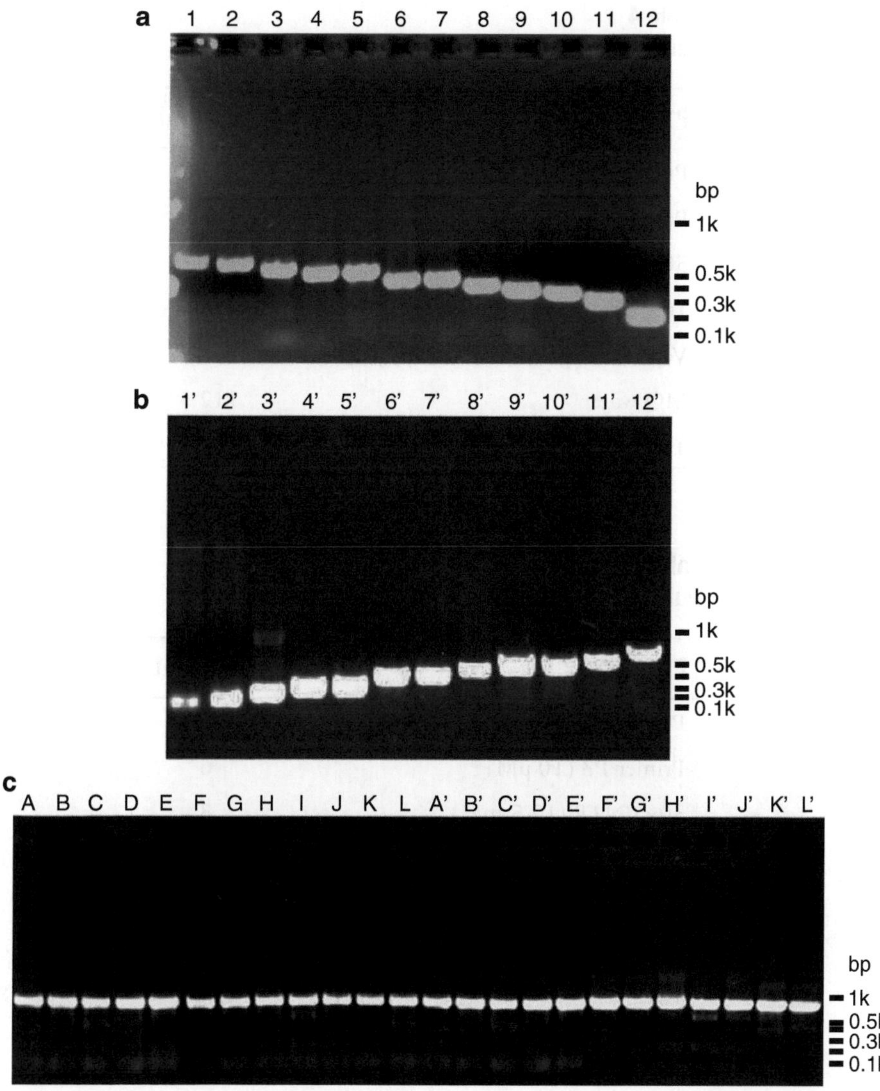

Fig. 2 Agarose gel electrophoresis of cell-free protein expression templates for 12 αCTS mutants generated by overlap/extension mutagenesis PCR. (**a**) The first-round PCR with each of the forward mutating primers 1–12 and the backward primer P4. The labeling on the top of the gel image represents the corresponding forward mutating primers 1–12, which also represents the 12 individual first-round PCR reactions hereafter. The amplified DNA bands were stained with ethidium bromide (EtBr) and visualized by UV light (254 nm). (**b**) The first-round PCR with each of the backward mutating primers 1′–12′ and the forward primer P3. The labeling on the top of the gel image represents the corresponding forward mutating primers 1′–12′, which also represents the 12 individual first-round PCR reactions hereafter. The amplified DNA bands were stained with EtBr and visualized by UV light (254 nm). (**c**) All the first-round PCR products shown in (**a**) and (**b**) were separately isolated from the gel to minimize contamination by the original plasmid template and the first-round PCR primers. The second-round PCR with primers P1 and P4 using 1:1 mixtures of the gel-purified first-round PCR products (i.e., 1 + 1′, 2 + 2′, 3 + 3′, and so on) are shown (labeled A–L, respectively). The second-round PCR with primers P2 and P3 using each of the 1:1 mixtures of the gel-purified first-round PCR products (i.e., 1 + 1′, 2 + 2′, 3 + 3′, and so on) are shown (labeled A′–L′, respectively). The 24 amplified DNA bands were stained with EtBr and visualized by UV light (254 nm)

Table 4
Full-length assembly PCR master mix (P1 + P4)

P1 + P4 mix	(×1) (μL)	(×13) (μL)
Primer P1 (10 μM)	6	78
Primer P4 (10 μM)	6	78
50× dNTP (12.5 mM)	3	39
10× ThermoPol Reaction Buffer	15	195
Vent DNA polymerase (2 U/μL)	1.5	19.5
MQ water	112.5	1,462.5
Total	144	1,872

Table 5
Full-length assembly PCR master mix (P2 + P3)

P2 + P3 mix	(×1) (μL)	(×13) (μL)
Primer P2 (10 μM)	6	78
Primer P3 (10 μM)	6	78
50× dNTP (12.5 mM)	3	39
10× ThermoPol Reaction Buffer	15	195
Vent DNA polymerase (2 U/μL)	1.5	19.5
MQ water	112.5	1,462.5
Total	144	1,872

6. Subject a 5 μL aliquot of the 24 PCR products (150 μL each) to 1.0 % AGE analysis, and stain the gel with an appropriate staining solution (e.g., EtBr, SYBR™ Green).
7. The AGE profiles from the second-round PCR are shown in Fig. 2c.

3.1.4 Purification of the Second-Round PCR Products and Annealing

1. Check that the DNA yields of the 24 PCR samples are similar and sufficient (*see* **Note 4**).
2. Using visually assessed band intensities as guidance mix the second-round DNA pairs (e.g., A and A′, B and B′, C and C′, and so on) together in an approximately 1:1 ratio.
3. The resulting 12 separate (1:1) DNA mixtures (about 300 μL each) from the second-round PCR are purified using the NucleoSpin Extract II kit following the manufacturer's protocol.

This procedure should result in 100 μL (50 μL per spin column × 2) each of 12 primer-free DNA solutions (*see* **Note 5**).

4. *DNA template re-annealing*: Put the resulting 12 DNA solutions onto a heat block (or thermal cycler) at 98 °C for 5 min, then turn off the heater, and let the samples cool down slowly to room temperature (around 25 °C) (*see* **Note 6**).
5. Determine the concentration of the 12 PCR DNA solutions (=PCR DNA templates for 12 Ile/Ala mutants of αCTS) individually by measuring absorbance at 260 nm using a NanoDrop spectrophotometer. Typical concentration of the 12 template PCR DNA is 0.15 mg/mL (in 100 μL).

3.2 Preparation of an E. coli Cell-Free S30 Extract

To prepare an *E. coli* cell-free extract, we usually use an *ompT-deficient E. coli* strain capable of expressing the T7 RNA polymerase gene, e.g., BL21 Star™ (DE3) (F⁻ *ompT hsdS*$_B$(r_B^-, m_B^-) *gal dcm rne*131 (λDE3)). The BL21 Star™ strains contain a mutation in the gene encoding RNaseE (*rne*131), which is one of the major sources of mRNA degradation. BL21 Star™ cells significantly improve the stability of mRNA transcripts and increase protein expression yields following transcription from the T7 ϕ10 promoter in the PCR DNA templates. The cell-free S30 extracts from *E. coli* BL21 Star™ (DE3) cells can be prepared from *E. coli* grown in a flask or a fermenter. Fermenter preparations have the advantage of high ribosome concentrations. The protocol described here is for a fermenter (BIOSTAT C, B. Braun Biotech International, Melsungen, Germany) preparation.

3.2.1 Day One: Streak Out E. coli BL21 Star™ (DE3) on an LBT Plate

1. Streak out *E. coli* BL21 Star™ (DE3) cells on an LBT plate, and then incubate it overnight at 37 °C.

3.2.2 Day Two: Buffer Preparation and Overnight Culture

1. Prepare and autoclave or filter sterilize stock solutions of glucose (2 M, 224 mL), thiamine (1 mg/mL, 200 mL), 10× S30 buffer (2 L), and 50 % PEG 8000 (2 L) as listed in Subheading 2 (*see* **Note 7**).
2. After autoclaving, store MQ water and 10× S30 buffer at 4 °C.
3. Pick a single colony of BL21 Star™ (DE3) with a sterile toothpick, inoculate it directly in 10 mL of LBT medium, and then let it grow overnight at 37 °C.

3.2.3 Day Three: Z-Medium Preparation and 1 L Overnight Culture

1. Prepare 5 L of 4× Z-medium.
2. Add 5 L of Z-medium concentrate, 4.5 mL of antifoam, and 14 L of MQ water into the fermenter. This gives a total volume of 19 L. When 1 L of overnight culture is added the following day, 20 L of Z-medium will result.
3. Sterilize the medium at 121 °C using the sterilization function of the fermenter.

4. Calibrate the pO2 electrode of the fermenter to 0 % after 25 min at 121 °C.
5. Inoculate 1 L of LBT medium prepared the previous day with the 10 mL overnight culture and incubate overnight at 37 °C.
6. Autoclave 2 L of 50 % PEG 8000 and store at room temperature.

3.2.4 Day Four: 20 L Culture and Processing of Cell Pellets

1. Equilibrate the fermenter at 37 °C, with 100 % air and mixing rate at 250 rpm.
2. Prior to addition of the overnight culture, remove a volume from the outlet at the base as a blank for later measurement of the absorbance of the culture at 600 nm (OD).
3. Add 224 mL of 2 M glucose and 200 mL of 1 mg/mL thiamine and then the 1 L overnight culture to the Z-medium *using a sterile glass funnel.*
4. Allow these to mix before removing a time zero volume for OD measurement.
5. Remove small volumes for OD measurements every 30 min until the culture reaches *an OD of 1.0*. This density should be achieved within 3 h.
6. Add the 40 mL of 0.5 M IPTG to the 20 L culture *to induce T7 RNA polymerase production in the cells.*
7. Remove small volumes for OD measurements every hour until the culture reaches *an OD of 3.0*. This density should be achieved within 3 h.
8. Fill a large sink or tub with a large amount of crushed ice in preparation for the next steps.
9. Drain the culture into 5 L conical flasks, and immediately place them on ice. *Weigh empty 1 L centrifuge tubes.*
10. Harvest the cells by serial centrifugation at 4 °C for 12 min at 10,000 × *g* in six 1 L tubes. Use pre-weighed tubes so that the pellet mass can be determined later.
11. Resuspend the cell pellets in a total of 400 mL of 1× S30 buffer α. Centrifuge the resuspended cells for 10 min at 10,000 × *g* and 4 °C.
12. Remove the supernatant, and pack the pellet by centrifugation for a further 5 min. Remove the remaining supernatant with a pipette.
13. Leave the cell pellets on ice in the cold room (or chamber at 4 °C) overnight, or snap freeze the pellets in liquid nitrogen, and then store them at –80 °C.

3.2.5 Day Five: Preparation of the S30 Extract

Optional: If the cell pellets were removed from the freezer, thaw them slowly on ice for at least 1 h prior to resuspension.

1. Resuspend the cell pellets (total of 100–120 g) in 400 mL of 1× S30 buffer α again and centrifuge for 12 min at 10,000 ×*g*.
2. Remove the supernatant, and pack the pellet by centrifugation for a further 5 min at 10,000 ×*g*. Remove the remaining supernatant with a pipette.
3. Resuspend the pellets in 1.3 mL of 1× S30 buffer α per gram of cells.
4. Subject the suspension to a single pass in a French press at a pressure of 6,000 psi (*see* **Note 8**).
5. Centrifuge the lysed cells in 500 mL tubes at 9,500 ×*g* and 4 °C for 15 min to remove any cell debris. Transfer the supernatant to clean tubes (50 mL).
6. Centrifuge the supernatant at 30,000 ×*g* at 4 °C for 1 h to produce the S30 fraction.
7. Prepare three batches of 3 L each of 1× S30 buffer β. Dialyze the S30 fraction in Spectra/Por #4 dialysis tubing against each 3 L batch of 1× S30 buffer β for 1 h at 4 °C, i.e., for 3 h in total.
8. Transfer the dialysis tubes containing the S30 fraction to a large evaporating dish *and dialyze against two batches of 1 L of* 50 % PEG 8000. Place the dish on a shaking tray to speed dialysis, and leave it for 2 h at 4 °C.
9. Change the 50 % PEG buffer, and repeat *until the volume of the extract has been reduced to approximately half the volume of the S30 extract (about 2 h)*. This can be assessed by visual comparison with dialysis tubing containing 60 mL of water.
10. The *concentrated* S30 fraction is then dialyzed against a further 3 L of 1× S30 buffer β for 15–30 min to remove the residual PEG 8000 buffer.
11. Measure the volume of the extract (*see* **Note 9**).
12. Dispense the extract into 1 mL aliquots in chilled 1.5 mL microcentrifuge tubes.
13. Snap freeze the aliquots in liquid nitrogen and store at −80 °C until further use.

For each new batch of S30 extract, the concentrations of magnesium acetate and the extract itself must be optimized. A series of reactions using 9–25 μL of S30 extract per 100 μL reaction mixture and 15–25 mM magnesium acetate should be sufficient to find the optimal concentrations of each.

3.3 Cell-Free Synthesis of the αCTS Ile/Ala Derivatives in Complex with τ_c16

Our target protein, αCTS (residues 964–1160 of the *E. coli* DNA polymerase III α subunit), is found to be insoluble when expressed alone but becomes soluble if expressed in the complex with its partner subunit Pol III τ_c16. The αCTS with a single Ile-to-Ala

Table 6
Composition of reaction mixture and outside buffer

Stock solution	Reaction mixture (μL)	Outside buffer (mL)
10× Protein synthesis buffer	150	1.5
1 M creatine phosphate	120	1.2
10 mg/mL creatine kinase	37.5	–
17.5 mg/mL tRNA	15	–
15 mM each of 20 amino acids (^{15}N-Ile)	99	0.99
1.07 M magnesium acetate	21	0.27
0.15 mg/mL template PCR DNA	100	–
Purified τ_c16 protein (8 mg/mL)	378	
MQ water	(Adjust)	(Adjust)
1 M KOH	–	(Adjust to pH 7.5)
E. coli S30 cell-free extract	300[a]	–
Total	1,500/sample	15/sample

[a]Needs prior optimization (*see* **Note 10**)

mutation at each of the 12 sequential locations can be synthesized in a soluble complex with τ_c16 by adding the following materials to an *E. coli* in vitro translation system. As template DNA, we use an expression vector (e.g., pETMCSI [8], pETMCSIII [8], or pET20b(+) (Novagen)) bearing a T7 promoter and terminator. However, PCR-generated templates are also highly recommended as no difference in final protein yield has been detected over many hundreds of reactions [6]. This PCR-directed CFPS approach is easily adapted for the high-throughput production of mutated proteins as more than ten different CFPS can be carried out in parallel and in high yield.

The protocol for CFPS *for each mutant*, i.e., an ^{15}N-Ile selectively labeled αCTS in complex with inlabeled τ_c16, is described below. The CFPS reaction mixture is originally based on the protocol of Kigawa et al. [1].

1. Prepare the reaction mixture and the outside buffer according to Table 6.
2. Place the reaction mixture (1.5 mL for each sample) into Spectra/Por #2 dialysis tubing, submerge it in the outside buffer (15 mL) (*see* **Note 11**), and then let it incubate at 30 °C for 7 h with gentle shaking.
3. Apply about 500 μL of each supernatant (1.5 mL) of the reaction mixture to a ProPur IMAC Mini spin column suitable to purify the His$_6$-tagged translation product (*see* **Note 12**).

3.4 Preparation of the Samples for NMR Analysis

1. Concentrate and wash the combined IMAC Mini spin column elution (3.9 mL per sample) as mentioned above with a desired NMR buffer (10 mM sodium phosphate, pH 6.8, 1 mM dithiothreitol, and 100 mM NaCl) repeatedly by ultrafiltration (e.g., Millipore Ultra-4, MWCO 1,0000) until the protein sample reaches the desired volume and buffer condition (a minimum 200 μL in volume and 50–100 μM in protein concentration are required for a typical ^{15}N-HSQC NMR measurements) (*see* **Note 13**).
2. Subject a 2 μL aliquot of the elution to 15 % SDS-PAGE analysis, and stain the gel with Coomassie brilliant blue.
3. The result of expression and purification of the protein complex is shown in Fig. 3.
4. Adjust the concentrated sample to a total volume of 200 μL by adding 20 μL of D_2O and the NMR buffer.
5. Transfer the sample to an NMR tube (3 mm in diameter) for ^{15}N-HSQC NMR measurement.
6. Record ^{15}N-HSQC NMR spectra. The results of ^{15}N-Ile-labeled αCTS (and its Ile-to-Ala mutants) in the complex with non-labeled τ_c16 are shown in Fig. 4.

4 Notes

1. *DO NOT* leave on ice, but leave *at room temperature* to prevent precipitation.
2. This gel purification is required to ensure that none of the vector is co-purified, thus avoiding unnecessary complications in the downstream NMR analysis due to the potential for a small amount of wild-type protein contamination.
3. If you can see the DNA band clearly as shown in Fig. 2a, b, then the quality is sufficiently high.
4. If you can see the DNA band clearly as shown in Fig. 2c, then the quality is sufficiently high.
5. This second-round PCR DNA purification is not necessary for the downstream CFPS, but the exact DNA concentration measurement is required.
6. This procedure is necessary to produce the self-circularizing PCR DNA template for CFPS [6].
7. Stir the mixture of 50 % PEG 8000 at room temperature overnight. Separately autoclave the stock solutions of glucose and 10× S30 buffer, 1 L of LBT medium, and 20 L of MQ water for diluting the 10× S30 buffer stock. Sterilization by filtration (0.2 μm) is advised for thiamine. Autoclave a large glass funnel

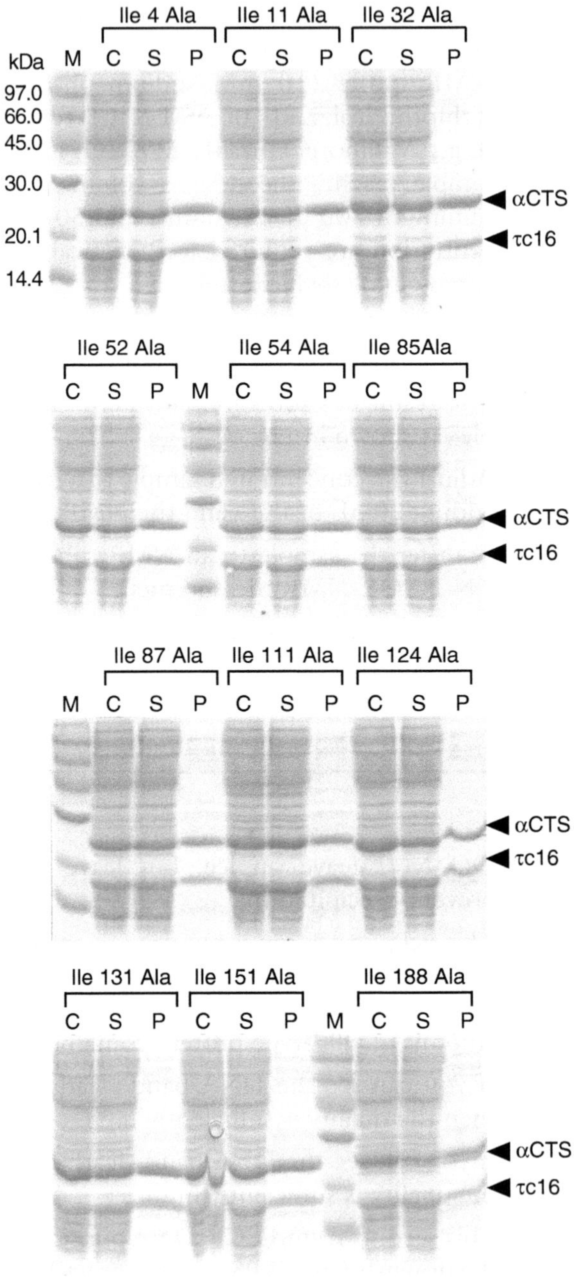

Fig. 3 Purification of 12 ^{15}N-Ile-labeled αCTS mutants in the complex with τ_c16. Purified complexes were analyzed using 15 % SDS-PAGE gels stained with Coomassie brilliant blue. M, Low molecular protein marker (molecular sizes are indicated on the left side of the top gel); C, crude cell-free extract; S, supernatant; P, purified His_6-αCTS (Ile-to-Ala substituted mutants):τ_c16 complexes. Ile4Ala, Ile11Ala, Ile32Ala, and so on represent the mutation sites in His_6-αCTS where Ile4, Ile11, Ile32, and so on were substituted with Ala. Twelve Ile-to-Ala-substituted His_6-αCTS mutants in the complex with τ_c16 (40 kDa in total) were separately and simultaneously expressed from the PCR-generated DNA templates and rapidly isolated as soluble complexes by single Ni-affinity column chromatography; protein yields were sufficient for downstream NMR measurements as shown in Fig. 4

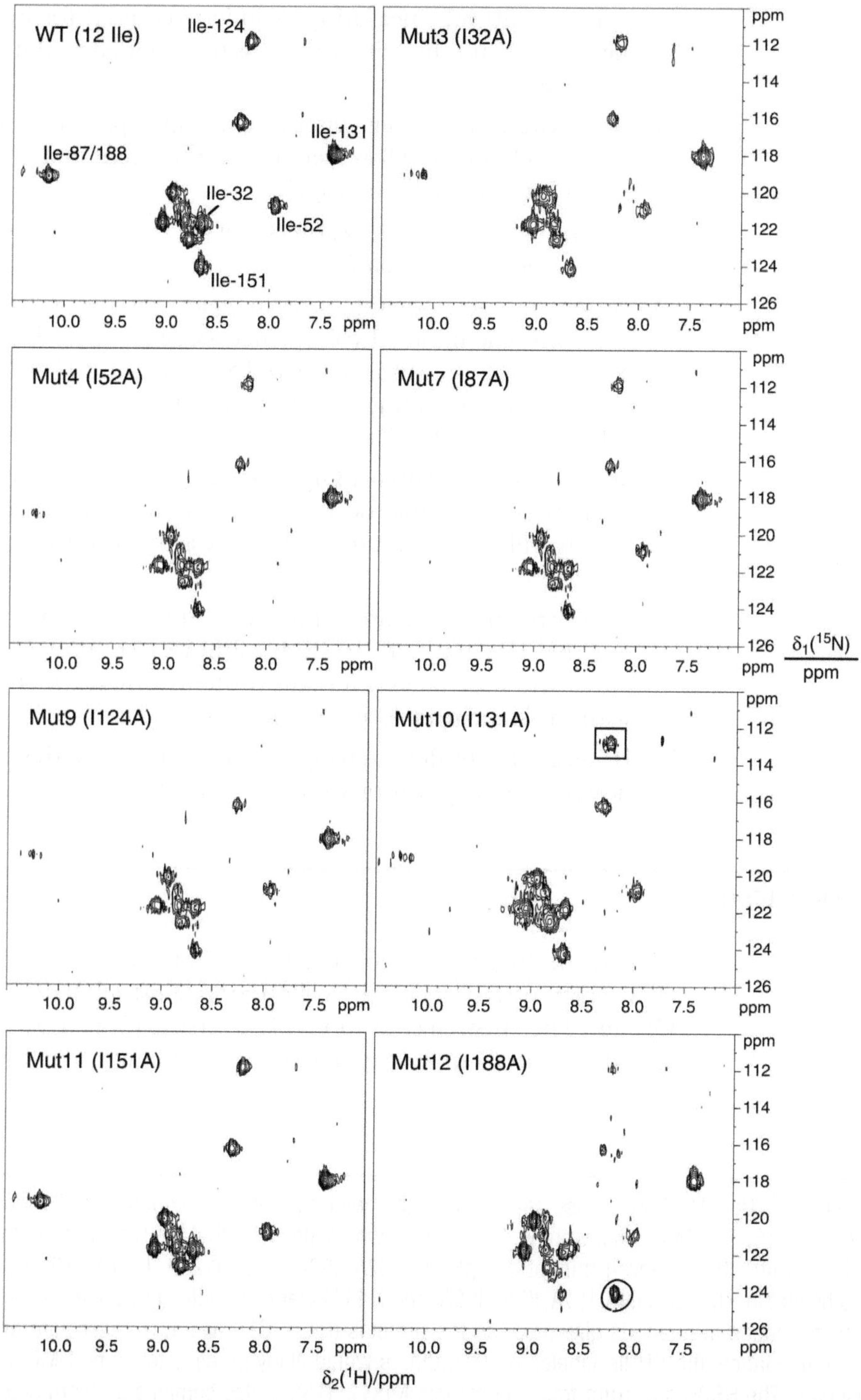

Fig. 4 ^{15}N-HSQC spectra of ^{15}N-Ile-labeled αCTS mutants in the complex with τ_c16. ^{15}N-HSQC spectra of several complexes of His$_6$-αCTS (^{15}N-Ile-labeled Ile-to-Ala mutants):τ_c16 (unlabeled) in a buffer of 10 mM sodium phosphate, pH 6.8, 1 mM dithiothreitol, and 100 mM NaCl. All NMR spectra were recorded at 25 °C using a Bruker 800 MHz or 600 MHz NMR spectrometer equipped with a cryoprobe. ^{15}N-HSQC spectra were recorded using 3 mm sample tubes, t_{1max} = 32 ms, t_{2max} = 102 ms, and total recording times between 22 and 24 h.

and two 250 mL measuring cylinders required for measuring the glucose and thiamine solutions to be added to the medium on day 3.

8. The lysate remains very viscous even after passing it through the French press. However, do not be tempted to pass it through the French press again.
9. Centrifuge (at 30,000 × *g* at 4 °C for 30 min) the S30 extract *if the level of precipitation is high.*
10. To give the best possible protein yield, it is very important to know first the minimal volume necessary for the cell-free reactions. Therefore, freshly prepared S30 extracts (T7-RNA polymerase induced) always need prior optimization of the input volume using a standard gene.
11. *See* Fig. 16.1 of Subheading 3.4.1 in ref. 10 to learn how to assemble the cut microcentrifuge tube in the knotted dialysis bag. It will enable you to readily access the reaction mixture in the dialysis bag.
12. Follow the manufacturer's protocol. Alternatively, you can also use the HisTrap™ FF (GE Healthcare) column (1 mL) to purify the His_6-tagged mutant protein complexes one by one with the standard procedure.
13. If necessary, buffer exchanges can be useful at this stage for downstream applications and/or storage.

Acknowledgements

This work has been supported by Profs. Nicholas E. Dixon (University of Wollongong, Wollongong, Australia) and Gottfried Otting (The Australian National University, Canberra, Australia) and the Australian Research Council, including a fellowship to

Fig. 4 (continued) The 12 NMR cross-peaks from the wild-type (WT) complex (^{15}N-Ile-labeled His_6-αCTS:unlabeled τ_c16; 40 kDa in total) were well dispersed, which indicates that the His_6-αCTS in the complex with τ_c16 is well structured. Seven mutant complexes (I32A, I52A, I87A, I124A, I131A, I151A, and I188A) appeared to be stable while five (I4A, I11A, I54A, I85A, and I111A) were unstable in solution. Perhaps the protein–protein interaction interface is disrupted, so the proteins (His_6-αCTS and τ_c16) associate only weakly. Under this circumstance, the ^{15}N-Ile-labeled αCTS mutants would likely precipitate, so no NMR cross-peaks can be observed. The HSQC spectrum from the sample Mut12 (I188A) also contains an extra cross-peak (at the ^{15}N-chemical shift of 124 ppm and ^{1}H-chemical shift of 8.15 ppm), probably from a C-terminal proteolysis product. The cross-peak of Ile-124 is shifted in the spectrum of the Mut10 (I131A) complex compared with the original location in the WT spectrum, presumably because they are close together. The cross-peak (at the ^{15}N-chemical shift of 119 ppm and the ^{1}H-chemical shift of 10.15 ppm) in the HSQC spectrum of the WT complex disappeared from spectra of both complexes I87A and I188A; therefore, it was not possible to assign

K.O. K.O. thanks Dr. Andrew V. Kralicek (Plant and Food Research, Auckland, New Zealand) for critical reading of an earlier version of the manuscript.

References

1. Kigawa T, Yabuki T, Yoshida Y et al (1999) Cell-free production and stable-isotope labeling of milligram quantities of proteins. FEBS Lett 442:15–19
2. Ozawa K, Headlam MJ, Schaeffer PM et al (2004) Optimization of an *Escherichia coli* system for cell-free synthesis of selectively ^{15}N-labelled proteins for rapid analysis by NMR spectroscopy. Eur J Biochem 271:4084–4093
3. Ozawa K, Dixon NE, Otting G (2005) Cell-free synthesis of ^{15}N-labeled proteins for NMR studies. IUBMB Life 57:615–622
4. Ozawa K, Wu PSC, Dixon NE et al (2006) ^{15}N-Labelled proteins by cell-free protein synthesis: strategies for high-throughput NMR studies of proteins and protein–ligand complexes. FEBS J 273:4154–4159
5. Morrison KL, Weiss GA (2001) Combinatorial alanine-scanning. Curr Opin Chem Biol 5: 302–307
6. Wu PSC, Ozawa K, Lim SP et al (2007) Cell-free transcription/translation from PCR amplified DNA for high-throughput NMR studies. Angew Chem Int Ed 46:3356–3358
7. Ozawa K, Jergic S, Park AY et al (2008) The proofreading exonuclease subunit ε of *Escherichia coli* DNA polymerase III is tethered to the polymerase subunit α via a flexible linker. Nucleic Acids Res 36:5074–5082
8. Neylon C, Brown SE, Kralicek AV et al (2000) Interaction of the *Escherichia coli* replication terminator protein (Tus) with DNA: a model derived from DNA-binding studies of mutant proteins by surface plasmon resonance. Biochemistry 39:11989–11999
9. Pratt JM (1984) Coupled transcription-translation in prokaryotic cell-free systems. In: Hames BD, Higgins SJ (eds) Transcription and translation. IRL Press, Oxford, pp 179–209
10. Apponyi M, Ozawa K, Dixon NE et al (2008) Cell-free protein synthesis for analysis by NMR spectroscopy. In: Kobe B, Guss M, Huber T (eds) Methods in molecular biology, vol 426, Structural proteomics: high-throughput methods. Humana, Totowa, NJ, pp 257–268
11. Jergic S, Ozawa K, Williams NK et al (2007) The unstructured C-terminus of the τ subunit of *Escherichia coli* DNA polymerase III holoenzyme is the site of interaction with the α subunit. Nucleic Acids Res 35:2813–2824

Chapter 12

Site-Specific Incorporation of Unnatural Amino Acids into Proteins by Cell-Free Protein Synthesis

Kiyoshi Ozawa and Choy Theng Loh

Abstract

Cell-free protein synthesis (CFPS) offers a fast and inexpensive means to incorporate unnatural amino acids (UAAs) site specifically into proteins. This enables engineering of proteins and allows production of protein-based probes for analysis of their interactions with other molecules. Using dialysis *Escherichia coli* CFPS system in combination with aminoacyl-tRNA synthetase and suppressor tRNA evolved from *Methanocaldococcus jannaschii* high expression yield of proteins with site specifically incorporated UAAs can be achieved. Typically the target protein can be prepared at concentrations of about 1 mg/mL, which is generally sufficient for subsequent applications.

Key words Cell-free protein synthesis, Suppressor tRNA, Aminoacyl-tRNA synthetase, Unnatural amino acids, Copper-free click chemistry

1 Introduction

Cell-free protein synthesis (CFPS) has generated great interest in structural biology since Kigawa et al. demonstrated the possibility of obtaining milligram per milliliter yields in a coupled cell-free transcription–translation system based on *Escherichia coli* S30 cell extract [1]. These amounts are sufficient for structure analysis by nuclear magnetic resonance (NMR) spectroscopy using modern high-field NMR spectrometers. It provides a convenient and rapid way to analyze proteins that would be difficult or costly to make otherwise, such as proteins that are toxic or sensitive to proteolysis in vivo [2], proteins enriched with unnatural amino acids (UAAs) [3, 4], and proteins requiring binding partners for expression in soluble form [5, 6], including membrane proteins ([7] for reviews *see* refs. [8–11]).

Since 2001, Schultz and co-workers have been developing a general method for site-specific introduction of UAAs into proteins in *E. coli* [12, 13] via genetic encoding, where an amber suppressor tRNA is enzymatically charged with the UAA. As opposed

Kirill Alexandrov and Wayne A. Johnston (eds.), *Cell-Free Protein Synthesis: Methods and Protocols*, Methods in Molecular Biology, vol. 1118, DOI 10.1007/978-1-62703-782-2_12, © Springer Science+Business Media, LLC 2014

to the former approach using in vitro chemically aminoacylated suppressor tRNAs, enzymatically aminoacylated tRNA can be continuously regenerated during protein expression, resulting in high protein yields. In general, modern CFPS systems produce more protein in a given volume of medium than in vivo systems.

In this chapter, we describe an efficient and rapid CFPS method to incorporate *p*-acetyl-phenylalanine (AcF) or *p*-azido-phenylalanine (AzF) at chosen sites of proteins by simply adding an optimized *M. jannaschii* amber suppressor $tRNA_{CUA}{}^{opt}$ [14], *p*-acetyl-phenylalanyl tRNA synthetase (AcF-RS) or *p*-azido-phenylalanyl tRNA synthetase (AzF-RS), and AcF or AzF into an *E. coli* cell-free translation system [15].

Similarly to the in vivo expression, we use an expression vector with T7 promoter and terminator in CFPS, such as pETMCSI [16], pETMCSIII [16], or pET20b(+) (Novagen). However, there is a big advantage in using PCR-generated templates in CFPS over in vivo expression given that no difference in final protein yield has been detected over many hundreds of experiments [17]. Therefore, this technique is highly recommended in CFPS (the detailed protocol is featured in accompanying chapter in this book).

Here, we describe a generic CFPS protocol that is used for production of the cytotoxic *E. coli* protein τ_c16 containing AcF at the 62nd, 126th, or 144th codon as well as the *S. aureus* sortase A containing AzF at the 40th codon. A template DNA encoding the target protein with one (or more) in frame amber codon(s) at desired position(s) is required [15].

In principle, AcF-substituted proteins can be labeled site specifically with hydroxylamine derivatives as tags in aqueous solution [18], and AzF-substituted proteins can be labeled with fluorescent tags to generate site-specific probes to facilitate the elucidation of molecular mechanisms, protein-to-protein networks, and protein–ligand or protein–protein interactions.

There is no naturally occurring azido group in normal proteins. Since AzF is site specifically introduced into a target protein by this method, an azido group on the protein can be readily modified using organic chemistry, for example using the Staudinger–Bertozzi ligation or the Huisgen 1,3-dipolar cycloaddition. The Huisgen 1,3-dipolar cycloaddition recently became very well known as "click chemistry," in which azido and alkyne groups make a 1,2,3-triazole ring in the presence of Cu(I). Therefore, we introduce one of the convenient methods using click chemistry to label the target protein containing AzF with a commercially available fluorescent alkyne derivative.

However, one major shortcoming of the copper-catalyzed click reaction is the fact that Cu(II) as well as the Cu(I) produced in the presence of ascorbate are highly cytotoxic and often cause damage to proteins [19]. To circumvent this problem, "copper-free click chemistry" has recently been developed which now allows azide-modified macromolecules to be labeled without metal catalysts. Here, we present the protocol to site selectively prepare fluorescently tagged proteins by copper-free click chemistry.

2 Materials

2.1 Preparation of *E. coli* S30 Extracts Expressing an Amber Suppressor $tRNA_{CUA}^{opt}$

1. Spectra/Por #2 (10 mm flat width corresponding to 6.4 mm diameter) and #4 (45 mm flat width corresponding to 29 mm diameter) dialysis tubing, 12,000–14,000 MWCO (Spectrum Laboratories Inc., Rancho Dominguez, CA, USA).
2. French press.
3. Centrifuges with rotors suitable for 50 mL, 500 mL, and 1 L tubes.
4. Optical densitometer.
5. 10 mL centrifugal ultrafiltration units, 10,000 MWCO, e.g., Millipore Ultra-4.
6. 10 and 50 mL disposable centrifuge tubes.
7. Benchtop orbital shaker.
8. SDS-PAGE apparatus.
9. 20 L capacity fermenter.
10. Vortex mixer.
11. Phenylmethanesulfonyl fluoride (PMSF).
12. Antifoam 204 (Sigma).
13. Milli-Q (MQ) water.
14. 1 mg/mL thiamine.
15. 2 M glucose.
16. 1 M Tris acetate buffer, pH 8.2.
17. 50 % PEG 8000 w/w: 1,000 g PEG 8000 and 1 L of 1× S30 buffer (total volume: ~2 L).
18. LBT medium: 10 g/L tryptone, 5 g/L yeast extract, 5 g/L sodium chloride, 2.5 mL/L of 1 M sodium hydroxide, 0.5 mg/L thymine. LBT-Cm medium contains additional 33 mg/L chloramphenicol, 1.5 % agar for plate use. LBT-AC medium contains additional 33 mg/L chloramphenicol, 100 mg/L ampicillin, and 1.5 % agar for plate use.
19. 4× Z-medium: 165 mM potassium dihydrogen phosphate, 664 mM dipotassium hydrogen phosphate, 40 g/L yeast extract [20].
20. 10× S30 buffer: 100 mM Tris acetate, initial pH 8.2, 160 mM potassium acetate, 140 mM magnesium acetate, pH 8.3 with potassium hydroxide.

21. S30 buffer α: 1× S30 buffer containing 0.5 mM PMSF, 1 mM dithiothreitol, and 7.2 mM β-mercaptoethanol.
22. S30 buffer β: 1× S30 buffer containing 1 mM dithiothreitol.
23. *E. coli* strain having T7 RNA polymerase gene (gene *1*) under transcriptional control of the *lacOP* operator/promoter, e.g., BL21 Star™::λDE3 (Invitrogen).
24. Isopropyl-β-D-1-thiogalactoside (IPTG).
25. A plasmid designated as pKO1474 [15] (*see* **Notes 1** and **19**).

2.2 Preparation of Optimized M. jannaschii Amber Suppressor $tRNA_{CUA}^{opt}$

1. A plasmid designated as pKO1474 [15] (*see* Subheading 2.1, **item 25**).
2. *E. coli* strain expressing the T7 RNA polymerase gene, e.g., BL21 Star™::λDE3 (Invitrogen).
3. 0.3 M potassium acetate (KOAc) buffer, pH 4.75 (500 mL).
4. Water-saturated phenol (200 mL).
5. Q-Sepharose High Performance (GE Healthcare) column (ø2.5 × 5 cm; 25 mL resin).
6. QA buffer: 20 mM Tris–HCl (pH 7.5), 10 mM $MgCl_2$, 0.2 M NaCl (1 L).
7. QB buffer: 20 mM Tris–HCl (pH 7.5), 10 mM $MgCl_2$, 1 M NaCl (500 mL).

2.3 Preparation of M. jannaschii-Derived Aminoacyl-tRNA Synthetase (RS)

The Schultz group has previously used directed molecular evolution of *M. jannaschii* tyrosyl-tRNA synthetase (RS) generating modified enzymes capable of incorporation of over 40 unnatural amino acids [13] (*see* also **Note 19**). We adapted and relocated these mutant RS genes individually into a T7-expression vector, pETMCSIII [16], to produce them in *E. coli* recombinantly with N-terminal His_6 tags [15].

1. Plasmids designated as pKO1420 and pKO1350 [15]: These plasmids encode N-terminal His_6-tagged ORFs for *p*-acetyl-phenylalanyl-tRNA synthetase (AcF-RS) and *p*-azido-phenylalanyl-tRNA synthetase (AzF-RS), respectively, placed under the control of the T7 promoter.
2. *E. coli* strain expressing the T7 RNA polymerase gene, e.g., BL21::λDE3/pLysS.
3. IPTG.
4. French press.
5. HisTrap™ FF (GE Healthcare) column (5 mL): All the buffers and the columns should be stored at 4 °C.
6. Buffer NA: 50 mM HEPES–KOH, pH 7.5, 300 mM NaCl, 20 mM imidazole, 5 % (v/v) glycerol (1 L).

7. Buffer NB: 50 mM HEPES–KOH, pH 7.5, 300 mM NaCl, 500 mM imidazole, 5 % (v/v) glycerol (500 mL).
8. Enzyme storage buffer: 50 mM Tris–HCl (pH 7.6), 2 mM dithiothreitol, 100 mM NaCl (200 mL).

2.4 Synthesis of p-Acetyl-Phenylalanine- or p-Azido-Phenylalanine-Containing Proteins Using Cell-Free Protein Synthesis

All buffers and solutions are stored at −30 or −80 °C.

1. 10× Protein synthesis buffer: 580 mM HEPES–KOH (pH 7.5), 17 mM dithiothreitol, 12 mM ATP, 8 mM GTP, 8 mM CTP, 8 mM UTP, 6.4 mM cyclic AMP, 0.68 mM folinic acid, 276 mM ammonium acetate, 2.08 M potassium glutamate (2.5 mL).
2. A mixture containing 15 mM each of 20 amino acids (1.5 mL) [21] (*see* **Note 2**).
3. 1.07 M magnesium acetate (0.5 mL).
4. 1 M creatine phosphate (Sigma) (2.0 mL).
5. 10 mg/mL creatine kinase (Roche) (0.1 mL): This enzyme is dissolved in MQ water.
6. 0.1–1.5 mg/mL of template plasmid DNA containing amber stop codon(s) (TAG) at the modification site in frame with the ORF of interest.
   (a) pETMCSIII-τ_c16 (Ser62TAG, Gln126TAG, and Arg144TAG); these T7 expression vectors [16] encode the *E. coli* τ_c16 ORF [22] having an amber stop codon (TAG) in frame at the 62nd, 126th, or 144th codon, resulting in N-terminally His$_6$-tagged τ_c16.
   (b) pETMCSI-SortA (Asn40TAG); this T7 expression vector [16] encodes *Staphylococcus aureus* sortase A having an amber stop codon (TAG) in frame at the 40th codon, resulting in C-terminally His$_6$-tagged sortase A. The plasmid DNA was usually purified with a Plasmid Maxi Kit (Qiagen).
7. *E. coli* cell-free S30 extract (*see* Subheading 3.1).
8. Total tRNA mixture including amber suppressor $tRNA_{CUA}^{opt}$: The amber suppressor $tRNA_{CUA}^{opt}$ (*see* Subheading 3.2) was diluted with MQ water to a final concentration of 17.5–35 mg/mL (Sup-$tRNA^{opt}$).
9. Evolved *M. jannaschii* tyrosyl-tRNA synthetase: AcF-RS and AzF-RS are described below [15] (*see* Subheading 3.3).
10. 50 mM AcF or AzF (dissolved in 0.1 M potassium hydroxide) (*see* **Note 3**).
11. ProPur IMAC Mini-spin columns (Nalgene Nunc International, USA).

2.5 Site-Specific Modification via Azide Group Incorporated into the Protein

1. 50 mM sodium phosphate (pH 8.0) (1 mL).
2. Click-iT™ tetramethylrhodamine (TAMRA) DIBO alkyne (Invitrogen) (*see* **Note 4**).

3 Methods

3.1 Preparation of an E. coli Cell-Free S30 Extract

To prepare an *E. coli* cell-free extract, we usually use an *ompT*-*deficient E. coli* strain expressing the T7 RNA polymerase gene, e.g., BL21 Star™::λDE3 (F^- *ompT hsdS*$_B$(r_B^-, m_B^-) *gal dcm rne*131 λDE3). The BL21 Star™ strains contain a mutation in the gene encoding RNaseE (*rne*131), which is one of the major causes of mRNA degradation. Use of BL21 Star™ cells significantly improves the stability of mRNA transcripts and increases protein expression yield from T7 promoter-based vectors. It is critical not to use *E. coli* strains for lysate production which have amber suppressor tRNA mutations, for example, JM109 (*endA*1 *glnV*44 *thi*-1 *relA*1 *gyrA*96 *recA*1 *mcrB*$^+$ Δ(*lac*-*pro*AB) e14- [F' *traD*36 *proAB*$^+$ *lacI*^{q} *lacZ*ΔM15] *hsdR*17($r_K^-m_K^+$)) or DH5α (F^- *endA*1 *glnV*44 *thi*-1 *recA*1 *relA*1 *gyrA*96 *deoR nupG* φ80d*lacZ*ΔM15 Δ(*lacZYA*-*argF*) U169, *hsdR*17(r_K^- m_K^+), $λ^-$). The cell-free S30 extracts from *E. coli* BL21 Star™::λDE3 cells can be prepared from *E. coli* grown in a flask or a fermenter. Fermenter preparations have the advantage of higher ribosome concentrations. Therefore, current protocol is adopted for the preparation of S30 extract using a fermenter (BIOSTAT C, B. Braun Biotech International, Melsungen, Germany) preparation.

3.1.1 Day One: Transformation of BL21 Star™::λDE3 with pKO1474

1. Transform an *E. coli* BL21 Star™::λDE3 with pKO1474: About 4 kb, Cm^r, and streak out on an LBT-Cm plate, and then incubate it overnight at 37 °C.

3.1.2 Day Two: Buffer Preparation and Overnight Culture

1. Prepare and autoclave or filter stock solutions of 2 M glucose, 224 mL; 1 mg/ml thiamine, 200 mL; 10× S30 buffer, 2 L; and 50 % PEG 8000, 2 L as listed in Subheading 2 (*see* **Note 5**).
2. After autoclaving, store MQ water and 10× S30 buffer at 4 °C.
3. Pick up a single colony of BL21 Star™::λDE3/pKO1474 transformant with a sterile toothpick, inoculate directly in a 10 mL LBT-Cm culture, and then let it grow overnight at 37 °C.

3.1.3 Day Three: Z-Medium Preparation and Overnight 1 L Culture

1. Sterilize the 20 L of 1× Z-medium in the fermenter.
2. Inoculate 1 L of LBT-Cm medium prepared the previous day with the 10 mL overnight culture and incubate overnight at 37 °C.
3. Autoclave 2 L of 50 % PEG 8000 and store at room temperature.

3.1.4 Day Four: 20 L Culture and Processing of Cell Pellets

1. Add 224 mL of 2 M glucose, 200 mL of 1 mg/mL thiamine, and 20 mL of 33 mg/mL of chloramphenicol followed by 1 L overnight culture into Z-medium with incubation at 37 °C.
2. Add 40 mL of 0.5 M IPTG to the 20 L culture when the culture reaches an A_{600} of 1.0 to induce T7 RNA polymerase expression.
3. Harvest the cells by centrifugation when the culture reaches an A_{600} of 3.0 (*see* **Note 6**).
4. Resuspend two-thirds of the cell pellets (80–100 g) in a total of 400 mL of 1× S30 buffer α. Centrifuge the resuspended cells for 10 min at 10,000 ×*g* at 4 °C (repeat this step twice).
5. At this stage the cell pellets can be stored on ice or in the cold room overnight (*see* **Note 7**).

3.1.5 Day Five: Preparation of the S30 Extract

1. Resuspend the pellets in 1.3 mL of 1× S30 buffer α per gram of biomass.
2. Subject the suspension to a single pass in a French press at a pressure of 6,000 psi (*see* **Note 8**).
3. Centrifuge the lysed cells at 30,000 ×*g* at 4 °C for 1 h to produce the S30 fraction.
4. Prepare three batches of 3 L each of 1× S30 buffer β. Dialyze the S30 fraction in Spectra/Por #4 dialysis tubing against each 3 L batch of 1× S30 buffer β for 1 h at 4 °C, i.e., for 3 h in total.
5. Transfer the dialysis tubes containing S30 fraction to a large evaporating dish and dialyze against two batches of 1 L of 50 % PEG 8000. Place the dish on a shaking tray to speed dialysis and leave for 2 h at 4 °C.
6. Change the 50 % PEG buffer, and repeat the operation until the volume of the extract has been reduced to approximately half the original volume of the S30 extract.
7. The concentrated S30 fraction is then dialyzed against a further 3 L of 1× S30 buffer β for 15–30 min to remove the traces of PEG 8000.
8. Measure the volume of the extract (*see* **Note 9**).
9. Dispense 1 mL aliquots of the extract into 1.5 mL microcentrifuge tubes.
10. Snap-freeze the aliquots in liquid nitrogen and store at −80 °C until further use.

For each new batch of S30 extract, the concentrations of magnesium acetate and the extract itself in the cell-free reactions must be optimized. Carry out a series of expression reactions using S30 extract and magnesium acetate, respectively, from 9 to 25 μL and from 15 to 25 mM ranges, which should be sufficient to find the optimal concentration of these components for CFPS.

3.2 Preparation of an Optimized Amber Suppressor $tRNA_{CUA}^{opt}$ Evolved from *M. jannaschii*

The isolation protocol for the total tRNA expressing an optimized amber suppressor $tRNA_{CUA}^{opt}$ is described below. Optimization of $tRNA_{CUA}^{opt}$ concentration in CFPS is critical to obtain high yields of proteins with UAAs [15].

1. Suspend 40 g of the cells, *E. coli* BL21 Star™::λDE3/pKO1474, harvested by centrifugation (*see* **Note 6**) with 200 mL of 0.3 M KOAc, pH 4.75 (*see* **Note 10**).
2. Add 200 mL of water-saturated phenol to the cell suspension, and shake it vigorously for more than 1 h. Allow the phases to settle, and remove the upper aqueous layer.
3. Add 200 mL of 0.3 M KOAc, pH 4.75, to the phenol layer, shake it, and recover 200 mL of the upper aqueous layer again.
4. Recover the total RNA fraction from 380 mL of the combined aqueous layers by ethanol precipitation.
5. Dissolve the total RNA fraction in 10 mL of QA buffer. The supernatant is clarified by centrifugation or filter and loaded on the Q-Sepharose High Performance column: ø2.5 cm × 5 cm; 25 mL resin; pre-equilibrated with QA buffer.
6. Elute bound total tRNA fraction including $tRNA_{CUA}^{opt}$ with QB buffer 500 mL, at a flow rate of 1.0 mL/min. Collect eluted fractions: 5 mL each of eluate per tube; and measure UV absorbance at 260 nm (A_{260}).
7. Pool the fractions with significant A_{260}, and recover the suppressor $tRNA_{CUA}^{opt}$ by ethanol precipitation.
8. Dissolve the total tRNAs including $tRNA_{CUA}^{opt}$:Sup-$tRNA^{opt}$ in an appropriate volume, less than 1 mL, of MQ water, and determine its concentration by measuring absorbance at $\lambda = 260$ nm (*see* **Note 11**).

3.3 Preparation of *M. jannaschii*-Derived Aminoacyl-tRNA Synthetase (RS)

The protocol for overexpression and purification of AcF-RS or AzF-RS is described below. BpyRS, BpaRS, HcoRS, and CNF-RS [15] can be also overexpressed and purified according to this protocol.

1. Transform *E. coli* BL21::λDE3/pLysS with pKO1420 or pKO1350, streak out on an LBT-AC plate, and incubate overnight at 30 °C.
2. Pick up a single colony with a toothpick and inoculate directly in 10 mL of LBT-AC medium followed by overnight incubation at 30 °C.
3. Inoculate 1 L of LBT-Cm medium prepared the previous day with the 10 mL overnight culture and incubate at 30 °C.
4. Cultivate until its A_{600} reaches between 0.5 and 1.0 at 30 °C.
5. To induce the expression of the AcF-RS add IPTG to a final concentration of 1.0 mM. Keep culture at 30 °C for 3 h: $A_{600} \approx 1.2$.

6. Suspend the cells, ≈3.55 g, harvested by centrifugation with 20 mL of buffer NA.
7. Disrupt cells with French press, 12,000 psi × 2 passes, keeping the suspension cool, and remove cell debris by centrifugation at 40,000 × *g* for 1 h at 4 °C in 50 mL tubes.
8. Load the supernatant onto a HisTrap™ FF column, 5 mL resin pre-equilibrated with NA buffer.
9. Elute proteins with a gradient of NA buffer, 50 mL to NB buffer, 50 mL at a flow rate of 1.0 mL/min. Collect eluted fractions.
10. Subject a 5 μL aliquot of each fraction to 15 % SDS-PAGE analysis, and stain the gel with Coomassie brilliant blue.
11. Pool the fractions containing AcF-RS or AzF-RS, 35.9 kDa, accordingly to the authentic protein marker.
12. Concentrate fractions down to 2 mL volume by ultrafiltration, e.g., Millipore Ultra-15.
13. Dialyze the concentrated sample against enzyme storage buffer, and determine its concentration by measuring UV absorption at 280 nm. The purified AcF-RS or AzF-RS 35.9 kDa has a molar extinction coefficient at 280 nm, $\varepsilon_{280} = 1.66 \times 10^4\ M^{-1}\ cm^{-1}$; absorbance at 280 nm of a 1 mg/mL enzyme solution equals 0.463 A_{280}. On average, 110–250 mg of AcF-RS or AzF-RS can be obtained with this protocol.

3.4 Cell-Free Synthesis of AcF- or AzF-Containing Proteins

1. Prepare the reaction mixture and the outside buffer according to Table 1.
2. Place the reaction mixture (500 μL) into Spectra/Por #2 dialysis tubing, submerge it in the buffer (5 mL) (*see* **Note 15**), and then incubate at 30 °C for 7 h or overnight with gentle shaking.
3. Apply the supernatant of the reaction mixture to a ProPur IMAC Mini-spin column to purify the His_6-tagged translation product (*see* **Note 16**).
4. Concentrate the eluate (1.3 mL) by ultrafiltration to desired volume (*see* **Note 17**).
5. Subject the 5 μL aliquot of elution to 15 % SDS-PAGE analysis, and stain the gel with Coomassie brilliant blue. A typical result of purification of AcF-containing protein is shown in Fig. 1.

3.5 Site-Selective Modification of the Azido Group Incorporated into the Protein

1. Prepare the azide-modified protein, 200 μM, in 50 mM sodium phosphate buffer, pH 8.0.
2. Mix 10 μL of azide-modified protein, 200 μM, with 1 μL of TAMRA DIBO alkyne (dissolved in DMSO), and 29 μL of MQ H_2O: 40 μL in total volume (*see* **Note 4**).

Table 1
Reaction mixture and the outside buffer recipe

Stock solution	Reaction mixture (μL)	Outside buffer (μL)
10× Protein synthesis buffer	50	500
1 M creatine phosphate	40	400
10 mg/mL creatine kinase	12.5	–
35 mg/mL Sup-tRNA^{opt}	2.5–10[a]	–
15 mM each of 20 amino acids	33	330
1.07 M magnesium acetate	7	90
50 mM AcF or AzF	10	100
1.0 mM AcF-RS or AzF-RS	1–20[a]	–
0.5 mg/mL template DNA(TAG)	16 (10 for PCR template)[b]	–
MQ H_2O	(Adjust)	(Adjust)
1 M KOH	–	(Adjust to pH 7.5)
E. coli S30 cell-free extract	5–200[c]	–
Total	500	5,000

[a]Needs prior optimization (*see* **Note 12**)
[b]*See* **Note 13**
[c]Needs prior optimization (*see* **Note 14**)

3. Incubate the "click" reaction mixture for 1 h at room temperature (approximately 20 °C).
4. Centrifuge the mixture at 16,000 ×*g* for 15 min to give a clear supernatant and pellet.
5. Subject an aliquot of supernatant and the pellet to 15 % SDS-PAGE analysis.
6. Detect fluorescence of TAMRA by taking a picture of the gel under UV light, 312 nm (*see* **Note 18**).
7. Stain the gel with Coomassie brilliant blue. A typical result of fluorescent labeling of an AzF-containing protein is shown in Fig. 2.

4 Notes

1. This plasmid carries a single copy of the recently optimized *M. jannaschii* amber suppressor tyrosyl-tRNA$_{CUA}$ (tRNA$_{CUA}^{opt}$) under a *proK* promoter that expressed constitutively in *E. coli* cells [14, 15].
2. DO NOT leave on ice, but leave at room temperature to prevent precipitation.

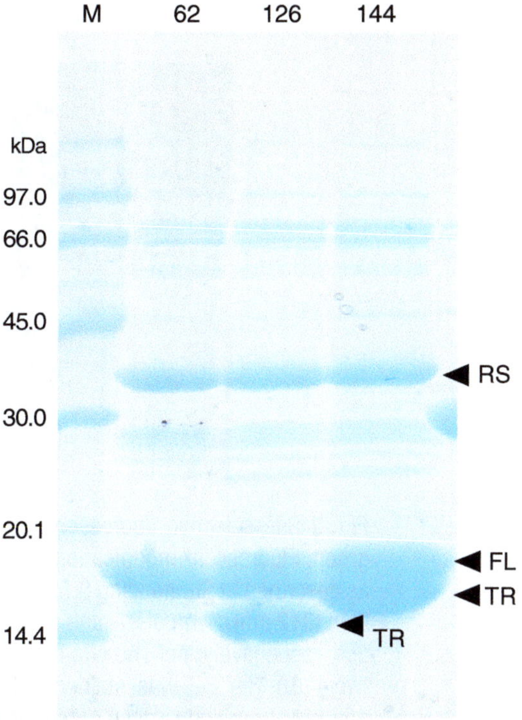

Fig. 1 Efficient incorporation of AcF into the τ_c16 amber mutants Ser62TAG, Asn126TAG, and Arg144TAG using AcF-RS (30 μM; 1.1 mg/mL) and $tRNA_{CUA}^{opt}$ (175 μg/mL) in the continuous exchange CF (CECF) system [1, 15]. N-terminally His_6-tagged products were purified by ProPur IMAC Mini-spin columns. M, low-molecular-weight marker; 62, Ser62TAG; 126, Asn126TAG; 144, Arg144TAG; RS, AcF-RS; FL, full-length product; TR, truncated product. The 15 % SDS-PAGE gel was stained with Coomassie brilliant blue. More than 2.0 mg/mL of full-length τ_c16 containing AcF were produced. (Compare the intensities of the FL proteins with that of AcF-RS (1.1 mg/mL) as an internal marker.) TR can be reduced if the concentration of amber suppressor $tRNA_{CUA}^{opt}$ is increased or a release factor 1 (RF-1)-free S30 extract [23] is used. TR can also be eliminated from the purified proteins by use of a C-terminal His_6 tag

3. The authors used reagents which were a kind gift from Mr. Ansis Maleckis. AcF and AzF are also commercially available from SynChem, Inc. (http://www.synchem.com). AzF is sensitive to UV light. Avoid long exposure of samples containing AzF to light.
4. 0.5 mg of TAMRA DIBO alkyne (Invitrogen) was dissolved in 70 μL of high-quality, anhydrous dimethylsulfoxide (DMSO), and the solution was stored at −80 °C after use.
5. Stir the mixture of 50 % PEG 8000 at room temperature overnight. Separately autoclave the stock solutions of glucose and 10× S30 buffer, 1 L of LBT medium, and 20 L of MQ water for diluting the 10× S30 buffer stock. Filter (0.2 μm) sterilization

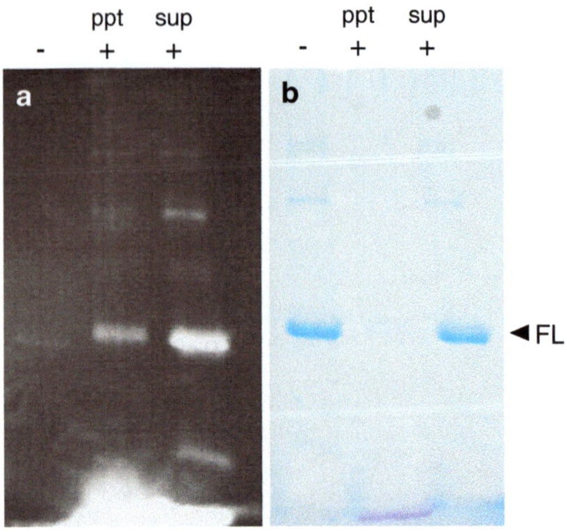

Fig. 2 Site-selective fluorescent labeling of sortase A(Asn40AzF) via the azido group. Fluorescent modification by copper-free click chemistry was carried out as described in Subheading 3.5. (**a**) The fluorescence image was taken under UV light (312 nm). Symbols, + and –, mean with and without DIBO TAMRA modification, respectively; ppt and sup represent the pellet and the supernatant, respectively. (**b**) The gel was stained with Coomassie brilliant blue. FL; Full-length product of purified sortase A(Asn40AzF) regardless of DIBO TAMRA modification

is recommended for thiamine. Autoclave a large glass funnel and two 250 mL measuring cylinders required on day 4 for measuring the glucose and thiamine solutions.

6. As an option, take one-third of cells (approximately 40 g) for a separate preparation of suppressor $tRNA_{CUA}^{opt}$ (*see* Subheading 3.2) before the following S30 processing step. Snap-freeze and store it at −80 °C. Use this cellular biomass later for the large-scale preparation of suppressor $tRNA_{CUA}^{opt}$.
7. Cells for S30 extract should be processed within the next day.
8. The lysate appears to be very viscous after passing it through the French press. Centrifuge the S30 extract at 30,000 × *g* at 4 °C for 30 min once again if the pellet appears to be too fluffy.
9. Centrifuge the S30 extract at 30,000 × *g* at 4 °C for 30 min *if the level of precipitation is high.*
10. Subjecting the cell suspension to double-pass through the French press at 12,000 psi can increase the final yield of suppressor $tRNA_{CUA}^{opt}$ as a part of total tRNA fraction.
11. Normally, there is no need to confirm tRNA integrity by gel analysis.
12. Optimization of Sup-$tRNA^{opt}$ and RS concentrations is essentially important for each DNA template containing amber mutations in the ORF; this often results in a significant increase in the final yields of full-length proteins [15].

13. Optimizing the DNA concentration can sometimes give a higher final yield of protein of interest. In the case of PCR-generated DNA template, the standard final DNA concentration is usually set to 10 μg/mL [17].
14. In order to obtain a highest protein yield, it is important to determine the minimal volume necessary for the cell-free reaction first. Therefore, freshly prepared S30 extracts (T7 RNA polymerase induced) always need a prior optimization of the input volume using a standard gene without any TAG mutations.
15. *See* Fig. 16.1 of Subheading 3.4.1 in ref. 21 if you would like to see the assembly of the cut microcentrifuge tube in the knotted dialysis bag. It will enable you to readily access the reaction mixture in the dialysis bag.
16. According to the manufacturer's protocol typical binding capacity for the mini-spin column is 1 mg His-tagged recombinant protein.
17. Typically lower than 500 μL would be required since the original reaction volume was 500 μL. If necessary, buffer exchange can be useful at this stage for downstream applications and/or storage.
18. A long exposure to UV light can result in a decrease of fluorescence.
19. Note added by editors: From analysis of the literature including this chapter that describes the construction of vectors for expression of *M. jannaschii* amber suppressor tRNA$_{\mathrm{CUA}}^{\mathrm{opt}}$ and engineered aminoacyl-tRNA synthetase (RS) it transpired to us that it is relatively difficult to extract the exact sequence of tRNA$_{\mathrm{CUA}}^{\mathrm{opt}}$ and RS using individual studies due to a large number of reported mutants and their combinations. While editing the chapter we combined the literature and experimental data to determine the optimal combinations of tRNA$_{\mathrm{CUA}}^{\mathrm{opt}}$ and engineered aminoacyl-tRNA synthetase. According to our analysis the optimal sequence of tRNA$_{\mathrm{CUA}}^{\mathrm{opt}}$ is *cggctaactaagcg-gcctgctgactttctcgccgatcaaaaggcattttgctattaagggattgacgag ggcg tatctgcgcagtaagatgcgccccGcattCCGGCCGTAGTTCAGC AGGGCAGAACGGCGGACTCTAAATCCGCATGGCAG GGGTTCAAATCCCCTCGGCCGGACCAaattcgaaaagcctg ctca acgagcaggctttttgcat*. tRNA-encoding unit is displayed in uppercase and is flanked by proK promoter and terminator sequences shown in italic; the consensus nucleotides of the promoter are underlined, and transcription initiating G is capitalized (based on [24]). The expression plasmid can contain up to three tRNA units. The orthogonal RS specific for both *p*-azido-L-phenylalanine and orthogonal tRNA (see above) has the sequence of MDEFEMIKRNPSEIISEEELREVLKKDEK SALIGFEPSGKIHLGHYLQIKKMIDLQNAGFDIIILLA DLHAYLNQKGELDEIRKIGDYNKKVFEAMGLKAKYV

YGSPFQLDKDYTLNVYRLALKTTLKRARRSMELIARE DENPKVAEVIYPIMQVNQIHSSGVDVAVGGMEQRK IHMLARELLPKKVVCIHNPVLTGLDGEGKMSSSK GNFIAVDDSPEEIRAKIKKAYCPAGVVEGNPIMEI AKYFLEYPLTIKRPEKFGGDLTVNSYEELESLFKN KELHPMRLKNAVAEELIKILEPIRKRL according to [25] and our experiments. The protein is typically N-terminally His_6 tagged to facilitate the purification of the enzyme.

Acknowledgements

This work has been supported by Profs. Gottfried Otting (Australian National University, Canberra, Australia) and Nicholas E. Dixon (University of Wollongong, Wollongong, Australia) and the Australian Research Council, including a fellowship to K.O. K.O. thanks Dr. Andrew V. Kralicek (Plant and Food Research, Auckland, New Zealand) for critical reading of an earlier version of the manuscript.

References

1. Kigawa T, Yabuki T, Yoshida Y et al (1999) Cell-free production and stable-isotope labeling of milligram quantities of proteins. FEBS Lett 442:15–19
2. Renesto P, Raoult D (2003) In vitro expression of rickettsial proteins. Ann N Y Acad Sci 990: 642–645
3. Kigawa T, Yamaguchi-Nunokawa E, Kodama K et al (2001) Selenomethionine incorporation into a protein by cell-free synthesis. J Struct Funct Genomics 2:29–35
4. Ozawa K, Headlam MJ, Mouradov D et al (2005) Translational incorporation of L-3,4-dihydroxyphenylalanine into proteins. FEBS J 272:3162–3171
5. Ozawa K, Jergic S, Crowther JA et al (2005) Cell-free protein synthesis in an autoinduction system for NMR studies of protein–protein interactions. J Biomol NMR 32:235–241
6. Ozawa K, Jergic S, Park AY et al (2008) The proofreading exonuclease subunit ε of *Escherichia coli* DNA polymerase III is tethered to the polymerase subunit α via a flexible linker. Nucleic Acids Res 36:5074–5082
7. Klammt C, Löhr F, Schäfer B et al (2004) High level cell-free expression and specific labeling of integral membrane proteins. Eur J Biochem 271:568–580
8. Klammt C, Schwarz D, Löhr F et al (2006) Cell-free expression as an emerging technique for the large scale production of integral membrane protein. FEBS J 273:4141–4153
9. Ozawa K, Dixon NE, Otting G (2005) Cell-free synthesis of ^{15}N-labeled proteins for NMR studies. IUBMB Life 57:615–622
10. Ozawa K, Wu PSC, Dixon NE et al (2006) ^{15}N-Labelled proteins by cell-free protein synthesis: strategies for high-throughput NMR studies of proteins and protein–ligand complexes. FEBS J 273:4154–4159
11. Shimizu Y, Kuruma Y, Ying BW et al (2006) Cell-free translation systems for protein engineering. FEBS J 273:4133–4140
12. Wang L, Schultz PG (2001) A general approach for the generation of orthogonal tRNAs. Chem Biol 8:883–890
13. Liu CC, Schultz PG (2010) Adding new chemistries to the genetic code. Annu Rev Biochem 79:413–444
14. Young TS, Ahmad I, Yin JA et al (2010) An enhanced system for unnatural amino acid mutagenesis in *E. coli*. J Mol Biol 395: 361–374
15. Ozawa K, Loscha KV, Kuppan KV et al (2012) High-yield cell-free protein synthesis for site-specific incorporation of unnatural amino acids at two sites. Biochem Biophys Res Commun 418:652–656
16. Neylon C, Brown SE, Kralicek AV et al (2000) Interaction of the *Escherichia coli* replication

terminator protein (Tus) with DNA: a model derived from DNA-binding studies of mutant proteins by surface plasmon resonance. Biochemistry 39:11989–11999

17. Wu PSC, Ozawa K, Lim SP et al (2007) Cell-free transcription/translation from PCR amplified DNA for high-throughput NMR studies. Angew Chem Int Ed 46:3356–3358
18. Fleissner MR, Brustad EM, Kálai T et al (2009) Site-directed spin labeling of a genetically encoded unnatural amino acid. Proc Natl Acad Sci U S A 106:21637–21642
19. Hong V, Presolski SI, Ma C et al (2009) Analysis and optimization of copper-catalyzed azide-alkyne cycloaddition for bioconjugation. Angew Chem Int Ed 48:9879–9883
20. Pratt JM (1984) Coupled transcription-translation in prokaryotic cell-free systems. In: Hames BD, Higgins SJ (eds) Transcription and translation. IRL Press, Oxford, pp 179–209
21. Apponyi M, Ozawa K, Dixon NE et al (2008) Cell-free protein synthesis for analysis by NMR spectroscopy. In: Kobe B, Guss M, Huber T (eds) Methods in molecular biology. Structural proteomics: high-throughput methods, vol 426. Humana, Totowa, NJ, pp 257–268
22. Jergic S, Ozawa K, Williams NK et al (2007) The unstructured C-terminus of the τ subunit of *Escherichia coli* DNA polymerase III holoenzyme is the site of interaction with the α subunit. Nucleic Acids Res 35:2813–2824
23. Loscha KV, Herlt AJ, Qi R, Huber T et al (2012) Multiple-site labeling of proteins with unnatural amino acids. Angew Chem Int Ed 51:2243–2246
24. Guo J, Melançon CE 3rd, Lee HS et al (2009) Evolution of amber suppressor tRNAs for efficient bacterial production of proteins containing nonnatural amino acids. Angew Chem Int Ed Engl 48(48):9148–9151
25. Chin JW, Santoro SW, Martin AB et al (2002) Addition of p-azido-L-phenylalanine to the genetic code of *Escherichia coli*. J Am Chem Soc 124(31):9026–9027

Chapter 13

In Vitro Translation of Papillomavirus Authentic and Codon-Modified L1 Capsid Gene mRNAs in Mouse Keratinocyte Cell-Free Lysate

Kong-Nan Zhao

Abstract

Keratinocytes are the major cell type in the epidermis responsible for constructing the protective barrier of mammalian skin. Primary keratinocytes cultured in vitro can mimic the in vivo differentiation process, providing an abundant source of pure keratinocytes used for investigating the regulatory mechanisms of gene expression associated with cell proliferation and differentiation. We developed a primary mouse keratinocyte cell-free lysate system for translation of papillomavirus authentic and codon-modified L1 capsid gene mRNAs in vitro. We demonstrated that the viral gene codon usage matched available aminoacyl-tRNAs to determine their translational efficiencies, which were associated with differentiation status of the keratinocytes used for preparing cell-free lysate. We revealed that a novel regulatory mechanism of gene expression is utilized by papillomavirus to direct viral capsid protein expression to the site of virion assembly in mature keratinocytes. Here, I describe the methods in details of how to establish primary mouse keratinocyte culture, to prepare cell-free lysate, to carry out in vitro translation of the viral gene mRNAs, and to detect the translated products using Western blotting analysis.

Key words Primary mouse keratinocytes, Cell-free lysate, Codon usage, Aminoacyl-tRNAs, Papillomavirus L1 gene mRNAs, In vitro translation

1 Introduction

Keratinocytes are the major cell type in the epidermis responsible for constructing the protective barrier of mammalian skin by undergoing a complex and carefully choreographed program of cell differentiation [1]. Proliferative keratinocytes in the basal layer periodically detach from an underlying basement membrane of extracellular matrix and move outward. Once in the suprabasal layer, keratinocytes stop dividing and enter a differentiation program. Terminally differentiated keratinocytes flatten and develop the cornified envelopes, which consist of a stabilized array of keratin filaments contained within a covalently cross-linked protein envelope and play a critical protection role in barrier function of the

Kirill Alexandrov and Wayne A. Johnston (eds.), *Cell-Free Protein Synthesis: Methods and Protocols*, Methods in Molecular Biology, vol. 1118, DOI 10.1007/978-1-62703-782-2_13, © Springer Science+Business Media, LLC 2014

epithelium [2]. The barrier function of the cornified keratinocytes at the outermost layer is to protect skin against environmental damage such as pathogens (bacteria, fungi, parasites, viruses), heat, UV radiation, and water loss [3]. Keratinization or cornification is part of the physical barrier formation in epidermis in which the keratinocytes produce more and more keratin and eventually undergo programmed cell death [3]. A number of structural proteins (filaggrin, keratin), enzymes (proteases), lipids, and antimicrobial peptides (defensins) contribute to maintain the important barrier function of the skin [4–8]. Once pathogens start to invade the upper layers of the epidermis, keratinocytes can react with the production of proinflammatory mediators and in particular chemokines such as CXCL10 and CCL2 which attract leukocytes to the site of pathogen invasion [9–12].

Keratinocytes cultured in vitro can mimic the in vivo differentiation process, which greatly improves our understanding of keratinocyte biology [13]. Keratinocyte cultures provide an abundant source of pure keratinocytes that can be used to test the effects of proliferation- and differentiation-regulating agents on cell function. The proliferation and differentiation capacity of cultured keratinocytes make the epidermal cells to be an ideal candidate for investigating the regulatory mechanisms of gene expression associated with cell proliferation and differentiation. By establishing mouse primary keratinocytes in culture, we observed that expression of the papillomavirus major capsid (L1) proteins by transient transfection of authentic (*Nat*) [14] or codon-modified (*Mod*) L1 gene expression plasmids was differentially associated with keratinocyte differentiation [15]. We demonstrate in vitro and in vivo that gene codon composition is in part responsible for differentiation-dependent expression of the L1 proteins in keratinocytes. We reveal a novel mechanism that keratinocytes substantially change their tRNA profiles upon differentiation to match the codon composition of the targeted genes to regulate their protein expression [15].

Different mammalian cell-free lysates as a complement to cell-based in vivo assays have been developed to study gene expression including DNA replication, RNA transcription, and protein translation [16–18]. The cytoplasmic extracts have been proven to be valuable tools to study the different regulatory mechanisms of translation from initiation, elongation, and termination to ribosome recycling, including the faithful recapitulation of the synergy between the 7-mGpppN cap structure and the poly(A) tail for translation initiation of mRNAs [19–21]. These cell-free systems allow the translation of exogenously added mRNAs in a specific and controlled manner independently of nuclear events [20–23].

Recently, we developed a cell-free lysate translation system from primary mouse keratinocyte cultures for in vitro translation of the papillomavirus L1 capsid gene mRNAs [24]. Keratinocytes are the host cells of papillomavirus infection that causes cervical cancer—a second common cancer in women worldwide and other cancers [25–28]. We thus chose papillomavirus L1 gene mRNAs with (*Mod*) or without codon modifications (*Nat*) as the target gene mRNAs to investigate whether and how codon usage of the viral genes determines the translation of their mRNAs in keratinocyte cell-free lysate. We proved further that the novel mechanism for regulation of gene expression is utilized by papillomavirus to direct viral capsid protein expression to the site of virion assembly in mature keratinocytes. We observed that the PV L1 mRNAs, whether they were transcribed in vitro from the *Nat* or *Mod* PV L1 genes, could be in vitro translated in the keratinocyte cell-free lysates when the L1 mRNAs were capped with m7 GpppG and the endogenous mRNAs in the cell-free lysates were completely hydrolyzed. We also observed that the translational efficiencies of the papillomavirus *Nat* and *Mod* L1 mRNAs were significantly different from the cell-free lysates prepared from the in vitro-grown keratinocytes with different differentiation status [24]. We confirmed that additional supplementation of the aminoacyl-tRNAs (aa-tRNAs) prepared from D0 and D8 mouse keratinocyte cultures could differentially enhance translation of the PV *Nat* and *Mod* L1 mRNAs in keratinocyte cell-free lysate. We believe that the developed keratinocyte cell-free lysate system is very useful for the mechanistic study of expression of mammalian gene mRNAs in vitro.

2 Materials

2.1 Animal

Newborn mouse pups at 1–3 days old, which are used for preparation of primary keratinocytes. The mouse pups can be stored at 4 °C until ready to proceed with the keratinocyte isolation procedure. Storage up to 4 h does not reduce the quantity or quality of the epidermal cells isolated.

2.2 Keratinocyte Culture Medium

3:1 medium. 500 ml of 3:1 medium contains 365 ml DMEM medium, 125 ml Ham's F12 medium, which was added with 2 mM glutamine, 100 U/ml penicillin, 100 U/ml streptomycin, 50 ml fetal bovine serum (FBS), 2.5 mg transferrin, 2.5 mg insulin, 4.2 μg cholera toxin, 0.12 mg hydrocortisone, 17 mg adenine, and 10 mg gentamicin (*see* **Note 1**).

Keratinocyte-SFM medium. Both calcium-free keratinocyte-SFM medium and low-calcium keratinocyte-SFM medium (Invitrogen, Australia) were used for the primary keratinocyte culture.

2.3 Target Gene Expression Plasmids

Four viral L1 capsid gene expression plasmids (pCDNA3HPV6b *Nat* L1, pCDNA3HPV6b *Mod* L1, pCDNA3BPV1 *Nat* L1, and pCDNA3BPV1 *Mod* L1) from two types of papillomaviruses (human papillomavirus 6b (HPV6b) and bovine papillomavirus 1 (BPV1)) were used in our experiments [24]. Both pCDNA3HPV6b *Nat* L1 and pCDNA3BPV1 *Nat* L1 plasmids contained wild-type L1 gene sequences, which are about 1.5 kb in length, encoding 500 amino acids, which show a strong codon usage bias among the degenerately encoded amino acids toward 18 codons mainly with T at the third position that are rarely used by mammalian genes [29]. Plasmids pCDNA3HPV6b *Mod* L1 and pCDNA3BPV1 *Mod* L1 expressed codon-modified L1 genes in which the L1 genes are substituted with codons preferentially used in the mammalian genome. 250 base substitutions were made in 250 codons rarely used in mammalian cells to produce unmodified L1 proteins encoded from the L1 gene with consensus codon usage [30]. We demonstrated that expression of the PV *Mod* and *Nat* L1 gene plasmids was differentially associated with cell differentiation in keratinocytes in vitro and in vivo due to that the gene codon composition is in part responsible for differentiation-dependent expression of L1 protein [15].

2.4 Enzymes

Different restriction enzymes can be used for linearization of target gene expression plasmids, which will be used for in vitro transcription of the target gene mRNAs. *Eco*RI, *Bam*HI, and *Hin*dIII were used to linearize the four papillomavirus L1 gene expression plasmids (pCDNA3HPV6b *Nat* L1, pCDNA3HPV6b *Mod* L1, pCDNA3BPV1 *Nat* L1, and pCDNA3BPV1 *Mod* L1) in our experiments.

2.5 Lysis Buffers

12 ml of lysis buffer contained 2.4 ml of 1 M HEPES–KOH (pH 7.4), 240 μl of 250 mM magnesium acetate (MgOAc) (pH 7.4), 576 μl of 5 M potassium acetate (KOAc), 24 μl of 1 M dithiothreitol (DTT), 120 μl of 0.5 M ATP, 480 μl of 50 mM GTP, 240 μl of 10 mM S-adenosylmethionine [31], 240 μl of 100 mM spermidine, 480 μl of 1 M creatine phosphate, 1,000 U of creatine phosphokinase, 24 μl of 1 mg/ml leupeptin, 24 μl of 1 mg/ml chymostatin, 24 μl of 1 mg/ml Pepstatin, 2.4 ml of 1 M sucrose, and 4.728 ml of H_2O.

2.6 Aminoacyl-tRNAs (aa-tRNAs) of Keratinocytes

Total tRNA fraction containing aminoacyl-tRNAs (aa-tRNAs) was extracted and purified from the primary keratinocyte cultures. The aa-tRNAs were used for supplementation of the keratinocyte cell-free lysates to demonstrate that available aa-tRNAs in cell-free lysates system match the codon composition of target genes to regulate translation of their mRNAs.

3 Methods

3.1 Primary Mouse Keratinocyte Isolation and Culture

1. Surface-sterilize the mouse pups (*see* **Note 2**).
2. Peel skin off from the mouse pups and cut the skins into even pieces in sterile phosphate-buffered saline (PBS) (*see* **Note 3**).
3. Place the skin pieces in PBS buffer containing 2.5 % Dispase (Gibco BRL) at 4 °C for 24 h and then for 1 h at room temperature before proceeding to separate epidermis from dermis (*see* **Note 4**).
4. Separate epidermis sheets from dermis with sterile forceps (*see* **Note 5**).
5. Place the epidermis sheets of 4 pups in 2 ml of 0.25 % trypsin–versene in 15 ml sterile centrifuge tubes and incubate the epidermis–trypsin mixture at 37 °C for 8 min (*see* **Note 6**).
6. Pipette the epidermis vigorously up and down for until the enzyme solution appears to be cloudy (*see* **Note 7**).
7. Add equal volume of 3:1 medium and 5 ml of PBS to each 15 ml sterile centrifuge tube.
8. Remove most of the lumpy bits that rise to the surface and spin the tube at 220 × *g* for 5 min.
9. Resuspend the keratinocytes in 10 ml of 3:1 medium after decanting carefully the supernatant.
10. Count cell numbers in a hemocytometer and centrifuge at 220 × *g* for 5 min (*see* **Note 8**).
11. Grow the primary keratinocytes in keratinocyte-SFM medium in collagen-coated T75 flask at a defined seeding density (*see* **Note 9**).
12. Change medium for the primary keratinocyte culture using the keratinocyte-SFM medium containing 0.2 mM calcium every other day until the cells will be harvested for keratinocyte cell-free lysate preparation (*see* **Note 10**). The newly purified keratinocytes (D0) showed significantly different cell morphology from the keratinocytes grown in vitro for 8 days (D8) (Fig. 1a). Expression of K14 protein (basal keratinocyte marker [32]) was significantly downregulated, and expression of involucrin (a terminal differentiation marker of the keratinocytes [33]) occurred in D8 keratinocytes (Fig. 1b).

3.2 Keratinocyte Cell-Free Lysate Preparation

1. Aspirate the medium and set the flask of the in vitro-grown keratinocytes on ice. Gently wash the keratinocytes with 10 ml of ice-cold phosphate-buffered saline (PBS).
2. Scrape gently the keratinocytes into 6 ml of ice-cold extraction buffer with a small-bladed cell scraper. Transfer the entire keratinocyte lysate from each flask to a separate 10 ml centrifuge tube.

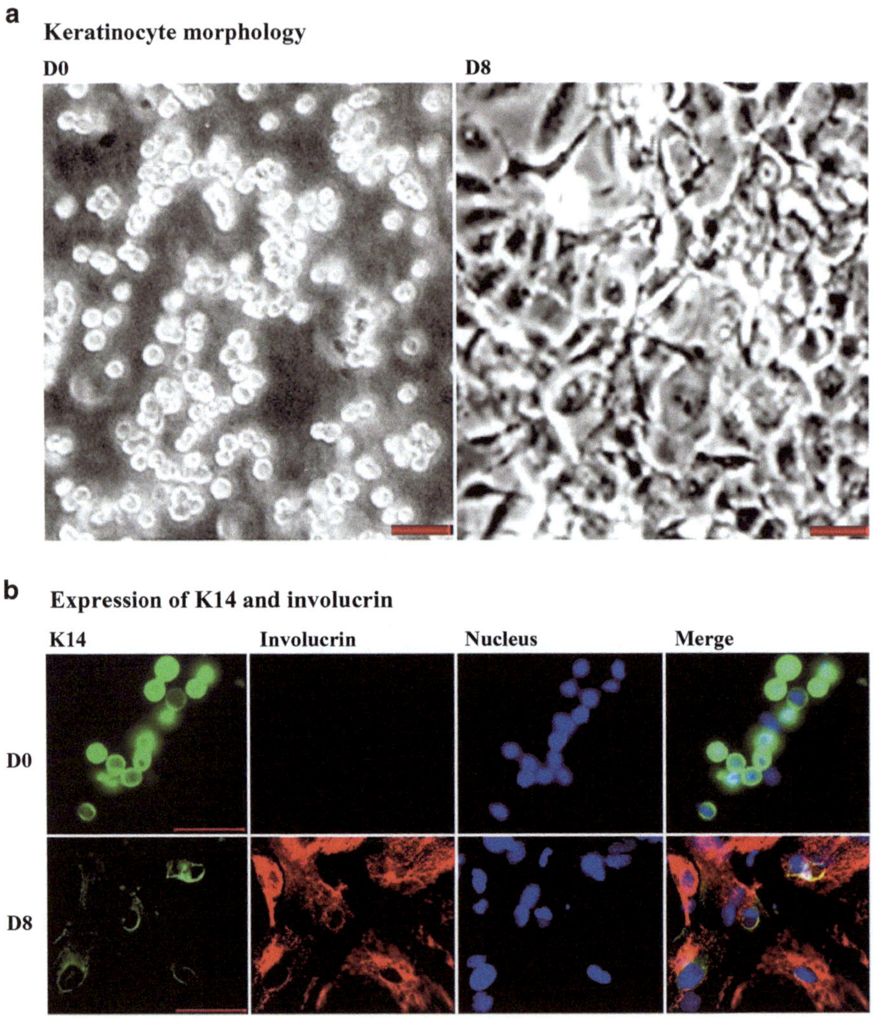

Fig. 1 Cell morphology and expression of K14 and involucrin in D0 and D8 mouse keratinocyte cultures. (**a**) Cell morphology. (**b**) Expression of K14 (basal keratinocyte marker) and involucrin (keratinocyte differentiation marker) by immunofluorescence staining. Scale bars = 20 μm

3. Pass the collected keratinocyte lysate through a 25-gauge needle ten times.
4. Centrifuge the lysate at 4 °C and 2,000 × *g* for 2 min.
5. Transfer the supernatant to a new 10 ml centrifuge tube to prepare mRNA-free keratinocyte cell-free lysate.
6. Hydrolyze the endogenous mRNAs in the prepared keratinocyte cell-free lysate by incubating the lysate at 20 °C for 10 min in the presence of 10 U/ml of micrococcal nuclease [34] and 1 mM calcium.
7. Stop the activity of enzyme by adding 2.5 mM ethylene glycol-bis(b-aminoethyl ether)-*N*,*N*,*N*′,*N*′-tetraacetic acid (EGTA, pH 7.0) following the hydrolysis of the endogenous mRNAs.

8. Store the keratinocyte cell-free lysate in small aliquots at –80 °C (*see* **Note 11**).
9. Supplement the cell-free lysate with 0.02 mM hemin before use (*see* **Note 12**).

3.3 In Vitro Transcription of Targeted Genes

1. Linearize target gene expression plasmids using appropriate restriction enzymes.
2. Transcribe the linearized target gene expression plasmids by T7 RNA polymerase (Promega, Australia) at 37 °C for 3 h (*see* **Note 13**).
3. Add 2 μl RNase-free DNase (2,000 U/ml, Promega, Australia) into each reaction to digest DNA for 1 h at 37 °C following transcription.
4. Purify the transcribed target mRNAs by phenol–chloroform extraction, followed by precipitation with 2 volumes of ethanol and washed twice with 70 % ethanol.
5. Resuspend the purified target mRNAs in RNase-free water used for in vitro translation (*see* **Note 14**).

3.4 In Vitro Translation of Targeted mRNAs in KC Cell-Free Lysate

1. Set up in vitro translation mixture in a 1.5 ml of microcentrifuge tube. The in vitro translation mixture with a volume of 25 μl consists of 17.5 μl of keratinocyte cell-free lysate, 2 μl of PV L1 mRNA (1 μg per reaction), 0.5 μl of RNase inhibitor, 0.5 μl of amino acid mixture (1 mM of each amino acid), and 4.5 μl of RNase-free H_2O (*see* **Note 15**).
2. Carry out in vitro translation reaction at 30 °C for various time lengths (*see* **Note 16**).
3. Stop the in vitro translation reaction by adding equal volume of 2× sodium dodecyl sulfate (SDS) sample buffer up to a volume of 50 μl.
4. Denature the in vitro translation mixture that contains 2× SDS sample buffer at 80 °C for 10 min (*see* **Note 17**).
5. Transfer 45 μl of supernatant of the in vitro-translated L1 product to a new tube after centrifugation for 5 min at 12,000 × *g* in a microcentrifuge.
6. Use the supernatant for detection of the in vitro-translated products using Western blotting analysis.

3.5 Detection of the In Vitro-Translated Proteins Using Western Blotting Analysis

1. Set up sodium dodecyl sulfate-polyacrylamide gel electrophoresis (SDS-PAGE).
2. Load 20 μl of the supernatant of the in vitro-translated proteins onto a well of SDS-PAGE.
3. Transfer the proteins to PVDF membranes.
4. Incubate the PVDF membranes with blocking buffer for 1 h or longer.

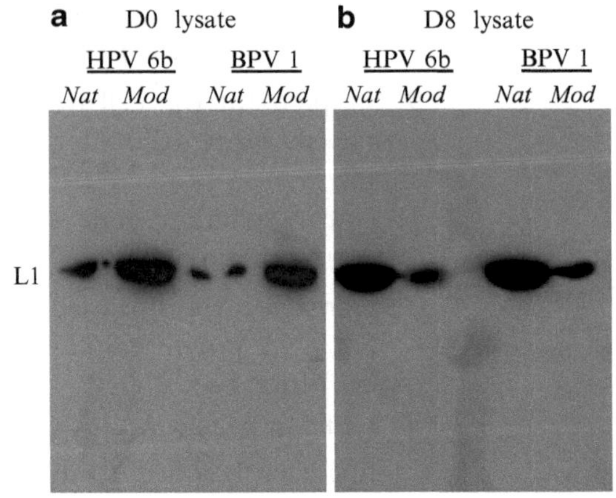

Fig. 2 Differentially translational expression of PV *Nat* and *Mod* L1 mRNAs in D0 (**a**) and D8 (**b**) keratinocyte cell-free lysates prepared from the primary mouse keratinocyte cultures. 1 μg of each of the PV L1 mRNAs was used for the in vitro translation reaction in keratinocyte lysates. Translation reaction was carried out at 30 °C for 3 h. The in vitro-translated L1 products were detected by Western blotting analysis

5. Label the PVDF membranes with first monoclonal antibody against the PV L1 protein at 4 °C overnight and then probe with horseradish peroxidase-conjugated anti-mouse secondary antibody for 2 h at room temperature.
6. Visualize the signals of the in vitro-translated proteins using a chemiluminescence system. We observed differentially translational expression of the PV *Nat* and *Mod* L1 mRNAs in D0 and D8 keratinocyte cell-free lysates (Fig. 2). The PV *Mod* L1 mRNAs prefer to translate the L1 proteins in D0 cell-free lysates that was prepared from less differentiated keratinocytes (Fig. 2a), while the PV *Nat* L1 mRNAs have the translational preference to the D8 cell-free lysate prepared from the differentiated mouse keratinocytes (Fig. 2b).

3.6 aa-tRNA-Regulation Studies

1. Extract and purify total tRNAs from the primary keratinocytes grown in vitro for a different time period (*see* **Note 18**).
2. Load 3 μg of tRNAs onto a well of 15 % denaturing polyacrylamide gel containing 8 M urea and 0.1 M sodium acetate (pH 5.2) [35].
3. Carry out tRNA gel electrophoresis in 0.1 M sodium acetate (pH 5.2) at 80 V for 5 h.
4. Stain the tRNA gel using ethidium bromide, after the electrophoresis is complete, to make tRNA bands visible under ultraviolet light.
5. Take gel photographs using Gel Doc (Biorad, Australia). Total tRNA profiles prepared from the newly prepared keratinocytes

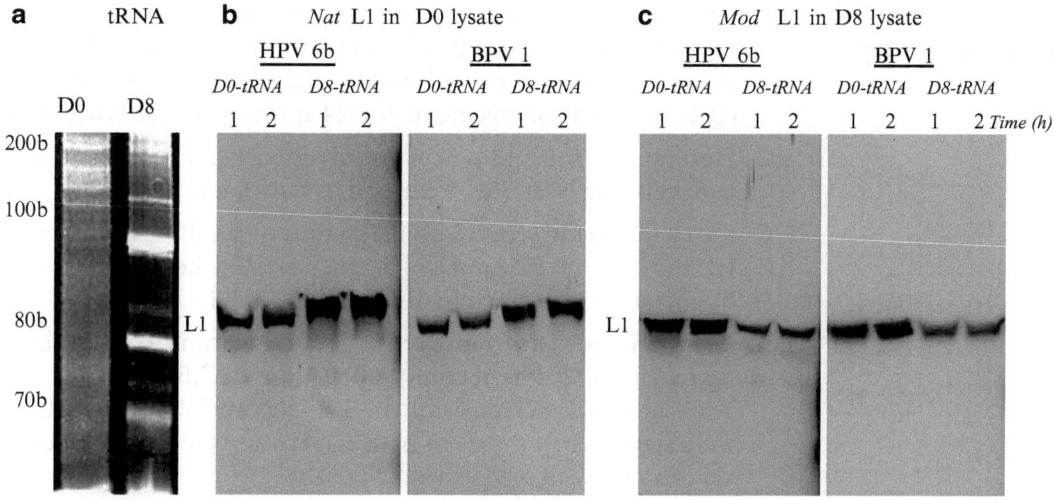

Fig. 3 aa-tRNAs from D0 and D8 mouse keratinocyte cultures differentially enhanced translation of the PV *Nat* and *Mod* L1 mRNAs in D0 and D8 keratinocyte cell-free lysates. (**a**) Profiles of total tRNAs prepared from D0 and D8 keratinocytes on 15 % denaturing polyacrylamide gel. Supplementation with 10^{-7} M of aa-tRNAs from D8 keratinocytes significantly enhanced L1 protein production from *Nat* in D0 keratinocyte cell-free lysate in vitro (**b**), while the L1 protein production from *Mod* L1 mRNAs in D8 keratinocyte cell-free lysate in vitro was significantly enhanced by supplementation with 10^{-7} M of aa-tRNA from D0 keratinocytes (**c**). 1 μg of each of the PV L1 mRNAs was used for the in vitro translation reaction. The in vitro-translated L1 products were detected by Western blotting analysis

(D0 keratinocytes) were significantly different from those prepared from the keratinocytes grown in vitro for 8 days (D8 keratinocytes) (Fig. 3a). While much more tRNA bands were observed in D0 keratinocytes, several extra tRNA bands in relative low molecular weights occurred in D8 keratinocytes, appearing that expression of these tRNA species is associated with keratinocyte differentiation (Fig. 3a) (*see* **Note 19**).

6. Produce aminoacyl-tRNAs (aa-tRNAs) from the total keratinocyte tRNA as previously reported [30] (*see* **Note 20**).
7. Supplement 10^{-7} M of D0 and D8 aa-tRNAs in D0 and D8 keratin cell-free lysates, respectively.
8. Carry out in vitro translation reaction of the PV *Nat* and *Mod* L1 mRNAs in D0 and D8 keratin cell-free lysates with different supplementation of the D0 and D8 aa-tRNAs at 30 °C for 1 and 2 h, respectively, as described in Subheading of 3.4.
9. Detect the in vitro-translated L1 proteins using Western blotting analysis as described in Subheading of 3.5. Supplementation with aa-tRNA from D8 cultured keratinocytes significantly enhanced L1 protein production from *Nat* L1 mRNAs in D0 keratinocyte lysate in vitro (Fig. 3b), while the L1 protein production from the *Mod* L1 mRNAs in D8 keratinocyte lysate in vitro was significantly enhanced by supplementation with 10^{-7} M aa-tRNA from D0 cultured keratinocytes (Fig. 3c).

3.7 Keratinocyte Cell Cycle Arrest Studies

1. Grow primary mouse keratinocytes in keratinocyte-SFM medium containing 0.2 mM calcium in vitro for 3 days and treat the cells with or without 10 μm of all-trans-retinoic acid (RA) or 3 mM of colchicine for 24 h (Fig. 4a) (*see* **Note 21**).
2. Harvest the RA and colchicine-arrested keratinocytes to prepare cell-free lysates as described in Subheading of 3.2.
3. Carry out in vitro translation reaction of the PV *Nat* and *Mod* L1 mRNAs in RA and colchicine-arrested keratinocyte cell-free lysate as described in Subheading of 3.4.
4. Detect the in vitro-translated L1 proteins using Western blotting analysis as described in Subheading of 3.5 (*see* **Note 22**). Translational efficiencies of the PV *Nat* and *Mod* L1 mRNAs were significantly different between RA and colchicine-arrested keratinocyte cell-free lysates (Fig. 4b).

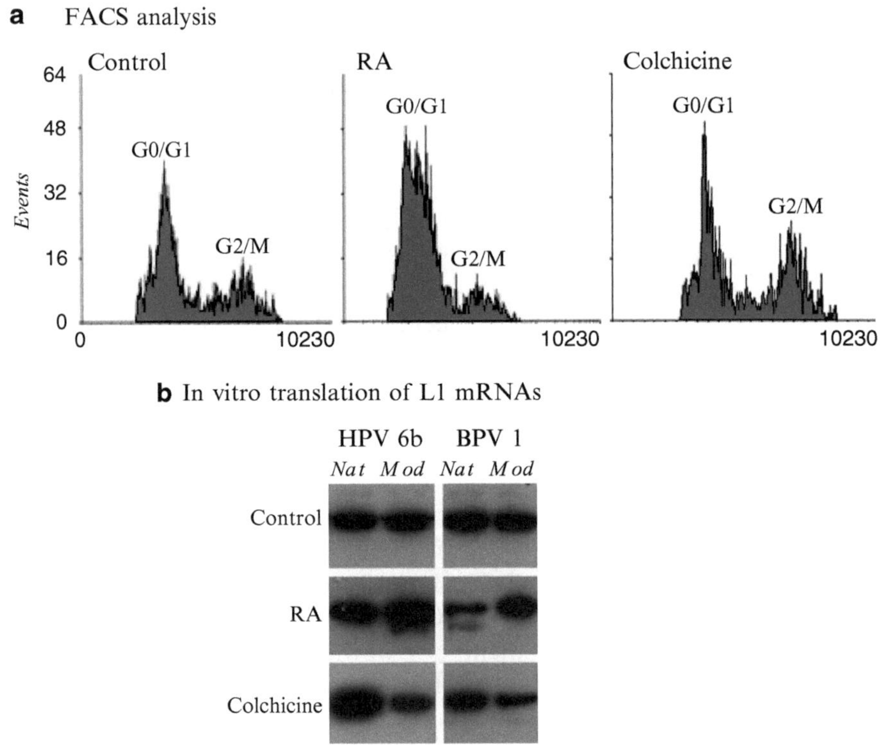

Fig. 4 Differential translation of PV *Nat* and *Mod* L1 mRNAs in cell-free lysates prepared from RA and colchicine-arrested keratinocytes. (**a**) FACS analysis of cell cycle events in control, RA, and colchicine-arrested keratinocytes [37]. (**b**) Translation of PV *Nat* and *Mod* L1 mRNAs in RA and colchicine-arrested keratinocyte cell-free lysates. 1 μg of each of the PV L1 mRNAs was used for the in vitro translation reaction. Translation reaction was carried out at 30 °C for 3 h. The in vitro-translated L1 products were detected by Western blotting analysis

4 Notes

1. Heat both adenine and insulin stock solutions at 100 °C for 10 min and then add them into the 3:1 medium. Filter-sterilize the 3:1 medium.
2. Use 70 % ethanol to surface-sterilize the mouse pups twice, with 15 s each time, and then rinse them with sterile PBS three times.
3. Peel skin off from the body wall of the mouse pups is like to remove a "skin jacket," pull arms through leaving paws attached the body. Avoid puncturing the abdominal cavity and gut as puncturing increases the likelihood of contamination of the cultures. Cut the skin of one pup into 6 even pieces.
4. Dispase is a protease that cleaves fibronectin, collagen IV, and to a lesser extent collagen I. Dispase is very useful for isolation of the primary keratinocytes from mouse skin with high quality and quantity.
5. Wash the epidermal sheets in sterile PBS three times to reduce fungal contamination. At this stage, the epidermis sheets appear opaque while the dermis appears moist and shiny.
6. The epidermis–trypsin mixture can be incubated at room temperature for 10 min, but the yield of keratinocytes is much lower.
7. Do not over pipette, which results in lysis of keratinocytes and releases cellular proteins so that solution looks like it contains cotton wool that will not pellet after centrifugation.
8. Typical keratinocyte yield is 4×10^6 cells per pup, but it varies from 10^6 to 10^7 cells per pup.
9. The primary keratinocytes grown at a cell density of 4.5×10^6cells/T75 flask in 12 ml of medium will be completely confluent within 5–6 days. 30 % of the cells show weak expression of involucrin in day 3 culture. Over 90 % of the cells show strong expression of involucrin, with organized involucrin filaments in day 8 culture (Fig. 1b).
10. 0.2 mM calcium can enhance differentiation of the keratinocytes in vitro culture.
11. The keratinocyte cell-free lysate stored at −20 °C is also very stable because of the presence of sucrose.
12. Hemin can prevent the inhibitory effect of protein synthesis by a cyclic AMP-independent protein kinase that catalyzes the phosphorylation of the small subunit of the protein synthesis initiation factor eIF-2 in the cell-free lysates [36].
13. The transcription reaction contained 5 mM m7 GpppG to synthesize capped PV L1 mRNAs.

14. Use denaturing gel electrophoresis to analyze the quality of each transcribed mRNA. Measure the yield of each transcribed mRNA. Generally, 1 μg of the plasmid DNA can produce 4 μg of mRNA.
15. [35S]-methionine labelling has traditionally been used as a sensitive method to detect the protein expression and synthesis from the targeted DNAs or mRNAs in different in vitro translation systems. But we found that immunoblotting analysis of the in vitro-translated protein was more sensitive than [35S]-methionine labelling.
16. The time length of in vitro translation reaction varies from 10 min to 24 h.
17. It is very critical to denature the in vitro-translated products at 80 °C for 10 min. High denaturation temperature (95–100 °C) or longer denaturation times will cause strong background of the blots.
18. We used a QIAGEN kit (QIAGEN) to extract and purify total tRNAs from the primary mouse keratinocyte cultured for a different time period.
19. We have previously demonstrated that keratinocytes substantially change their tRNA profiles upon differentiation in both bovine and mouse epidermises [15].
20. Production of aminoacyl-tRNAs (aa-tRNAs). tRNAs ($2.5 \times 1^{0-4}$ M) were added to a 20 μl reaction mixture containing 10 mM Tris-acetate (pH 7.8), 44 mM KCl, 12 mM MgC_{l2}, 9 mM β-mercaptoethanol, 38 mM ATP, 0.25 mM GTP, and 7 μl of keratinocyte-free lysate. The reaction was carried out at 25 °C for 20 min, and 30 μl of H_2O was added to the reaction to dilute the tRNAs to 1^{0-4} M. The in vitro-produced aa-tRNAs were aliquoted and stored at −70 °C.
21. Both RA and colchicine enhance keratinocyte differentiation. But we observed that both RA and colchicine treatments lead to the changes of total tRNA profiles in mouse keratinocytes, indicating that tRNA profile changes link to a cell cycle phase in mammalian cells [37].
22. The blots were examined for the PV L1 proteins by immunoblotting analysis using the monoclonal antibody against L1 proteins as in Subheading of 3.5.

Acknowledgements

This work was partially supported by a National Health and Medical Research Council of Australia Industry Research Fellowship (301256 to K.N.Z.) and grants from Cancer Council of Queensland (401623 and 511210 to K.N.Z.).

References

1. Toulza E, Mattiuzzo NR, Galliano MF et al (2007) Large-scale identification of human genes implicated in epidermal barrier function. Genome Biol 8:R107
2. Fuchs E, Raghavan S (2002) Getting under the skin of epidermal morphogenesis. Nat Rev Genet 3:199–209
3. Eckert RL, Crish JF, Robinson NA (1997) The epidermal keratinocyte as a model for the study of gene regulation and cell differentiation. Physiol Rev 77:397–424
4. Jensen JM, Pfeiffer S, Akaki T et al (2007) Barrier function, epidermal differentiation, and human beta-defensin 2 expression in tinea corporis. J Invest Dermatol 127:1720–1727
5. O'Regan GM, Sandilands A, McLean WH et al (2009) Filaggrin in atopic dermatitis. J Allergy Clin Immunol 124:R2–R6
6. Sandilands A, Sutherland C, Irvine AD et al (2009) Filaggrin in the frontline: role in skin barrier function and disease. J Cell Sci 122:1285–1294
7. Spazierer D, Fuchs P, Reipert S et al (2006) Epiplakin is dispensable for skin barrier function and for integrity of keratin network cytoarchitecture in simple and stratified epithelia. Mol Cell Biol 26:559–568
8. Jensen JM, Schutze S, Neumann C et al (2000) Impaired cutaneous permeability barrier function, skin hydration, and sphingomyelinase activity in keratin 10 deficient mice. J Invest Dermatol 115:708–713
9. Lyons JG, Patel V, Roue NC et al (2008) Snail up-regulates proinflammatory mediators and inhibits differentiation in oral keratinocytes. Cancer Res 68:4525–4530
10. Southall MD, Li T, Gharibova LS et al (2003) Activation of epidermal vanilloid receptor-1 induces release of proinflammatory mediators in human keratinocytes. J Pharmacol Exp Ther 304:217–222
11. Wilmer JL, Burleson FG, Kayama F et al (1994) Cytokine induction in human epidermal keratinocytes exposed to contact irritants and its relation to chemical-induced inflammation in mouse skin. J Invest Dermatol 102:915–922
12. Morita E, Yamada S, Kimura I et al (1993) Interleukin-8 production from cultured human dermal fibroblasts by stimulation with supernatant of cultured human epidermal cells. Skin Pharmacol 6:161–169
13. Leask A, Rosenberg M, Vassar R et al (1990) Regulation of a human epidermal keratin gene: sequences and nuclear factors involved in keratinocyte-specific transcription. Genes Dev 4:1985–1998
14. Su D, Smith SM, Preti M et al (2009) Stathmin and tubulin expression and survival of ovarian cancer patients receiving platinum treatment with and without paclitaxel. Cancer 115:2453–2463
15. Zhao KN, Gu W, Fang NX et al (2005) Gene codon composition determines differentiation-dependent expression of a viral capsid gene in keratinocytes in vitro and in vivo. Mol Cell Biol 25:8643–8655
16. Sokoloski KJ, Wilusz J, Wilusz CJ (2008) The preparation and applications of cytoplasmic extracts from mammalian cells for studying aspects of mRNA decay. Methods Enzymol 448:139–163
17. Lee KA, Bindereif A, Green MR (1988) A small-scale procedure for preparation of nuclear extracts that support efficient transcription and pre-mRNA splicing. Gene Anal Tech 5:22–31
18. Wakiyama M, Takimoto K, Ohara O et al (2007) Let-7 microRNA-mediated mRNA deadenylation and translational repression in a mammalian cell-free system. Genes Dev 21:1857–1862
19. Pisarev AV, Skabkin MA, Pisareva VP et al (2010) The role of ABCE1 in eukaryotic post-termination ribosomal recycling. Mol Cell 37:196–210
20. Svitkin YV, Sonenberg N (2004) An efficient system for cap- and poly(A)-dependent translation in vitro. Methods Mol Biol 257:155–170
21. Bergamini G, Preiss T, Hentze MW (2000) Picornavirus IRESes and the poly(A) tail jointly promote cap-independent translation in a mammalian cell-free system. RNA 6:1781–1790
22. Witherell G (2000) AG-7088 Pfizer. Curr Opin Investig Drugs 1:297–302
23. Thoma C, Ostareck-Lederer A, Hentze MW (2004) A poly(A) tail-responsive in vitro system for cap- or IRES-driven translation from HeLa cells. Methods Mol Biol 257:171–180
24. Gu W, Ding J, Wang X et al (2007) Generalized substitution of isoencoding codons shortens the duration of papillomavirus L1 protein expression in transiently gene-transfected keratinocytes due to cell differentiation. Nucleic Acids Res 35:4820–4832
25. Orth G, Favre M, Jablonska S et al (1978) Viral sequences related to a human skin papillomavirus in genital warts. Nature 275:334–336

26. Riou G, Favre M, Jeannel D et al (1990) Association between poor prognosis in early-stage invasive cervical carcinomas and non-detection of HPV DNA. Lancet 335:1171–1174
27. Durst M, Gissmann L, Ikenberg H et al (1983) A papillomavirus DNA from a cervical carcinoma and its prevalence in cancer biopsy samples from different geographic regions. Proc Natl Acad Sci USA 80:3812–3815
28. Gissmann L, Wolnik L, Ikenberg H et al (1983) Human papillomavirus types 6 and 11 DNA sequences in genital and laryngeal papillomas and in some cervical cancers. Proc Natl Acad Sci USA 80:560–563
29. Zhao KN, Liu WJ, Frazer IH (2003) Codon usage bias and A+T content variation in human papillomavirus genomes. Virus Res 98: 95–104
30. Zhou J, Liu WJ, Peng SW et al (1999) Papillomavirus capsid protein expression level depends on the match between codon usage and tRNA availability. J Virol 73: 4972–4982
31. Samorski R, Gissmann L, Osen W (2006) Codon optimized expression of HPV 16 E6 renders target cells susceptible to E6-specific CTL recognition. Immunol Lett 107:41–49
32. Merritt AJ, Berika MY, Zhai W et al (2002) Suprabasal desmoglein 3 expression in the epidermis of transgenic mice results in hyperproliferation and abnormal differentiation. Mol Cell Biol 22:5846–5858
33. Dover R, Watt FM (1987) Measurement of the rate of epidermal terminal differentiation: expression of involucrin by S-phase keratinocytes in culture and in psoriatic plaques. J Invest Dermatol 89:349–352
34. Pelham HR, Jackson RJ (1976) An efficient mRNA-dependent translation system from reticulocyte lysates. Eur J Biochem 67:247–256
35. Gu W, Li M, Zhao WM et al (2004) tRNASer(CGA) differentially regulates expression of wild-type and codon-modified papillomavirus L1 genes. Nucleic Acids Res 32:4448–4461
36. Ranu RS, Levin DH, Delaunay J et al (1976) Regulation of protein synthesis in rabbit reticulocyte lysates: characteristics of inhibition of protein synthesis by a translational inhibitor from heme-deficient lysates and its relationship to the initiation factor which binds Met-tRNAf. Proc Natl Acad Sci USA 73: 2720–2724
37. Ding J, Doorbar J, Li B et al (2010) Expression of papillomavirus L1 proteins regulated by authentic gene codon usage is favoured in G2/M-like cells in differentiating keratinocytes. Virology 399:46–58

Chapter 14

An Optimized Yeast Cell-Free Lysate System for In Vitro Translation of Human Virus mRNA

Xiao Wang, Liang Zhao, and Kong-Nan Zhao

Abstract

Yeast (*Saccharomyces cerevisiae*) as a model organism has long been established to study various aspects of eukaryotic cellular and molecular biology. Cell-free lysates prepared from different yeast strains have been used as a powerful tool to study eukaryotic protein expression in vitro. Recently, we established a yeast cell-free lysate system for in vitro translation long and short L1 capsid gene mRNAs of human papillomavirus type 58. We were able to significantly improve the translation efficiencies of the viral mRNAs in the established system by optimizing the concentrations of potassium and magnesium and controlling the physiological status of the yeast cells used for lysate preparation. We proved that a single specific amino acid can be rate limiting for translation of a target mRNA. Here, we describe the materials, method, and technique used for the development of an efficient yeast cell-free translation system.

Key words Yeast, Cell-free lysate, In vitro translation, In vitro transcription, Papillomavirus, DNA, RNA

1 Introduction

Cell-free translation systems prepared from different organisms have widely been used to produce eukaryotic proteins in vitro. The reported systems include *Escherichia coli* cell extract [1], wheat germ cell extract (WGE), and rabbit reticulocyte lysate (RRL) [2]. A key goal for the use of these cell-free translation systems is to synthesize biologically active proteins [3]. However, these systems have inherent limitations, primarily regarding protein folding and posttranslational modifications. For example, the *E. coli* system has a much greater tendency to cause aggregation of the protein products due to the prokaryotic nature of its translation and folding mechanisms. Posttranslational modifications such as phosphorylation, acetylation, and glycosylation widely occur during mammalian gene expression in order to produce functional proteins. However, glycosylation, which is the most

Kirill Alexandrov and Wayne A. Johnston (eds.), *Cell-Free Protein Synthesis: Methods and Protocols*, Methods in Molecular Biology, vol. 1118, DOI 10.1007/978-1-62703-782-2_14, © Springer Science+Business Media, LLC 2014

widespread and complex form of posttranslational modification in eukaryotes [4], does not take place in the WGE system (http://www.piercenet.com/browse.cfm). In addition, WGE frequently gives rise to truncated proteins due to premature translational termination [5]. RRL, another popular cell-free system, is able to produce proteins folded correctly and with appropriate posttranslational modifications. However, the developmental specialization together with the unique codon usage of reticulocytes limits RRL's application in synthesizing preparative amount of proteins other than globins [6].

Budding yeast (*Saccharomyces cerevisiae*), a simple unicellular eukaryote, has long been established as a model organism to study various aspects of eukaryotic cellular and molecular biology, including translational regulation [7]. The cell-free translation system from *S. cerevisiae* has been developed to study eukaryotic protein expression [8–10]. Derived from eukaryotic cells, this system can faithfully translate exogenous mRNAs and produce correctly folded proteins [10]. Compared with other eukaryotic cell-free systems such as WGE and RRL, it offers two main advantages. Firstly, large amounts of starting material are readily available at low cost, and secondly, the yeast lysate exhibits high stability on long term storage [11]. However, the yeast cell-free system, like other eukaryotic in vitro translation systems, suffers from low translational efficiency and productivity. To circumvent this issue, we recently have developed a yeast cell-free lysate system for efficient translation of human papillomavirus (HPV) type 58 long and short L1 gene mRNAs in vitro [12, 13]. Our rationale for developing this cell-free lysate system was threefold. Firstly, the codon usage of HPV genes resembles that of yeast genes more closely rather than that of the mammals [14]. Secondly, bovine papillomavirus type 1 (BPV1) virions can infect yeast, and BPV-1 episomes can replicate in the infected yeast cells [15, 16]. Thirdly, HPV L1 capsid protein expressed in yeast can self-assemble into viruslike particles (VLPs) [17]. We showed that the translational efficiencies of the targeted HPV L1 mRNAs were significantly improved using the yeast cell-free lysates we prepared. The established yeast cell-free lysate system also provides a unique opportunity to reveal one of the regulatory mechanisms of protein synthesis by using lysates prepared from genetically modified yeast strains [13].

Although developed and verified for HPV L1 and GFP genes, this cell-free lysate system can be useful for the efficient expression of the exogenous mRNAs from different organisms. In this chapter, we describe the materials, method, and technique in details used for developing the cell-free lysate system that we have reported previously [12, 13].

2 Materials

2.1 Yeast Strains and Culture

1. Suitable yeast strains such as Y303 and M2915 [12, 13].
2. YEPD yeast cell culture medium: 1 % of yeast extract, 1 % of bactopeptone, and 2 % of glucose. Part A: Weigh 10 g of yeast extract and 10 g of bactopeptone; add water to a volume of 900 mL; autoclave at 121 °C for 20 min. Part B: Weigh 20 g glucose, and add water to 100 mL; autoclave at 115 °C for 15 min. Store at 4 °C. Mix Parts A and B thoroughly before use (*see* **Note 1**).

2.2 Lysis Buffer

1. *Stock solution* (*see* **Note 2**).
   (a) 1 M HEPES–KOH (pH 7.4): Dissolve 11.915 g HEPES (MW 238.3, free acid, *N*-2-hydroxyethylpiperazine-*N*′-2-ethanesulfonic acid) in 50 mL of water, and adjust pH to 7.4 with 1 M KOH. Store at 4 °C.
   (b) 1 M Dithiothreitol (DTT): Dissolve 1.542 g DTT in 10 mL of water.
   (c) 5 M Potassium acetate (KOAc, pH 7.4): Dissolve 4.907 g KOAc in 10 mL of water, and adjust pH to 7.4 with glacier acetic acid. Store at 4 °C (*see* **Note 3**).
   (d) 250 mM magnesium acetate (MgOAc): Store at 4 °C.
   (e) 0.5 M ATP.
   (f) 50 mM GTP.
   (g) 10 mM *S*-Adenosyl-L-methionine [18].
   (h) 100 mM Spermidine.
   (i) 1 M Sucrose: Store at 4 °C.
   (j) 1 M Creatine phosphate.
   (k) Creatine phosphokinase.
   (l) 1 mg/mL Leupeptin.
   (m) 1 mg/mL Pepstatin.
   (n) 1 mg/mL Chymostatin.
2. *2× lysis buffer.*
   Prepare 2× lysis buffer as shown in Table 1 (*see* **Note 4**), and store single-use aliquots at –70 °C. Avoid repeated freeze–thaw cycles. Dilute to 1× buffer with water before use.

2.3 Other Materials for Making Yeast Lysate

1. Micrococcal nuclease (from *Staphylococcus aureus*, 100–300 U/mg protein, Sigma).
2. 2 mM Hemin.
3. 0.5 M EGTA: Dissolve 19.02 g EGTA (sodium salt) in 100 mL of water, and adjust pH to 7.0 with KOH. Store at 4 °C.

Table 1
2× lysis buffer components and volume

Components	Volume	Final concentration
1 M HEPES–KOH, pH 7.4	1.2 mL	200 mM
1 M Sucrose	1.2 mL	200 mM
250 mM MgOAc	120 µl	5 mM
5 M KOAc	288 µl	240 mM
1 M DTT	12 µl	2 mM
0.5 M ATP	60 µl	5 mM
50 mM GTP	240 µl	2 mM
10 mM SAM	120 µl	200 µM
100 mM Spermidine	120 µl	2 mM
1 M Creatine phosphate	240 µl	40 mM
Creatine phosphokinase	500 U	83.4 U
1 mg/mL Leupeptin	12 µl	2 µg/mL
1 mg/mL Pepstatin	12 µl	2 µg/mL
1 mg/mL Chymostatin	12 µl	2 µg/mL
H_2O	2.364 mL	
Total volume	6 mL	

4. β-Mercaptoethanol.
5. SCE buffer (1 M sorbitol, 10 mmol/L sodium citrate, 10 mmol/L EDTA): Store at 4 °C.
6. STC buffer (1 M sorbitol, 10 mM $CaCl_2$, 10 mM Tris–HCl, pH 7.5): Store at 4 °C.
7. Lyticase (partially purified grade, Sigma, USA).
8. 40 mL Dounce glass/glass homogenizer (Jencons Scientific Ltd., Leighton Buzzard, UK): Bake at 230 °C for at least 4 h before use.
9. Sterile 40 mL screw-cap centrifuge tubes and Eppendorf tubes.
10. 10 % (w/v) Sodium dodecyl sulfate (SDS).

2.4 *In Vitro* Transcription Components

1. Target gene expression plasmids (*see* **Note 5**).
2. DNA extraction kit (#51304, Qiagen).
3. RNA Transcription kit (# E2030, New England Biolabs).
4. Phenol, chloroform, and ethanol (100 and 75 %).
5. 0.1 M 7-methyl-diguanosine triphosphate (7mGpppGor 7-mG).

2.5 In Vitro Translation Components

1. Amino acid mixture (1 mM of each amino acid, Sigma, Australia).
2. RNase inhibitor (Promega, Australia).

2.6 Materials for Western Blotting Analysis

1. 2× SDS sample buffer.
2. 1.5 M Tris–HCl buffer pH 8.8.
3. 0.5 M Tris–HCl buffer pH 6.8.
4. 10 % (w/v) SDS solution.
5. 30 % Acrylamide/Bis solution (37.5:1).
6. 10 % Ammonium persulfate solution.
7. *N*,*N*,*N*′,*N*′-Tetramethylethylenediamine (TEMED).
8. Mini protein gel apparatus.

3 Methods

3.1 Yeast Cell Culture

1. Grow a starter yeast culture aerobically to stationary phase (ca. 2×10^8 cells/mL) in 50 mL of YEPD medium. Incubate the culture at 28–30 °C with vigorous agitation (150 rpm) for 24 h. The starter yeast culture can be stored at 4 °C up to 1 month without significant loss of viability.
2. Inoculate 100 μL of the starter yeast culture into 500 mL of YEPD medium in a 2 L flask. Incubate the culture overnight with vigorous agitation (150 rpm) at 28–30 °C until a cell density of $1–2 \times 10^7$ cells/mL (~OD600 = 1) is reached. This would take 14–18 h, depending upon the strain and age of the starter culture (*see* **Note 6**).

3.2 Preparation of Yeast Cell-Free Lysate

1. Harvest yeast cells by centrifugation at 500 × *g* for 10 min at 4 °C, and discard the supernatant.
2. Resuspend the yeast pellet in 200 mL cold sterile water.
3. Wash the yeast suspension by centrifugation at 500 × *g* for 10 min at 4 °C, and discard the supernatant.
4. Measure the wet weight of the cell pellet (usually ~1 g from a 500 mL culture).
5. Resuspend the cell pellet in 40 mL of 1 M sorbitol, and wash the yeast suspension again by centrifugation as in **step 3**.
6. Resuspend the cell pellet in 20 mL of SCE buffer containing 20,000 U lyticase and 0.2 mM β-mercaptoethanol added just before use.
7. Incubate the yeast suspension with gentle agitation (~40 rpm) to digest yeast cell wall in the dark for 3 h at 25 °C, and monitor spheroplast formation every 30 min (*see* **Note 7**).

8. Harvest yeast spheroplasts by centrifugation at 700×*g* for 15 min at 4 °C, and discard the supernatant.
9. Wash twice with 40 mL cold sterile STC buffer by centrifugation at 850×*g* for 5 min at 4 °C (*see* **Note 8**).
10. Resuspend spheroplast pellet in 0.5 mL of 1× ice-cold lysis buffer per gram of original wet-weighted pellet on ice, and transfer the spheroplast suspension into a prechilled Dounce glass/tight-fitting glass homogenizer (*see* **Note 9**). Dounce the spheroplasts to lyse them about 25–150 strokes. Monitor the lysis process every 25 strokes by microscopic examination.
11. Transfer the crude lysate to a prechilled Eppendorf tube and centrifuge at 2,000×*g* for 10 min at 4 °C.
12. Transfer the lysate using a cold sterile pipette tip to a new sterile Eppendorf tube, carefully avoiding the upper lipid layer and the flocculent material above the cell pellet.
13. Incubate the lysate by adding 10 U/mL of micrococcal nuclease and 1 mM $CaCl_2$ to hydrolyze the endogenous mRNAs at 20 °C in a water bath for 10 min.
14. Terminate the reaction by adding 0.5 M EGTA, pH 7.0 to a final concentration of 2.5 mM.
15. Aliquot the lysate, snap freeze the aliquots in a dry ice/ethanol bath, and store at −70 °C. The lysate can be thawed and refrozen once without significant loss of activity (*see* **Note 10**).
16. Supplement the yeast cell-free lysate with 0.02 mM hemin before use.

3.3 In Vitro Transcription

1. Linearize a targeted gene plasmid by cutting the plasmid DNA using an appropriate restriction enzyme (*see* **Notes 11–13**).
2. Confirm the enzyme-digested plasmid DNA by agarose gel electrophoresis, and estimate its quantity by comparing the band intensity to a quantitative DNA marker. Small amounts (<5 %) of undigested plasmid DNA present have little effect on RNA yield (*see* **Note 14**).
3. Extract the linearized plasmid DNA using standard phenol/chloroform extraction protocol (*see* **Note 15**). Dissolve the purified plasmid DNAs in nuclease-free water.
4. Carry out in vitro transcription for the linearized target gene expression plasmid DNA by T7 RNA polymerase. Thaw both 10× transcription buffer and 20× NTP mix at room temperature, until just thawed (do not thaw at 37 °C). If precipitant is evident following thawing, vortex briefly to redissolve. Keep the 20× high-molecular-weight (HMW) component mix and

T7 RNA polymerase at –20 °C until needed. Take precautions to avoid RNase contamination (*see* **Note 16**).

5. Set up transcription reaction. Add the transcription reagents into a reaction tube for one plasmid in the following order: 27 μl RNase-free H_2O, 4 μl 10× transcription buffer, 2 μl 20× ribonucleotide solution mix, 2 μl 20× HMW mix, 1 μl 0.1 M 7-mG, 2 μl T7 RNA polymerase (500 U/μl), and 2 μl plasmid DNA template(s) (1–2 μg). Total reaction volume is 40 μl. An additional in vitro transcription reaction is set up using the linearized basal plasmid DNA as a control.
6. Incubate at 42 °C for 2–4 h. This transcription setup is for transcribing templates larger than 0.3 kb (not for shorter templates, *see* **Note 17**).
7. Take a 1 μl aliquot of the transcription reaction for detecting the quality of the synthesized mRNAs, and roughly estimate the mRNA amounts by 1 % agarose gel electrophoresis.
8. Add 2 μl of RNase-free DNase I (2,000 U/mL, Promega) into the reaction tube to terminate the transcription reaction at 37 °C for 1 h if both quality and quantity of the synthesized RNAs are satisfactory.
9. Purify the transcribed mRNA by standard phenol/chloroform extraction, followed by precipitation with 2 volumes of ethanol, and wash twice with 70 % ethanol.
10. Resuspend the purified mRNA in nuclease-free H_2O, and measure its amount for in vitro translation by NanoDrop 1000 Spectrophotometer.

3.4 In Vitro Translation

1. Set up in vitro translation reaction by adding the reagents in the following order: 17.5 μl yeast cell-free lysate, 2 μl mRNA (1 μg per reaction), 0.5 μl RNase inhibitor, 0.5 μl amino acid mixture (1 mM of each amino acid), and 4.5 μl RNase-free H_2O, with a volume of 25 μl (*see* **Notes 18–20**).
2. Perform the translation reaction at 30 °C for 3–4 h (*see* **Note 21**).

3.5 Western Blotting Assay

1. Stop the in vitro translation reaction by adding equal volume (25 μl) of 2× SDS sample buffer. Vortex, and keep on ice.
2. Denature the samples by incubating at 80 °C for 10 min (*see* **Note 22**). Chill on ice. Samples can either be analyzed by SDS-PAGE immediately or can be stored at –70 °C.
3. Detect the protein products of the targeted gene mRNAs by Western blotting assay. Isotope labeling is an alternative method for detecting protein expression (*see* **Note 23**).

4 Notes

1. The Parts A and B of YEPD medium should be autoclaved separately to avoid potential chemical reactions between yeast extract, bactopeptone, and glucose, which may have detrimental effects on yeast growth.
2. Prepare all solutions using ultrapure nuclease-free water and analytical grade reagents. All the solutions listed hereafter should be sterilized by passing through a 0.45 μm syringe filter and stored in single-use aliquots at –20 °C unless otherwise indicated.
3. KOAc instead of KCl as the source of K^+ was used to significantly increase protein synthesis in the in vitro translation system, because Cl^- in KCl has an inhibitory effect [10].
4. We have reported that magnesium and potassium cations and sucrose affected the translation efficiency of HPV58 L1 gene mRNAs in the yeast cell-free lysate [12]. The optimal concentrations of magnesium, potassium, and sucrose may vary in different gene mRNAs. Systematic optimization of their concentrations for individual gene mRNAs is required.
5. We used three viral gene plasmids [12, 13] and one GFP expression plasmid [19] as the target gene expression plasmids. The GFP plasmid was only used in preliminary experiments. The three viral gene plasmids included (1) a pcDNA3-BPV1 L1 plasmid that expressed BPV L1 protein and (2) two pcDNA3-HPV58 L1 plasmids pcDNA3-HPV58L1-L and pcDNA3-HPV58L1-S that expressed both long and short L1 proteins, respectively [12, 13]. The target gene expression plasmids were constructed to express protein products using their coding sequences by standard molecular cloning techniques. The target gene expression plasmids must contain a promoter, such as T7, T3, or SP6 promoter, which can drive transcription of the target gene in vitro.
6. The physiological status of yeast cells used for lysate preparation will greatly affect the translational activity of the system [12]. Do not use overgrown yeast cells to prepare cell-free lysates. The cell-free lysates prepared from the yeast cells harvested in early log-phase growth (OD600) will have the maximal translational activity.
7. Start to monitor spheroplast formation in enzymatic digestion of yeast cells every 30 min. Different methods, (a) microscopy and (b) OD600 measurement, have been used for monitoring spheroplast formation. (a) Place 20 μl of enzyme–yeast cell suspension at one end of a glass slide and 20 μl of yeast cell suspension only at the other end. Cover the glass slide with cover slips, and observe under the microscope. Unlike the undigested

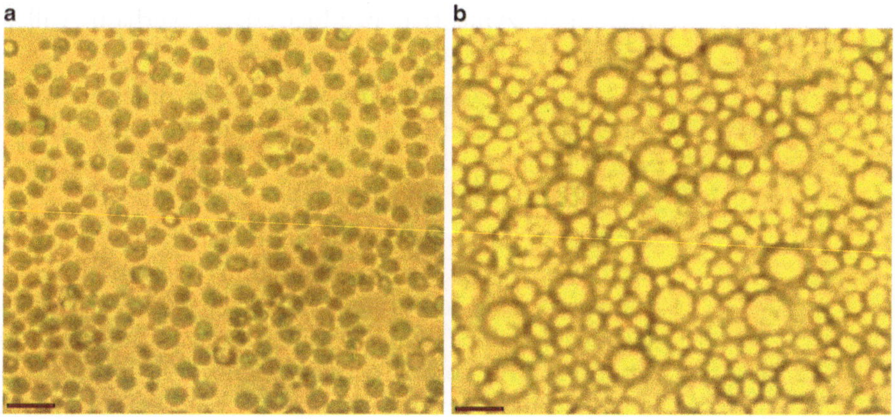

Fig. 1 Morphological differences between yeast cells without enzyme digestion (**a**) and 1-h enzyme-digested yeast cells (**b**). High percentage of the enzyme-digested yeast cells formed spheroplasts. Scale bars = 10 μM

yeast cells that have a thick cell wall and are small in size, the spheroplasts have a thinner cell wall and appear to be larger (Fig. 1). Spheroplasting is sufficient when less than 5 % of the cells remain intact in the visual field with the indicating solution, relative to the control solution. (b) Transfer 20 μL of the enzyme–yeast cell suspension into 1 mL of 0.1 % (w/v) SDS and another 20 μL of the enzyme–yeast cell suspension into 1 mL of 1 M sorbitol. Measure the OD600 of the two preparations using H_2O as the blank. If the OD600 value in the SDS preparation is less than 30 % of that in the sorbitol preparation, the spheroplast formation is considered to be completed.

8. The spheroplast pellet at this stage can be snap frozen in a dry ice/ethanol bath and stored at −70 °C for 1–2 days.
9. Prechill the homogenizer, tubes, and pipette tips on ice or in a fridge. Keep the homogenizer on ice to reduce the heat accumulated during homogenization. Keep samples on ice all the times.
10. Generally, yeast cell-free lysate is remarkably stable at −70 °C when compared with other cell-free systems. We found that the yeast cell-free lysates (containing sucrose) were very stable at −20 °C for at least 6 months.
11. We linearized the four plasmid DNAs described in **Note 5** and used the linearized plasmid DNAs for mRNA transcription in vitro. All the four plasmids were efficiently expressed in the yeast cell-free translation system. Thus, this cell-free translation system should be suitable for expression of other gene plasmid DNAs.
12. The restriction enzyme used to linearize the plasmids should leave blunt ends or 5′ overhangs (linearization of template

with an enzyme that produces a 3′ overhang will result in aberrant transcripts).

13. 30–40 μg of plasmid DNA should be digested for 1 mL of transcription reaction. 1 μg of plasmid DNA should be digested in a volume of 10 μl using the recommended buffer and temperature for each enzyme.
14. Linearized DNA from the digestion reaction can be used directly if desired, with only a slight reduction of yield (<10 %). However, the unpurified linearized DNA must be heat-inactivated at 65 °C for 20 min prior to adding to transcription reaction. The volume of the unpurified linearized template DNA should not constitute more than 10 % of the total transcription reaction.
15. We recommend using a commercial Miniprep kit to purify plasmid DNA. If the crude alkaline lysis method is used, the prepared plasmid DNA often contains unacceptable level of ribonuclease that negatively impacts in vitro transcription.
16. Take great care to avoid inadvertently introducing RNases into the transcription reaction. The following precautions should be taken.
    (a) Always wear gloves when working with RNA.
    (b) Use either a dedicated set of pipettes for RNA work or aerosol-resistant (barrier) pipette tips.
    (c) Use ultrapure water (Milli-Q or equivalent) to make up solutions, and autoclave all solutions if possible. It is not necessary to treat solutions and equipment with diethyl pyrocarbonate (DEPC). DEPC can inhibit subsequent reactions if not completely inactivated.
    (d) Use disposable plasticware instead of glassware whenever possible.
    (e) The addition of 7-mG helps to protect the newly synthesized mRNA from degradation.
17. It is recommended to increase the amounts of plasmid DNA template, polymerase, and reaction time for efficient transcription of short templates (50–300 nt), since transcription of small RNAs is largely dependent on the number of initiation events. Total reaction volume consists of 25 μl RNase-free H_2O, 4 μl 10× transcription buffer, 2 μl 20× ribonucleotide solution mix, 2 μl 20× HMW mix, 1 μl 0.1 M 7-mG, 2 μl T7 RNA polymerase (500 U/μl), and 4 μl plasmid DNA template(s) (2–4 μg). Total reaction volume is 40 μl. Incubate at 42 °C for a minimum of 3–4 h (up to overnight). Longer incubation time generally results in higher yields.
18. If a protein to be synthesized contains a higher proportion of one or two specific amino acids, the amino acids can be the

rate-liming factor that affects the protein yield [12, 13]. In this case, additional supplementation of the specific amino acids is required to increase the yield.

19. Two negative (template) controls are required in a translation experiment, with one using H_2O only and other one using linearized control plasmid DNA.
20. The optimal amount of template mRNA can be determined experimentally by assaying a range of amounts. We usually use 1 μg mRNA in 25 μl of translation reaction.
21. Longer incubation time of in vitro translation does not always result in higher yields. We have examined the translation efficiencies of HPV58 long and short L1 mRNAs in our established yeast cell-free system and found that they have quite different translation efficiencies [12, 13]. The reaction of long L1 mRNA can last for 3 h; however, the reaction of the short one was decreased. So different mRNAs might have different optimal reaction time length. If you use the yeast cell-free system for protein synthesis, various reaction time lengths should be tried to determine the optimum.
22. The temperature used to denature the in vitro-translated products is very critical. High denaturing temperature (95–100 °C) will cause strong background in the blots.
23. We initially used [^{35}S]-methionine labeling to detect the protein expression of the two HPV 58 L1 gene expression plasmid DNAs in RRL with additional supplementation of yeast aa-tRNAs by autoradiography [12, 13] but found that this traditional method is more time consuming and less sensitive than immunoblotting assay. Thus, we recommend using immunoblotting assay to detect the in vitro translational products, which is very convenient and safe.

Acknowledgments

This work was partially supported by a National Health and Medical Research Council of Australia Industry Research Fellowship (301256 to KNZ) and grants from Cancer Council of Queensland (401623 to KNZ).

References

1. Adhikary G, Chew YC, Reece EA et al (2010) PKC-delta and -eta, MEKK-1, MEK-6, MEK-3, and p38-delta are essential mediators of the response of normal human epidermal keratinocytes to differentiating agents. J Invest Dermatol 130:2017–2030
2. Endo Y, Sawasaki T (2006) Cell-free expression systems for eukaryotic protein production. Curr Opin Biotechnol 17:373–380
3. Katzen F, Chang G, Kudlicki W (2005) The past, present and future of cell-free protein synthesis. Trends Biotechnol 23:150–156

4. Lowe JB, Marth JD (2003) A genetic approach to Mammalian glycan function. Annu Rev Biochem 72:643–691
5. Abraham AK, Olsnes S, Pihl A (1979) Fidelity of protein synthesis in vitro is increased in the presence of spermidine. FEBS Lett 101:93–96
6. Lodish HF, Desalu O (1973) Regulation of synthesis of non-globin proteins in cell-free extracts of rabbit reticulocytes. J Biol Chem 248:3420–3427
7. Botstein D, Chervitz SA, Cherry JM (1997) Yeast as a model organism. Science 277:1259–1260
8. Gasior E, Herrera F, Sadnik I et al (1979) The preparation and characterization of a cell-free system from *Saccharomyces cerevisiae* that translates natural messenger ribonucleic acid. J Biol Chem 254:3965–3969
9. Szczesna E, Filipowicz W (1980) Faithful and efficient translation of viral and cellular eukaryotic mRNAs in a cell-free S-27 extract of *Saccharomyces cerevisiae*. Biochem Biophys Res Commun 92:563–569
10. Tuite MF, Plesset J, Moldave K et al (1980) Faithful and efficient translation of homologous and heterologous mRNAs in an mRNA-dependent cell-free system from *Saccharomyces cerevisiae*. J Biol Chem 255:8761–8766
11. Tuite MF, Plesset J (1986) mRNA-dependent yeast cell-free translation systems: theory and practice. Yeast 2:35–52
12. Wang X, Liu J, Zheng Y et al (2008) An optimized yeast cell-free system: sufficient for translation of human papillomavirus 58 L1 mRNA and assembly of virus-like particles. J Biosci Bioeng 106:8–15
13. Wang X, Liu J, Zhao WM et al (2010) Translational comparison of HPV58 long and short L1 mRNAs in yeast (*Saccharomyces cerevisiae*) cell-free system. J Biosci Bioeng 110: 58–65
14. Zhao KN, Liu WJ, Frazer IH (2003) Codon usage bias and A+T content variation in human papillomavirus genomes. Virus Res 98:95–104
15. Zhao KN, Frazer IH (2002) *Saccharomyces cerevisiae* is permissive for replication of bovine papillomavirus type 1. J Virol 76:12265–12273
16. Zhao KN, Frazer IH (2002) Replication of bovine papillomavirus type 1 (BPV-1) DNA in *Saccharomyces cerevisiae* following infection with BPV-1 virions. J Virol 76:3359–3364
17. Lowe RS, Brown DR, Bryan JT et al (1997) Human papillomavirus type 11 (HPV-11) neutralizing antibodies in the serum and genital mucosal secretions of African green monkeys immunized with HPV-11 virus-like particles expressed in yeast. J Infect Dis 176: 1141–1145
18. Wilson VG, West M, Woytek K et al (2002) Papillomavirus E1 proteins: form, function, and features. Virus Genes 24:275–290
19. Zhao KN, Tomlinson L, Liu WJ et al (2003) Effects of additional sequences directly downstream from the AUG on the expression of GFP gene. Biochim Biophys Acta 1630:84–95

Chapter 15

In Vitro Translation-Based Protein Kinase Substrate Identification

Szilvia K. Nagy and Tamás Mészáros

Abstract

Identification of a particular protein as a physiological substrate towards kinases of interest is an extremely complex process. Under physiological conditions kinases and their putative substrates are in low abundance, and production of active eukaryotic kinases with standard overexpressing methods is an arduous task. Herein, we describe a cell-free in vitro protein translation procedure combined with fluorescent phosphoprotein staining as a simple and rapid method for identification of putative kinase substrates.

Key words Protein kinase, Substrate identification, Phosphorylation assay, In vitro translation, Phosphoprotein staining

1 Introduction

Reversible protein phosphorylation is one of the key mechanisms by which the cellular processes are controlled. It plays a role in numerous signal transduction pathways; thus, its proper functioning is critical for almost all phenomena of life including cell growth and death regulation, metabolic control, inflammation, immune response, and even learning and memory. The central role of kinases is further exemplified by their great number in all eukaryotic organisms. The analysis of human genome revealed over 500 kinases [1], and this enzyme family is even more extended in plants with the Arabidopsis genome encoding more than 1,000 protein kinases [2]. Despite the increasing number of annotated kinases, the majority of the kinase substrates have not yet been described. Considering the central role of protein kinases in numerous signal transduction pathways, the importance of delineation of protein kinase-substrate networks hardly can be overestimated.

Definitely depicting a particular protein as a physiological substrate of the kinase in question is a complex procedure involving diverse biochemical methods and is a challenging task for two reasons.

Kirill Alexandrov and Wayne A. Johnston (eds.), *Cell-Free Protein Synthesis: Methods and Protocols*, Methods in Molecular Biology, vol. 1118, DOI 10.1007/978-1-62703-782-2_15,

Firstly, most of the kinases and putative substrates are low abundance proteins under physiological conditions. The problem is further exacerbated by the weak affinity of kinases for their substrates. Due to these inherent limitations, isolation of sufficient amount of either the kinase–substrate complex or in many cases even the separated form of the proteins of interest is often impractical. Secondly, most of the eukaryotic kinases and their substrates cannot be efficiently produced in *Escherichia coli*-based protein overexpressing systems as the multidomain eukaryotic proteins do not fold properly and tend to form aggregates in prokaryotic cells.

Despite the varied improvements to prokaryotic cell-based protein synthesis and the use of several eukaryotic cell types for routine protein production, cell-free in vitro translation is the rational choice of method for alleviating the paucity of functional recombinant kinases. Various sources of translation machinery have been explored for construction of in vitro translation systems, but due to its low cost and capacity for producing properly folded, multidomain eukaryotic proteins wheat germ protein extracts appear particularly suited for the kinase production [3]. The wheat germ-based translation systems are ideally suited for production of milligram quantities of proteins and parallel synthesis of a panel of proteins [4]. Furthermore, simple and rapid generation of required DNA templates for in vitro transcription is ensured by the availability of optimized vectors and PCR protocols and transition from a gene to a purified protein can be accomplished in few days [4, 5].

To identify a protein as a *bona fide* substrate of a kinase-of-interest some criteria have to be fulfilled, such as the demonstration of in vitro phosphorylation of the putative substrate by the kinase. Traditional detection methods of protein phosphorylation rely on utilization of either radioactively labeled ATP or anti-phospho amino acid-specific antibodies. Both approaches are well established and highly sensitive; however, the former is hazardous and requires a dedicated radioactivity laboratory, the latter depends on antibody quality and is time consuming. Invention of phosphoprotein-specific dye technology has provided an attractive alternative for identification of phosphorylated proteins. The commercially available dye technology is capable of the fluorescent detection of phosphoserine-, phosphothreonine-, and phosphotyrosine-containing proteins directly in sodium dodecyl sulfate (SDS)-polyacrylamide gels [6]. The staining protocol is rapid in comparison to the traditional methods and its sensitivity is sufficient for unambiguous detection of in vitro phosphorylated proteins.

Here, we describe a cell-free in vitro protein translation and fluorescent phosphoprotein staining-based protocol for identification of putative substrates of protein kinases.

2 Materials

2.1 Cloning and Purification of In Vitro Translation Vectors

1. Primers:
   pEU3-NIIfor 5′-TACTTCCAATCCAATGCA*ATG*... *18–22 gene specific nucleotide*-3′
   pEUrev 5′-TTATCCACTTCCAATG*TCA*... *18–22 gene specific nucleotide*-3′ (*see* **Note 1**).
2. Vectors: with T7 promoter [5].
   pEU3-NII-HLICNot (6× His tag).
   pEU3-NII-GLICNot (GST tag).
3. iProof High-Fidelity DNA Polymerase with the relevant buffers (5× HF or 5× GC buffer) (Bio-Rad).
4. 10 mM CleanAmp dNTP mix (TriLink).
5. 50× TAE buffer: 2 M Tris, 1 M acetic acid, 0.05 M EDTA.
6. Agarose.
7. Agarose gel staining solution such as GelGreen (Biotium).
8. PCR Clean Up Buffer: 26 % (w/v) PEG 8000, 6.5 mM $MgCl_2$, 0.6 M sodium acetate, pH 6–7.
9. FastDigest *Ssp*I and *Not*I restriction enzymes (Fermentas).
10. PCR MasterMix (Fermentas).
11. Gel extraction kit such as PureLink™ Quick Gel Extraction Kit (Invitrogen).
12. 100 mM dGTP and 100 mM dCTP solutions (Bioline).
13. 10× NEBuffer 2 (New England BioLabs) + 1 % BSA.
14. T4 DNA Polymerase (New England BioLabs).
15. 50 mM EDTA stock solution.
16. α-Select Chemically Competent Cell (Bioline).
17. LB agar plate containing 100 μg/mL carbenicillin.
18. SOC medium: 0.5 % (w/v) yeast extract, 2 % (w/v) tryptone, 10 mM NaCl, 2.5 mM KCl, 10 mM $MgCl_2$, 10 mM $MgSO_4$, 20 mM glucose.
19. Plasmid miniprep kit such as PureLink™ Quick Plasmid Miniprep Kit (Invitrogen).
20. Nuclease-free water.
21. Barrier tips.
22. Bench top microcentrifuge.
23. Spectrophotometer.

2.2 In Vitro Transcription and Translation

1. TranscriptAid T7 High Yield Transcription Kit (Fermentas).
2. 7.5 M ammonium acetate.

3. ENDEXT® Technology Wheat Germ Cell-Free Expression System (including WEPRO® and 40× SUB-AMIX® solution) (Cell-Free Sciences).
4. 1 μg/μL creatine kinase stock solution (Roche).
5. 10× Laemmli buffer.
6. 5× SDS Loading Dye: 0.312 M Tris, 10 % (w/v) SDS, 250 mM DTT, 50 % (v/v) glycerol, 0.01 % Bromophenol blue, pH 6.8.
7. Protein staining solution such as PageBlue™ protein staining solution (Fermentas).
8. Prestained and unstained protein markers such as ProSieve™ QuadColor™ Protein Marker, ProSieve Unstained Protein Marker II (Lonza).
9. 1.5 mL Protein LoBind Eppendorf tubes (Eppendorf).
10. Sterile barrier tips.

2.3 Protein Affinity Purification, Kinase Assay

1. Pierce Glutathione Magnetic Beads (Pierce).
2. GST Wash & Binding Buffer: 25 mM Tris, 150 mM NaCl, pH 8.0.
3. MagneHis™ Ni-Particles (Promega).
4. His Wash & Binding Buffer: PBS, 300 mM NaCl, 0.01 % Tween-20.
5. 5 M NaCl.
6. PBS Buffer: 137 mM NaCl, 2.7 mM KCl, 10 mM Na_2HPO_4, 2 mM KH_2PO_4, pH 7.4.
7. 600 ng/μL TEV protease in 50 % glycerol [7].
8. 20× TEV Protease Buffer: 1 M Tris–HCl, 10 mM EDTA, pH 8.0.
9. Kinase buffer: 20 mM HEPES, 15 mM $MgCl_2$, 5 mM EGTA, 50 μM ATP, 10 mM Na_3VO_4, 100 mM NaF, 20 mM β-glycerophosphate, 1 mM DTT, pH 7.5.
10. Lambda Protein Phosphatase (New England BioLabs).
11. Pro-Q® Diamond Phosphoprotein Gel Stain (Invitrogen).
12. Gel scanner.
13. Magnetic stand.
14. Sterile, RNase-free 96-well, flat bottom plate.

3 Methods

3.1 Amplification of Protein Coding cDNA

Amplify gene of interest from cDNA library or other template with iProof enzyme in a 50 μL PCR reaction with the following conditions:

10 μL HF Buffer.

1 μL 10 mM CleanAmp dNTP mix.

2.5 μL 10 μM forward primer.

2.5 μL 10 μM reverse primer.

1 μL template (5–50 pg plasmid or 50–500 ng cDNA library).

0.2 μL iProof enzyme.

Made up to 50 μL with nuclease-free water.

Set up the PCR program according to the melting temperature of the primers and the length of the expected product (*see* **Note 2**).

Analyze 5 μL of PCR reaction by agarose gel electrophoresis before precipitation. Primers and dNTPs can be removed by PCR Clean Up solution. Complete the PCR reaction's volume to 100 μL with nuclease-free water and add 100 μL of PCR Clean Up buffer. Following 20 min incubation at room temperature (RT), centrifuge at 13,000 rpm for 10 min with bench top microcentrifuge. Discard the supernatant and wash the pellet with 1 mL of cold absolute ethanol, centrifuge at 13,000 rpm for 10 min, and repeat this step (*see* **Note 3**). Finally, dry the pellet by air or vacuum centrifugation, dissolve it in 10 μL of nuclease-free water, and measure the DNA concentration at OD_{260} with spectrophotometer.

3.2 Ligation Independent Cloning

Digest 3 μg of pEU3-NII-HLICNot or pEU3-NII-GLICNot vector with 30 U FastDigest *Ssp*I enzyme for 30 min at 37 °C (*see* **Note 4**) and separate the digested DNA with agarose gel electrophoresis at 60 V. Stain gel with GelGreen for 20 min. Cut out the linearized vector, purify it with Gel Extraction Kit, elute DNA from column with water, and measure the DNA concentration (*see* **Note 5**).

To generate the cohesive ends, set up T4 DNA polymerase treatment by mixing 500 ng of linearized plasmid construct, 2 μL 10 mM dGTP, 2 μL 10× NEB2 + 1 % BSA, 0.2 μL T4 DNA polymerase and bring it to 20 μL with water. In case of insert, use dCTP instead of dGTP. Incubate the reaction mixtures for 10 min at 37 °C, then for 20 min at 75 °C to inactivate T4 DNA polymerase.

Mix approximately 50 ng of T4-treated vector and the required amount of PCR product to ensure 1:3 molar vector:insert ratio. Add 1.25 μL of 50 mM EDTA and complete the reaction mixture with water to 12.5 μL and incubate for 20 min at RT. Thaw competent cells on ice for 20 min. Add 5 μL of the ligation mixture to 50 μL of the cell suspension and mix it gently. Leave the tube on ice for 20 min. Heat shock the cells at 42 °C for 1 min and cool tube on ice for 2 min. Finally, add 500 μL of ambient temperature SOC medium and incubate the cells for 1 h at 37 °C, shaking at 220 rpm. Centrifuge for 5 min at 4,000 rpm and discard approximately 400 μL of the supernatant, resuspend the cells and spread them onto LB + carbenicillin plate and incubate overnight at 37 °C.

Check the colonies with PCR using pEU3-NIIfor/pEUrev primers (*see* **Note 6**). Purify the plasmid from the positive clones by growing 5 mL LB + carbenicillin liquid cultures overnight at 37 °C and using Plasmid Miniprep Kit. Determine DNA concentration of plasmid preparation.

3.3 Linearization of Plasmid Construct

Digest 3 μg of the purified plasmid construct with 30 U of FastDigest *Not*I enzyme for 5 min at 37 °C in 1× FastDigest buffer (*see* **Note 7**). Precipitate the DNA as described in Subheading 3.1 and dissolve in 10 μL nuclease-free water.

3.4 In Vitro Transcription

Set up the reaction mixture on ice and ensure RNase-free conditions by using nuclease-free equipment and wearing gloves.

Prepare transcription mixture by adding 1 μg linearized plasmid, 8 μL NTP mix, 4 μL of 5× TranscriptAid™ Reaction Buffer, 2 μL TranscriptAid™ Enzyme mix, complete to 20 μL with provided DEPC-treated water. Incubate the reaction for 2 h at 37 °C (*see* **Note 8**).

Precipitate the mRNA by adding 55 μL 7.5 M ammonium acetate, 875 μL prechilled absolute ethanol, and 330 μL of nuclease-free water. Vortex briefly and leave the tube on ice for 15 min. Collect the pellet by centrifugation for 15 min at 4 °C and 13,000 rpm. Discard the supernatant above the compact white pellet and add 1 mL prechilled absolute ethanol. Invert the tube few times and centrifuge again for 15 min at 4 °C and 13,000 rpm. After discarding ethanol spin it shortly again to remove the remaining ethanol and air-dry the pellet for 5 min. Dissolve the pellet in 21 μL of 1× SUB-AMIX. Optionally, you can analyze the template mRNA (*see* **Note 9**) by running 1 μL of preparation on 0.8 % agarose gel (Fig. 1).

3.5 In Vitro Translation

Thaw the reaction components and keep them on ice ensuring RNase-free conditions.

Prepare 20.8 μL of translation mixture by the addition of 5 μL (approximately 15 μg) of mRNA, 10 μL appropriate WEPRO® solution, 0.8 μL creatine kinase, and 5 μL 1× SUB-AMIX solution and mix gently by pipetting (*see* **Note 10**). Avoid bubble formation. Underlay the solution to 206 μL SUB-AMIX solution in a sterile 96-well plate carefully (*see* **Note 11**). Seal the wells with a plate sealer sticker to avoid evaporation. Incubate the translation mixture for 20 h at 20 °C (*see* **Note 12**).

Remove the plate sealer, transfer the reaction mixture into a suitably sized eppendorf tube and store it on ice until further analysis (*see* **Note 13**). Separate 5 μL of translation reaction mixture on standard SDS-PAGE and visualize the proteins by Coomassie staining (Fig. 2). In most cases, the translated exogenous protein is clearly visible as an extra band in comparison to protein components of control wheat germ extract (*see* **Note 14**).

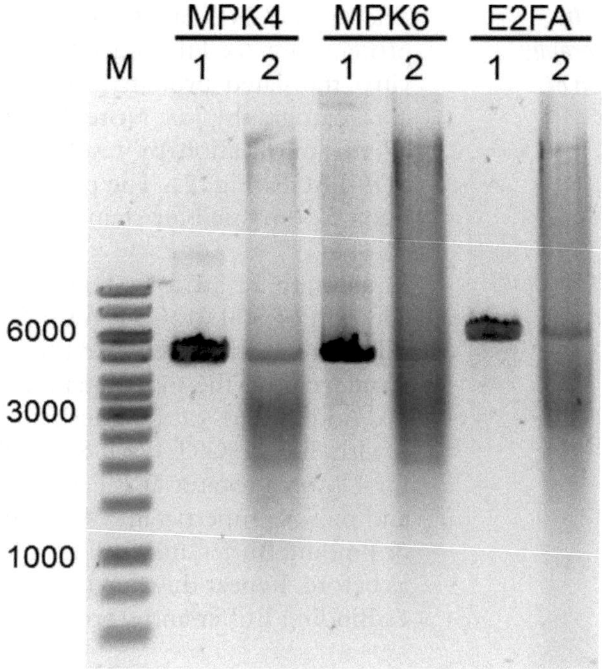

Fig. 1 Agarose gel analysis of linearized plasmid constructs and in vitro transcribed mRNAs. Arabidopsis thaliana kinases (MPK4, MPK6) and their putative substrate (E2FA) were cloned into pEU3-NII-HLICNot and pEU3-NII-GLICNot translation vectors, respectively. Following the linearization of vectors, in vitro transcription was performed. (*M*) DNA ladder, (*1*) linearized plasmid constructs, (*2*) in vitro-transcribed mRNAs

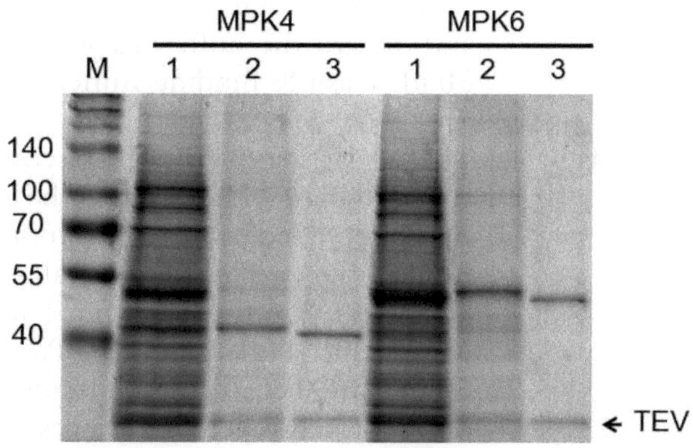

Fig. 2 SDS-PAGE analysis of in vitro translated and affinity purified proteins. His-tagged Arabidopsis thaliana MPK4 (43 kDa), MPK6 (45 kDa) were translated and purified with MagneHis™ Ni-Particles. The purified proteins were cleaved by TEV protease. The proteins were separated on 10 % SDS-PAGE and visualized by PageBlue™ Protein Staining Solution. (*M*) protein ladder, (*1*) total translation mixture, (*2*) affinity purified kinase, (*3*) protein cleaved by TEV protease

3.6 Affinity Chromatography Purification of Proteins

The pEU3-NII-HLICNot- and pEU3-NII-GLICNot-coded proteins are 6× His and GST tagged, respectively; hence the in vitro translated proteins can be effectively separated by affinity chromatography (*see* **Note 15**). Optionally, check the success of affinity purification by resolving 1/10 of total bead volume on SDS-PAGE (Fig. 2). The purified protein should be clearly visible upon Coomassie blue staining.

3.6.1 GST Tag-Based Purification

Equilibrate 25 μL of Glutathione Magnetic Bead slurry by the addition of 300 μL GST Wash & Binding Buffer in an eppendorf tube. Invert the tube several times, collect beads with a magnetic stand, remove the supernatant, and repeat this step twice.

Add 200 μL of the translation mixture to the beads; complete it with 300 μL GST Wash & Binding Buffer. Incubate the samples for 1 h on a rotator at RT. Collect the beads with magnetic stand and remove supernatant. Wash the beads with 300 μL GST Wash & Binding Buffer, mix well for a few minutes, and collect the beads as before. Repeat this step twice. After washing, add 50 μL of Wash & Binding Buffer and store at 4 °C.

3.6.2 6× His Tag-Based Purification

Equilibrate 25 μL of the MagneHis™ Ni-Particles slurry by the addition of 300 μL His Wash & Binding Buffer. Rotate the tubes for a minute and collect the beads using a magnetic stand. Discard the supernatant and repeat the equilibration step twice.

Complete the translation mixture to 700 μL by the addition of 500 μL His Wash & Binding Buffer and 7 μL 5 M NaCl to the beads. Incubate for 30 min on a rotator at RT. Place the tubes into a magnetic stand to collect the beads and remove the supernatant. Add 700 μL His Wash & Binding Buffer and place the tubes on rotator for 5 min at RT. Collect the beads with a magnetic stand and discard the buffer. Repeat the washing three times then add 50 μL Wash & Binding Buffer and store at 4 °C (*see* **Note 16**).

3.7 Phosphatase Treatment

The wheat germ extract is a concentrated protein solution with various kinases; thus the synthesized proteins of interest may be phosphorylated by the endogenous kinases. To avoid false-negative results, the produced substrates are dephosphorylated prior to kinase assay. Add 44 μL 1× Lambda Phosphatase Buffer, 5 μL $MnCl_2$ and 1 μL Lambda PPase to the beads and incubate for 30 min at 30 °C. Wash the beads with 250 μL of PBS for 5 min on a rotator and repeat the process three times. Store the samples in an appropriate buffer or proceed to TEV cleavage (*see* **Note 17**).

3.8 TEV Protease Cleavage

In an in vitro kinase assay, either the kinase or the substrate has to be in solution (*see* **Note 18**). The most convenient method to free the purified proteins from the beads is by TEV protease cleavage (*see* **Note 19**). Transfer the beads into a Protein LoBind tube, discard the buffer, and add 30 µL 1× TEV Buffer with 1 mM DTT and 20 µL of TEV protease.

Leave the sample at RT for 20 min, followed by incubation at 4 °C overnight. The following day, place the tubes into a magnetic stand to remove the beads and transfer the supernatant into a Protein LoBind tube. To eliminate TEV protease, add 10 µL of the equilibrated MagneHis™ Ni-Particles to the supernatant and incubate them on a rotator at RT for 10 min. Place the tube in a magnetic stand, collect the supernatant, and store the protein according to your optimized protocol (*see* **Note 17**).

Check the success of protease treatment by analyzing 5 µL of supernatant on SDS-PAGE (Fig. 2). Following Coomassie staining, you can estimate the concentration of purified protein with densitometry measurement in relation to protein bands with known concentration. The expected yield is approximately 300 ng/µL.

3.9 Performing In Vitro Kinase Assay

In most of the cases, 10–100 ng of kinase is sufficient for in vitro kinase assay, and 1 µg of phosphorylated substrate is clearly detectable with Pro-Q phosphoprotein-specific fluorescent staining.

Take the required volume of the slurry of kinase or dephosphorylated substrate coated beads, collect the beads with a magnetic stand, and discard the supernatant. Add the required volume of TEV cleaved substrate or kinase to the beads, bring it with kinase buffer (*see* **Note 20**) to 16 µL and incubate for 30–60 min at the temperature which ensures maximal activity of the studied kinase (*see* **Note 21**). Stop the reaction by adding 4 µL of 5× SDS Loading Dye.

3.10 PRO-Q Staining

Separate the samples, including non-kinase treated substrate and prestained or fluorescent protein marker on SDS-PAGE.

Immerse the gel into 100 mL fix solution, and incubate with gentle agitation at RT for 30 min. Repeat this step. Alternatively, you can leave the gel in fixing solution at 4 °C overnight. Wash the gel in 100 mL ultrapure water with agitation at RT for 10 min to remove all of the methanol and acetic acid from the gel, repeating the step three times. Use a dark box for incubation in the following steps. Incubate the gel in 50 mL Pro-Q solution for 60–90 min at RT with gentle agitation for staining. Wash the gel with 100 mL of destain solution for 30 min at RT with agitation, repeating this step three times. Remove the destain solution from the gel by incubating the gel for 5 min in 100 mL water. Repeat this step. Transilluminate the gel using an imaging instrument with light sources and filters that match the 555 nm excitation and 580 nm emission maximum of Pro-Q Stain (Fig. 3) (*see* **Note 22**).

3.11 Total Protein Staining

After staining with Pro-Q stain, the gel can be stained with SYPRO or Coomassie blue to perform total protein stain (*see* **Note 23**).

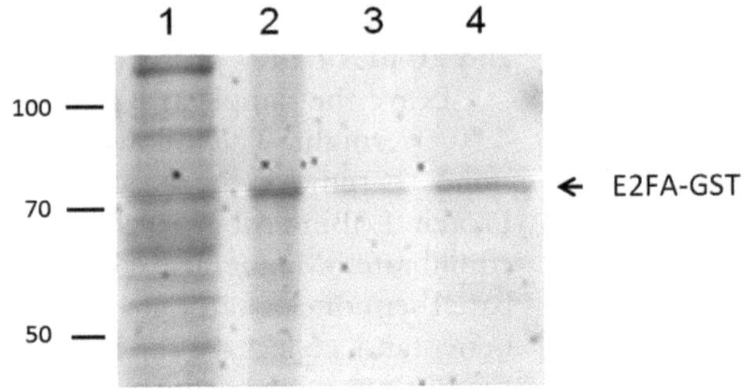

Fig. 3 Pro-Q phosphoprotein staining of the in vitro kinase assay. His-tagged Arabidopsis thaliana MPK6 and its putative substrate E2FA-GST (78 kDa) were translated and purified with Pierce Glutathione Magnetic Beads. The purified MPK6 was cleaved by TEV protease, and the purified substrate was treated on the beads with Lambda PPase. After implementation of in vitro kinase reaction, the proteins were separated on 10 % SDS-PAGE and visualized by Pro-Q Diamond phosphoprotein gel stain. (*1*) Total translation mixture of substrate, (*2*) affinity purified substrate without PPase treatment, (*3*) PPase treated affinity purified substrate (*4*) dephosphorylated substrate following in vitro kinase treatment with MPK6

4 Notes

1. The start and stop codons of the gene of interest are indicated in italics.
2. The preferred high-fidelity polymerase of your laboratory can also be used. Standard dNTP mixture is applicable if you use hot start DNA polymerase. Conditions of PCR reaction should be optimized according to the insert.
3. If you cannot see a pellet, leave a small amount of fluid in the tube to avoid losing the DNA by pipetting.
4. Vectors containing SP6 promoter are also available: pEU0-NII-HLICNot (6× His tag) and pEU0-NII-GLICNot (GST tag) [5]. Complete digestion of vector is a crucial step during the cloning procedure.
5. For visualization of the stained agarose gel LED light transmission is preferable to UV light source, because UV radiation could cause DNA damage. The use of TE buffer is to be avoided during the elution step of Gel Extraction of the linearized vector, because EDTA binds Mg ions, which is the cofactor of T4 DNA polymerase.
6. The expected number of colonies is 20–30. Efficiency of ligation independent cloning is nearly 100 %, therefore checking four colonies from each construct should be sufficient for identification of positive clones.
7. Linearization of template DNA of in vitro transcription generally increases the efficiency of the reaction. In case the insert

possesses *Not*I restriction site, the following restriction endonucleases can be used: *Sca*I, *Fsp*I.

8. Success of transcription is indicated by formation of white precipitate (magnesium pyrophosphate) after 2 h.
9. An example of high-quality mRNA is shown in Fig. 1. Smeared or ladder like pattern of RNA is normal, but occurrence of RNA below 1 kb is a sign of RNA degradation. Avoid using degraded mRNA in the following step.
10. Dedicated, pre-purified wheat germ extracts are available for synthesis of His-tagged and GST-tagged proteins. Store the wheat germ extract and creatine kinase solutions in aliquots at −80 °C, and avoid freeze–thaw cycles. The 2× SUB-AMIX can be stored at −20 °C for 2 weeks.
11. The process of underlaying the translation mixture needs a lot of practice and this step is crucial for successful protein production. You can make yourself familiar with this technique by practicing with underlaying 5× SDS Loading Dye to water.
12. Plates should be handled with special attention to avoid mixing of the two phases. The translation process can be shortened by incubating at 26 °C for 8–16 h, but generally higher activity, solubility, and productivity are expected at lower temperature.
13. During the overnight incubation, the two phases of the in vitro translation mixture are mixed by diffusion, and the solution becomes homogenous. Some proteins are sensitive to freezing; therefore, generally it is safer to keep the samples at 4 °C for up to 1 week.
14. The wheat germ extract is a highly concentrated protein solution; thus, the molecular weight of exogenously translated protein may overlap with abundant endogenous proteins. In this case, the success of in vitro translation can be detected by Western blot analysis.
15. Considering the expected yield of the described in vitro translation reactions, the target proteins can be efficiently purified by a few microliters of magnetic beads with high protein binding capacity. Addition of the affinity matrix in excess increases binding of nonspecific proteins which decreases the efficiency of purification. Working with such volumes is an arduous task with traditional affinity matrixes, therefore application of magnetic particles is recommended.
16. Formation of tiny clumps of beads after affinity purification is a good indication of protein binding.
17. The most effective storage conditions of purified proteins should be evaluated on a case by case basis. However, assuming that bacterial and fungal contamination is averted (0.01 % NaN_3), most of the proteins can be safely stored at 4 °C for up

to a few weeks. On the other hand, some proteins can be kept in 30 % glycerol at −20 °C for months. The purified proteins also can be stored in bead bound form.

18. TEV cleavage is superior to the elution of proteins by imidazole or reduced glutathione for two reasons. First, the TEV cleavage site is located between the affinity tag and the protein of interest; thus, the purified protein will possesses only one extra amino acid on its N terminus. Second, the TEV cleavage generally produces purer protein than elution, and the TEV can be eliminated via affinity purification.
19. The TEV protease cleavage of purified proteins should be rationally planned. If more substrates are to be tested with the same kinase, the kinase should be cleaved, and vice versa. The GST tag is a 21 kDa protein with putative phosphorylation sites; therefore, the His tag is the preferred labeling option if the substrate is not to be cleaved.
20. The provided kinase assay buffer composition should be handled as guideline. Modify it according to your needs. DTT has to be added freshly to kinase buffer. If you use the GST-tagged substrate without TEV cleavage, GST should be included as negative control.
21. The majority of protein kinases produced by wheat germ in vitro translations possess high kinase activity, but it can be increased further by co-translation with the relevant activators. In cases when the kinase of interest is activated by phosphorylation, the mRNAs should be added in a 10:1 kinase:activator ratio. If the kinase activity depends on complex formation, the ratio of the mRNA of the kinase and the regulatory subunit should be kept at 1:1.
22. Fix buffer for Pro-Q should be prepared fresh before each use. The provided staining protocol is suitable for mini gel staining. The volume of solutions should be increased proportionally to the gel size. Check the details of Pro-Q staining in the provided manual.
23. Determining the ratio of Pro-Q dye to total protein dye signal intensities provides a measure of the phosphorylation level normalized to the total amount of protein.

Acknowledgments

This work was supported by the Hungarian Scientific Research Fund and National Office for Research and Technology grant K69187 and K68160. T.M. gratefully acknowledges the support of Bolyai János fellowship.

References

1. Manning G, Whyte DB, Martinez R et al (2002) The protein kinase complement of the human genome. Science 298:1912–1934
2. Champion A, Kreis M, Mockaitis K et al (2004) Arabidopsis kinome: after the casting. Funct Integr Genomics 4:163–187
3. Endo Y, Sawasaki T (2006) Cell-free expression systems for eukaryotic protein production. Curr Opin Biotechnol 17:373–380
4. Goshima N, Kawamura Y, Fukumoto A et al (2008) Human protein factory for converting the transcriptome into an in vitro-expressed proteome. Nat Methods 5:1011–1017
5. Bardoczy V, Geczi V, Sawasaki T et al (2008) A set of ligation-independent in vitro translation vectors for eukaryotic protein production. BMC Biotechnol 8:32
6. Otani M, Taniguchi T, Sakai A et al (2011) Phosphoproteome profiling using a fluorescent phosphosensor dye in two-dimensional polyacrylamide gel electrophoresis. Appl Biochem Biotechnol 164:804–818
7. van den Berg S, Lofdahl PA, Hard T et al (2006) Improved solubility of TEV protease by directed evolution. J Biotechnol 121: 291–298

Chapter 16

Preparation of Protein Arrays Using Cell-Free Protein Expression

Elizabeth A. Cook and Mingyue He

Abstract

Protein microarrays are a miniaturized format for assaying functional protein interactions and proteomics using a high-throughput, multiplexed approach. The primary limitations of this technology include the time and cost involved in the production of highly purified individual proteins for arraying and also the limited stability of functional proteins once arrayed. In light of these difficulties, cell-free protein expression systems are being increasingly used to generate protein arrays in situ from coding DNA sequences. This chapter describes the method DNA array to protein array (DAPA) allowing the repeated "printing" of protein arrays directly from a DNA array template using cell-free protein expression. Once the DNA templates have been spotted, the generation of the protein array involves only simple handling procedures and is relatively time and cost-efficient, and protein arrays can easily be produced "on demand" as and when required. The resultant protein array may be used for any downstream applications as for conventionally spotted protein arrays.

Key words Protein array, Protein microarray, DNA microarray, Cell-free protein synthesis, Protein immobilization

1 Introduction

Protein microarrays have emerged as an extremely useful technological platform for the large-scale proteomics and functional protein analysis. As miniaturized assays, they enable the display of many proteins on a single slide, allowing high-throughput analysis of proteins and the screening of many interactions in parallel. They may be used to assay protein interactions with different molecules, including proteins, antibodies, nucleic acids, peptides, small molecules, or lipids. They are increasingly used in including drug target discovery, identification of disease biomarkers, profiling of protein expression, and protein interaction studies [1, 2].

Kirill Alexandrov and Wayne A. Johnston (eds.), *Cell-Free Protein Synthesis: Methods and Protocols*, Methods in Molecular Biology, vol. 1118, DOI 10.1007/978-1-62703-782-2_16, © Springer Science+Business Media, LLC 2014

A variety of approaches have been described that use cell-free protein synthesis as a means of producing and simultaneously arraying multiple proteins in parallel [3–8]. Cell-free protein expression systems comprise homogenized cell lysates which can transcribe and translate proteins from PCR-generated linear or circular plasmid DNA templates. These methods allow the generation of protein arrays "on demand" directly from DNA constructs, offering a number of advantages over directly spotted protein arrays. DNA arrays are more stable compared to protein arrays and can be stored for extended periods of time with no degradation. This allows for generation of the protein array quickly as and when required and reduces the risk of the proteins becoming functionally inactive upon storage of the array. In addition, as the proteins are synthesized directly from DNA templates on the array, this removes the need for expensive and time-consuming expression and purification of individual proteins for arraying since the production of DNA coding for individual protein sequences is relatively inexpensive and the DNA constructs can be readily amplified by PCR.

We have previously described the generation of protein arrays through cell-free protein expression using the DNA array to protein array (DAPA) method, involving the direct "printing" of a protein array from a corresponding DNA array (as shown in Fig. 1) [2].

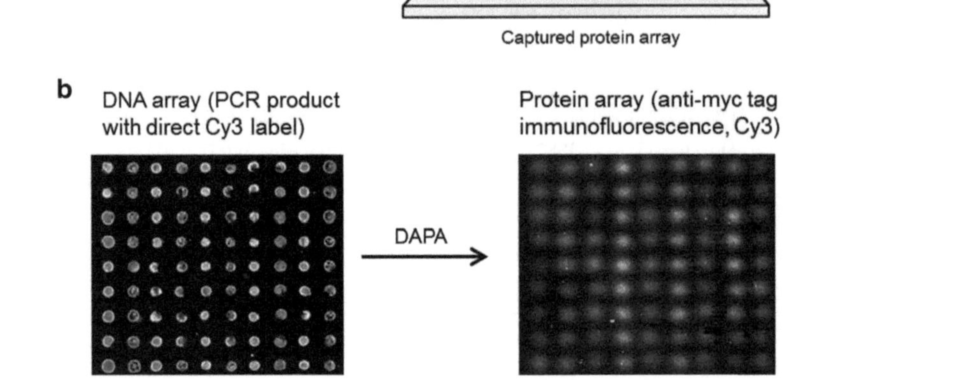

Fig. 1 (**a**) Schematic of the DAPA procedure. (**b**) Sample results from DAPA, showing a DNA microarray template slide (*left*) and resultant protein microarray (*right*)

In brief, a DNA slide is printed with PCR-generated constructs encoding proteins with a double hexa-histidine [2(His)$_6$] tag at the C-terminus. This slide is assembled face-to-face with a protein-capture Ni–NTA-functionalized slide. A membrane soaked with an *E. coli* cell-free lysate is placed between the slides. During the DAPA process, the DNA immobilized to the microarray is transcribed and translated and the synthesized proteins diffuse through the membrane filter to the protein-capture slide, producing a protein array corresponding to the DNA array template. The protein array typically has spots with well-defined geometry, with the intensity of each spot following a Gaussian profile due to diffusion of the proteins through the membrane (Fig. 1) [3].

In addition to the advantages mentioned above, DNA template slides can also be used more than once, permitting many protein slides to be printed by repeating the DAPA procedure from a single reuseable DNA template. This further reduces the need for time-consuming production of separate protein arrays. We have previously demonstrated that a single DNA array template can be used to print up to 20 copies of the protein array [3].

2 Materials

2.1 Template DNAs

DNA constructs for DAPA-generated protein arrays are in a format to allow expression using an *E. coli* cell-free system and then subsequent immobilization and detection of the expressed proteins containing designed tags (Fig. 2). They are generally produced by PCR assembly of three fragments, namely, an upstream fragment, gene of interest, and downstream fragment (Figs. 2 and 3). These fragments can be cloned into plasmid vectors (which may be generated by standard molecular biology techniques) for easy amplification by PCR (Fig. 3):

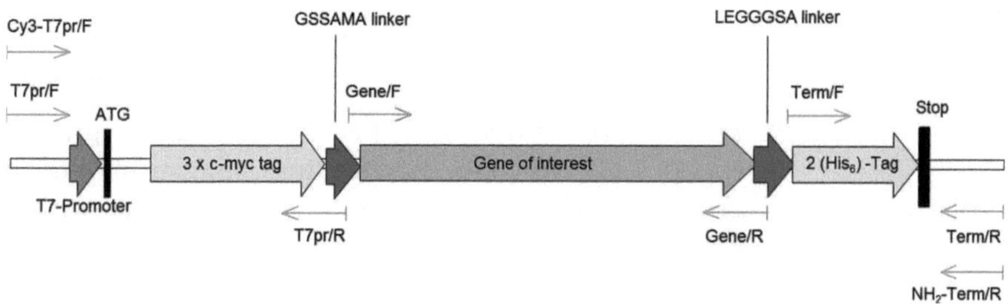

Fig. 2 Representation of a full-length DNA construct for DAPA, showing the important features of the construct and positions of all primers

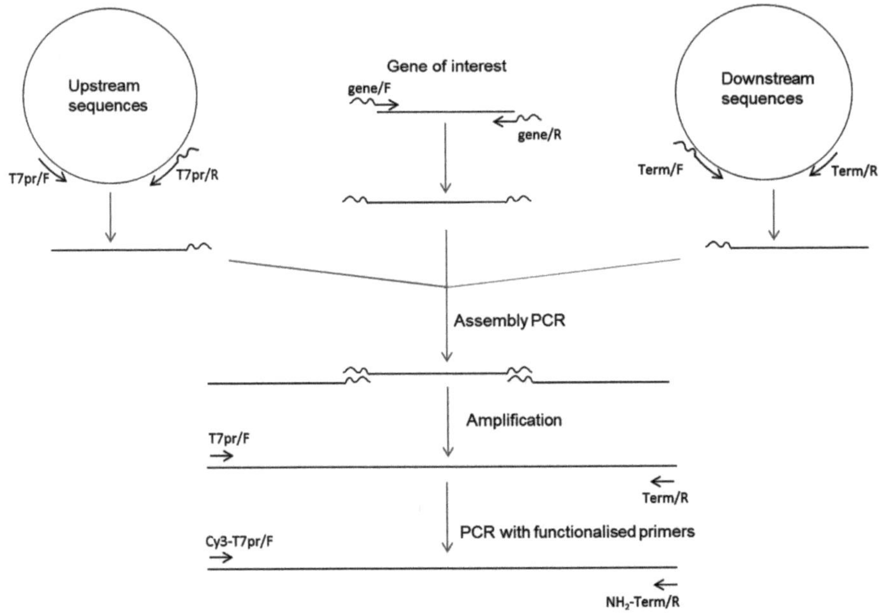

Fig. 3 Overview of PCR-based construction strategy for generation of full-length DNA for DAPA. The primers used to generate individual fragments are indicated

1. *Upstream fragment.* The upstream fragment should contain the T7 promoter, the ribosome binding site, the start codon ATG, and an N-terminal tag. The following sequence shows the upstream fragment we use:

   5′GATCTCGATCCCGCGAAAT*TAATACGACTCACTATA*GGGAGACCACAAC
   GGTTTCCCTCTAGAAATAATTTTGTTTAACTTTAA*GAAGGAG*ATATCCACC
   M E Q K L I S E E D L N G E Q K L
   ATGGAGCAAAAGCTCATTTCTGAAGAGGACTTGAATGGAGAACAGAAATTG
   I S E E D L N G E Q K L I S E E D
   ATCAGTGAGGAAGACCTCAACGGTGAGCAGAAGTTAATATCCGAGGAGGAT
   L **G S S A M A**
   CTT**GGCTCTAGTGCCATGGCT-**3′

   The T7 promoter sequence and ribosome binding site are in italics. A triple c-myc tag is underlined (*see* **Note 1**). A linker sequence and the initiation codon ATG are in bold. The encoded amino acids are shown above the DNA sequence.

2. *Gene of interest.* Any type of DNA encoding gene of interest.

3. *Downstream fragment.* The downstream fragment we use encodes a 2(His)$_6$ tag, used for immobilization of the expressed protein, separated from the gene of interest with a flexible linker (see below) (*see* **Note 2**):

   L E G G G S A H H H H H H S R A W R H
   5′-**CTCGAGGGTGGCGGTAGC**GCACATCACCATCACCA
   TCACTCTAGAGCTTGGCGTCAC
   P Q F G G H H H H H H . .
   CCGCAGTTCGGTGGTCACCACCACCACCACCACTAATAA
   (A)$_{28}$CCGCTGAGCAATAACTAGC

ATAACCCCTTGGGGCCTCTAAACGGGTCTTGAGGGGT
*TTTTTG*CTGAAAGGAGGAACTATAT
CCGGA-3′

The linker sequence is in bold, followed by the 2(His)$_6$ tag (underlined) and the transcription termination region in italics. The encoded amino acids are shown above the DNA sequence.

2.2 Primers

Individual sets of primers are required for generation of the three separate fragments listed above and include overlapping sequences to allow for the assembly of full-length constructs (Fig. 3). All primer sequences are given 5′ → 3′:

1. Primers for generating the upstream fragment:
   (a) Forward primer T7pr/F: GATCTCGATCCCGCG
   (b) Reverse primer T7pr/R: AGCCATGGCACTAGAGCC AAGATCCTCCTCGGATATTAAC
2. Primers for generating the gene of interest:
   (a) Forward primer Gene/F: GGCTCTAGTGCCATGGCT $(N)_{15–25}$
   (b) Reverse primer Gene/R: GCTACCGCCACCCTCGAG $(N)_{15–25}$

   Where $(N)_{15–25}$ are the coding nucleotides from the 5′ and 3′ end of the gene of interest with the correct reading frame. 15–25 is the number of nucleotides. The final stop codon is removed (Gene/R). Sequences underlined overlap with the upstream fragment (*see* Subheading 1) and downstream fragment (*see* Subheading 3) to allow for PCR assembly of the full-length construct.
3. Primers for generating the downstream fragment:
   (a) Forward primer Term/F: CTCGAGGGTGGCGGTAGC ACATCACCATCACCAT
   (b) Reverse primer Term/R: TCCGGATATAGTTCCTCC
4. Cy3- and NH_2-coupled primers for labelling of the full-length DNA construct (*see* Subheading 3.1 below).
   (a) Forward primer Cy3-T7pr/F: Cy3-GATCTCGATCC CGCG
   (b) Reverse primer NH_2-Term/R: NH_2-TCCGGATAAGTT CCTCC

2.3 General Molecular Biology Reagents

1. Deoxy nucleotides.
2. High-fidelity polymerase for PCR, for example, Expand High Fidelity PCR system (Roche) (*see* **Note 3**).
3. PCR cleanup and gel extraction kit, for example, NucleoSpin Extract II (Macherey-Nagel).
4. RTS *E. coli* HY cell-free protein expression kit (5 PRIME, Hamburg, Germany).

2.4 Solutions and Buffers

1. 6× spotting buffer: 300 mM sodium phosphate (pH 8.5).
2. 0.1 % Tween-20 (v/v).
3. 1 mM HCl.
4. 100 mM KCl.
5. Quenching buffer: 0.1 M Tris–HCl (pH 9.0) containing 50 mM ethanolamine (added immediately before use).
6. Saturated NaCl solution (for humidification chamber).
7. Wash buffer: PBS, 0.05 % Tween-20 (PBST).
8. Blocking buffer: PBST, 1 % BSA.
9. ddH_2O.

2.5 Slides, Spotting Plate, and the Membrane Filter

1. Epoxy-coated slide for immobilization of DNA, for example, Nexterion slide E (Schott Nexterion).
2. 0.22 μm membrane filter, for example, Millipore membrane.
3. Ni–NTA-coated slide (Xenopore).
4. 384-well plate (Nunc).
5. Parafilm.
6. Source of pressurized air.
7. Microarray scanner.

3 Methods

In this section, we describe the following procedures in order: (1) generation of PCR DNA constructs for DAPA; (2) immobilization of the DNA constructs on an epoxy-functionalized slide, creating a DNA array template; (3) making a protein array by the DAPA method; and (4) detection of arrayed proteins on the slide.

3.1 Generation of PCR DNA Constructs for DAPA

Full-length templates for DAPA consist of three components. The detailed final construct with the positions of all primers as well as a simplified construction strategy is shown in Figs. 2 and 3. Each of these three fragments is generated using a separate PCR reaction, which also introduce overlapping sequences at each end of the gene of interest (*see* Subheading 2.2). The full-length construct is then assembled by overlapping PCR consisting of the three fragments. Finally, this construct is amplified and labelled by PCR using Cy3- and NH_2-coupled primers for both efficient immobilization (an NH_2 group at the 3′ end) and detection (a Cy3 fluorophore at the 5′ end).

3.1.1 Standard PCR Reactions

1. Set up a 50 μl PCR reaction containing the following components:

   10× PCR buffer (containing $MgCl_2$) 5 μl.

dNTP mix (2.5 mM each) 4 μl.
Forward primer (16 μM) 2 μl.
Reverse primer (16 μM) 2 μl.
DNA template 1–10 ng.
High-fidelity DNA polymerase 2.6 units.
H_2O to final 50 μl.
PCR program: 30 cycles of thermal cycling (94 °C, 30 s; 54 °C, 30 s; 72 °C, 120 s; then 72 °C, for 8 min). Finally, hold at 4 °C.

2. Run the PCR product on a 1 % agarose gel to confirm correct size and isolate the fragment from the gel using gel extraction kit when needed.
3. Measure the concentration and purity of the PCR products by absorption at 260/280 nm.

3.1.2 Assembly PCR Reaction Using the Fragments Generated

1. Set up the assembly reaction as follows:
10× buffer (containing $MgCl_2$) 1 μl.
dNTP mix (2.5 mM each) 1 μl.
Equimolar ratio mix of DNA fragments (upstream fragment, the gene of interest, and downstream fragment) 10–50 ng.
High-fidelity DNA polymerase 1 unit.
H_2O to final 10 μl.
PCR program: 8 cycles of thermal cycling (94 °C, 30 s; 54 °C, 60 s; 72 °C, 120 s); then hold at 4 °C.
2. Amplify the assembled full-length product by adding 1 μl of the above mixture into a second 50 μl PCR reaction as described in Subheading 3.1.1, but using the primers T7pr/F and Term/R.
3. Analyze the PCR product on a 1 % agarose gel to confirm correct size and purity, and purify it using gel excision kit (*see* **Note 4**) for cell-free protein synthesis (*see* **Note 5**).

3.1.3 Labelling of the Full-Length Construct by PCR

1. Set up a standard 50 μl PCR reaction (*see* Subheading 3.1.1), but using the primers Cy3-T7pr/F and NH_2-Term/R.
2. Analyze the product on a 1 % agarose gel to confirm correct size and purity.
3. Purify the PCR product using a PCR purification kit.

3.2 Immobilization of the PCR DNA Constructs on Epoxy-Coated Slides

1. Add 1 volume 6× spotting buffer to 5 volumes labelled PCR DNA construct and transfer to a 384-well plate.
2. Spot 1–10 nl each DNA construct onto an epoxy-silane slide, using a spot-to-spot distance of 0.6–1 mm (*see* **Note 6**).
3. Transfer slides to a humidified chamber (sealed container with saturated NaCl solution) immediately after spotting and incubate for 12–20 h at RT (or longer at 4 °C) (*see* **Note 7**).
4. Bake slides at 60 °C for 30 min (*see* **Note 8**).

5. Wash slides while rocking at RT as follows:
   (a) 0.1 % Tween-20, 1 × 5 min.
   (b) 1 mM HCl, 2 × 2 min.
   (c) 100 mM KCl, 1 × 10 min.
   (d) Rinse with ddH_2O.
6. Block any unreacted epoxy groups on the slide surface with 1× quenching buffer for 15 min at 50 °C.
7. Rinse slides in ddH_2O and dry using pressurized air.
8. Scan slides to confirm correct immobilization of DNA using a microarray scanner, detecting the Cy3 fluorophore. DNA slides can now be stored at 4 °C until needed (see below).

3.3 Making a Protein Array by the DAPA Method

1. Prepare the individual components for the DAPA:
   (a) Cut a piece of 0.22 μm membrane filter to the size required to cover the full spotted DNA array.
   (b) Prepare 10 μl *E. coli* cell-free lysate for each 1 cm^2 of membrane, plus an extra 10 μl for the positive and negative controls (*see* **Notes 9** and **10**).
2. Assemble the DNA array slide from Subheading 3.2 and protein-capture slide (Ni–NTA coated) with the cell-free mix in the order as follows (doing this as quickly as possible to avoid drying out of the membrane):
   (a) Protein-capture slide.
   (b) Pipette the cell-free expression mixture onto the center of the protein-capture slide, taking care to avoid bubbles.
   (c) Place the membrane over the protein-capture slide (in an area equivalent to the spotted DNA array), ensuring the entire membrane is wet.
   (d) Place the DNA array slide on top (DNA template facing downwards), aligning with the protein slide.
   The assembled slide sandwich needs to be sealed to prevent drying out of the membrane during DAPA, which may be done with a layer of parafilm covering each slide. Figure 4 shows a schematic of the full DAPA assembly. Constant pressure must be exerted on the DAPA sandwich for the duration of the process, which is done using a slide holder such as shown in Fig. 4 (*see* **Note 11**).
3. Incubate the DAPA sandwich (along with positive and negative controls, *see* **Note 10**) at 30 °C for 4 h.
4. Disassemble the DAPA sandwich and immerse both slides in PBST buffer while separating the slides.

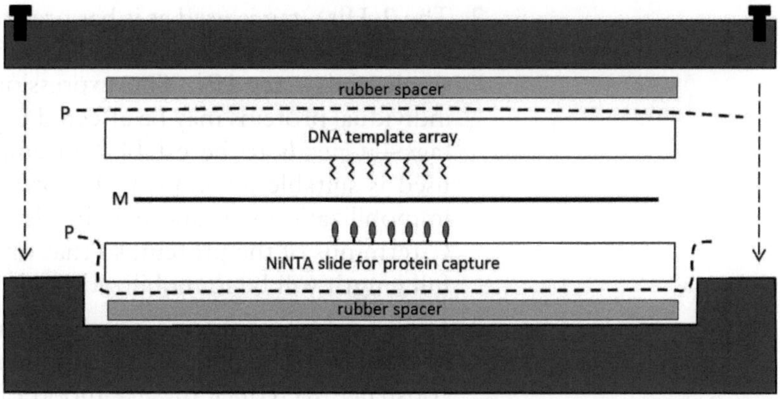

Fig. 4 Schematic cross section of the prototype apparatus used for DAPA procedure, showing the assembly of the DAPA sandwich. P = parafilm, M = membrane soaked in cell-free expression lysate

5. Wash the DNA array slide in ddH_2O and dry with pressurized air. The slide can be stored at 4 °C for repeated use.
6. Wash the protein-capture slide three times in PBST for 5 min. The slide is now ready for detection of the immobilized proteins (see below).

3.4 Detection of Arrayed Proteins on the Slide

The immobilized proteins can be detected with an antibody as follows:

1. Block the protein-capture slide in 1 % BSA in PBST for 1 h with gentle shaking at RT.
2. Add a suitable primary antibody into the blocking buffer for 1 h at RT (*see* **Notes 1** and **12**). If the primary antibody is already labelled by a fluorophore, go to **step 3** below. Otherwise, incubate with a fluorophore-labelled secondary antibody (in the blocking buffer) for an additional 1 h at RT.
3. Wash the protein slide in PBST 3 × 5 min, rinse in ddH_2O, and dry with pressurized air before scanning. Now the slide is ready for detection using microarray scanner.

4 Notes

1. The use of an epitope tag for detection of proteins allows the expression of all proteins to be detected using a single anti-tag antibody and removes the requirement for individual antibodies specific for each protein. It also facilitates the comparison of relative protein expression level on the slide.

2. The 2$(His)_6$ tag is used as it has previously been shown to have an improved affinity for Ni–NTA surfaces than a conventional single (His_6) tag [9]. The expression/solubility/function of individual proteins may be affected by the presence of different tags—it needs to be established empirically whether the tag used is suitable for individual proteins. It is ideal to have the immobilization tag (and may be also the detection tag) at the C-terminus of the protein, so that only those proteins that are full length will be immobilized (and detected).
3. High-fidelity polymerase is recommended for the amplification of the constructs, due to the number of PCR cycles for each construct, to reduce the likelihood of introducing PCR-based mutations.
4. PCR products for cell-free protein expression should be purified via spin column and eluted in water. Direct isolation of PCR fragments with trace agarose contamination may inhibit cell-free protein expression.
5. Correct expression of the protein from the assembled PCR construct can be verified by a small-scale cell-free protein expression from the PCR product, followed by Western blotting detection.
6. Slides can be labelled using a diamond pen for ease of identification and to help with correct orientation.
7. Once the labelled PCR products have been generated, care should be taken to avoid prolonged exposure to light that result in fluorophore bleaching.
8. This is an extra step in processing of the DNA slides, which has been found to significantly improve immobilization of the DNA.
9. As in the manufacturer's instructions, a repeated freeze-thawing of the *E. coli* cell-free lysate should be avoided by splitting into convenient aliquots. Aliquots should be stored at –20 °C and thawed on ice just before use.
10. It is useful to set up both positive and negative control reactions in tubes (10 μl lysate each) at the same time to check the cell-free protein expression. The positive control should contain 0.5 μl GFP vector (provided in the RTS *E. coli* HY kit), and the negative control 0.5 μl ddH_2O. Expression of GFP is monitored under a UV light to visualize the green fluorescence.
11. The slide holder shown in Fig. 4 is a prototype device. Any similar setup which allows constant pressure and prevents the drying out of the membrane between the slides may also be suitable. This could, for example, be achieved using an alternative slide holder or by placing a heavy weight on top of the slide sandwich.

12. The primary antibody used may be specific either for each protein expressed or for a designed epitope tag encoded by the DNA construct. If primary antibodies are labelled with different fluorophores (or have different secondary antibodies), then multiple antibodies may be incubated at once, for the detection of different proteins.

Acknowledgements

The Protein Technology Group at Babraham Bioscience Technologies is a partner in the EC FP7 222635 AffinityProteome and 241481 AFFINOMICS projects. Research at Babraham institute is supported by BBSRC, UK.

References

1. Berrade L, Garcia AE, Camarero JA (2011) Protein microarrays: novel developments and applications. Pharm Res 28:1480–1499
2. Hall D, Ptacek J, Synder M (2007) Protein microarray technology. Mech Ageing Devel 128:161–167
3. He M, Stoevesandt O, Palmer EA et al (2008) Printing protein arrays from DNA arrays. Nat Methods 5:175–177, Printing protein arrays from DNA arrays
4. Stoevesandt O, Vetter M, Kastelic D et al (2011) Cell free expression put on the spot: advances in repeatable protein arraying from DNA (DAPA). N Biotechnol 28:282–290
5. He M, Taussig MJ (2001) Single step generation of protein arrays from DNA by cell-free expression and in situ immobilisation (PISA method). Nucleic Acid Res 29:e73
6. Angenendt P, Kreutzberger J, Glökler J et al (2006) Generation of high-density protein microarrays by cell-free in situ expression of unpurified PCR products. Mol Cell Proteomics 5:1658–1666
7. Ramachandran N, Hainsworth E, Bhullar B et al (2004) Self-assembling protein microarrays. Science 305:86–90
8. Ramachandran N, Raphael JV, Hainsworth E et al (2008) Next-generation high-density self-assembling functional protein arrays. Nat Methods 5:535–538
9. Khan F, He M, Taussig MJ (2006) A double-Hid tag with high affinity binding for protein immobilisation, purification and detection on Ni-NTA surfaces. Anal Chem 78: 3072–3079

Chapter 17

Posttranscriptional Control of Protein Synthesis in *Drosophila* S2 Cell-Free System

Motoaki Wakiyama and Shigeyuki Yokoyama

Abstract

Protein synthesis is regulated transcriptionally and posttranscriptionally, with the latter including both the translation and mRNA degradation steps. Eukaryotic mRNAs have a characteristic 7-methyl-G cap structure at their 5′ ends and a polyadenylated tail at their 3′ ends. These structures, and the sequences of the untranslated regions (UTR) flanking the coding region on the 5′ and 3′ sides, are recognized by various RNA-binding proteins and determine translational efficiency and mRNA stability. RNA interference is a sequence-specific inhibition of protein synthesis triggered by double-stranded RNA (dsRNA). This process is mediated by RNA-binding proteins named Argonaute. Argonautes incorporate dsRNAs of 21–22 nucleotides (termed short-interfering RNAs or siRNAs) and cleave mRNAs containing sequences complementary to siRNAs. In this chapter, we describe a cell-free translation system from *Drosophila* Schneider line 2 (S2) cells that recapitulates RNA interference. This system can be programmed with multiple RNA transcripts, a target and a control, and chemically synthesized short-interfering RNA (siRNA). The production of the target protein is reduced in the presence of the target-specific siRNA, in a dose-dependent manner. We also describe a coupled transcription and translation system using the S2 cell lysate.

Key words *Drosophila* S2 cells, Translation, 5′-Untranslated region, siRNA, RNAi

1 Introduction

During the past half-century, cell-free systems have played an important role in the analyses of posttranscriptional control of protein synthesis. Rabbit reticulocyte lysate (RRL) and wheat germ extract (WGE) are the most popular eukaryotic systems and have been used in many studies of translation initiation, elongation, and mRNA degradation.

Tuschl and coworkers employed a lysate prepared from embryos of the fruit fly *Drosophila melanogaster* to recapitulate RNA interference (RNAi) [1]. They showed that gene-specific inhibition by double-stranded RNA (dsRNA) was induced in the *Drosophila* embryo lysate, but not in WGE or RRL. Thus, the *Drosophila* embryo lysate became a valuable tool in dissecting the

Kirill Alexandrov and Wayne A. Johnston (eds.), *Cell-Free Protein Synthesis: Methods and Protocols*, Methods in Molecular Biology, vol. 1118, DOI 10.1007/978-1-62703-782-2_17, © Springer Science+Business Media, LLC 2014

molecular basis of RNAi. Indeed, using this system, Tuschl and coworkers demonstrated that 21- and 22-nt RNA fragments are the sequence-specific mediators of RNAi [2]. Their discovery was a breakthrough for understanding of RNAi in mammals [3].

Although useful as a system for RNAi investigation, preparation of a *Drosophila* embryo lysate is laborious. In this chapter, we describe a cell-free protein synthesis system using *Drosophila* Schneider line 2 (S2) cells [4, 5]. The S2 cell was established from a primary culture of 20- to 24-h-old *Drosophila melanogaster* embryos [6]. This cell line is easy to grow in a monolayer in a culture dish and also in suspension, using an Erlenmeyer flask at 25–27 °C (without requiring a CO_2 supply). Thus, it is very convenient for large-scale preparations of cell lysates.

The key to obtaining a lysate with high activity is the choice of cell disruption method. The Dounce homogenizer disrupts cells by pushing the sample between the sides of the tube and the pestle. Although this technique is used in many cases, it is difficult to obtain reproducible results. It often causes the nuclei breakage, which results in the leakage of chromosomal DNA and nucleases into the cell lysate. Instead, we have successfully employed the nitrogen cavitation method [4, 5]. This technique involves the equilibration of nitrogen gas with a cell suspension under high pressure, followed by sudden decompression. The process is easy to control, and the nuclei remain intact if an appropriate lysis solution is used.

In this chapter, we describe preparation of the *Drosophila* S2 cell lysate and subsequent constitution of the S2 cell-free protein synthesis system. This cell-free system is a useful tool for research into the posttranscriptional control of protein synthesis.

2 Materials

2.1 Cell Culture

1. *Drosophila* S2 cells (Invitrogen; *see* **Note 1**).
2. Express Five SFM (Invitrogen; *see* **Note 1**). To 1,000 mL of the medium, add 90 mL of 200 mM L-glutamine solution.
3. Polycarbonate Erlenmeyer flask with vent cap, 500 mL, 3,000 mL (Corning, USA).
4. Incubator/shaker suitable for 110 rpm, 25 °C.
5. Vented cap tissue culture flasks (25 mL fill volume).

2.2 Preparation of Cell Lysate

1. Dulbecco's phosphate-buffered saline (D-PBS).
2. Cell lysis solution: 40 mM HEPES-KOH (pH 8.0), 100 mM potassium acetate, 1 mM magnesium acetate, and 1 mM dithiothreitol (DTT).
3. Cell disruption bomb 4639 (Parr Instrument Company, USA; *see* Fig. 1).
4. Compressed nitrogen gas (grade 1).

Fig. 1 Cell disruption bomb 4639 (Parr Instrument Company). The discharge valve is indicated by an *arrow*

2.3 *In Vitro* Transcription and Translation

1. T7 RNA transcription kit, such as mMESSAGE mMACHINE T7 Kit (Ambion).
2. RNA purification kit, such as MEGAclear Kit (Ambion).
3. Ribonucleotides: ATP, CTP, UTP, and GTP (100 mM).
4. Creatine phosphokinase (Roche Applied Science): 10 mg/mL. Dissolve 100 mg in 10 mL of 20 mM HEPES-KOH, pH 7.6, 50 %(v/v) glycerol; store at −80 °C in 50–100 μL aliquots.
5. Amino acids: RTS Amino Acid Sampler (5 PRIME, Hamburg, Germany).
6. FluoroTect™ $Green_{Lys}$ in vitro Translation Labeling System (Promega, L5001).
7. Translation master mix: 30 mM HEPES-KOH, pH 8.0, 100 mM potassium acetate, 4.5 mM magnesium acetate, 40 mM creatine phosphate, 2.5 mM spermidine, 5 mM ATP, 1 mM GTP, and 100 μM amino acid mixture (as above).
8. TnT master mix: 30 mM HEPES-KOH (pH 8.0), 100 mM potassium acetate, 12 mM magnesium acetate, 150 mM creatine phosphate, 2.5 mM spermidine, 5 mM ATP, 2.5 mM GTP, 2.5 mM CTP, 2.5 mM UTP, and 100 μM amino acid mixture.

2.4 Recapitulation of RNA Interference (RNAi)

1. Luciferase GL2 siRNA (QIAGEN SI03650353).
2. Luciferase GL3 siRNA (QIAGEN SI03650360).
3. Plasmids encoding firefly luciferase GL3 (*see* **Note 2**).
4. Plasmids encoding *Renilla* luciferase (*see* **Note 2**).

3 Methods

3.1 Preparing S2 Cell Lysate

Drosophila S2 cells (previously adapted to Express Five SFM medium supplemented with glutamine) are maintained in a tissue culture flask:

1. Dilute S2 cells with fresh medium to 2–4 × 10^6 cells/mL, and dispense 500 mL into a 3 L polycarbonate Erlenmeyer flask with vent cap (*see* **Note 3**).
2. Culture cells in an incubator/shaker at 25 °C with 110 rpm rotation (*see* Fig. 2).
3. Grow cells until the culture reaches the stationary phase. The cell density may approach around 4 × 10^7 cells/mL (*see* **Note 4**).
4. Sediment the cells by centrifugation (400 g, 10 min at 4 °C) and resuspend them in ice-cold D-PBS.
5. Sediment the cells by centrifugation as above. Repeat **steps 4** and **5** three times.
6. Suspend the cell pellet in ice-cold cell lysis solution (at 1 × 10^9 cells/mL; *see* **Note 5**).
7. Place the cell suspension in an ice-cold cell disruption bomb 4639 (*see* **Note 6**).

Fig. 2 Culturing S2 cells using a 3 L Erlenmeyer flask

8. Apply 100 psi of nitrogen gas to the cell suspension for 60 min on ice (*see* **Note 7**).
9. Collect the cell lysate in a 50 mL tube placed in ice (*see* **Note 8**).
10. Centrifuge the cell lysate at 1,000 × *g*, 4 °C for 10 min.
11. Transfer the supernatants to fresh micro-tubes.
12. Centrifuge the cell lysate at 16,000 × *g*, 4 °C for 10 min.
13. Transfer the supernatants to fresh micro-tubes (*see* **Note 9**).
14. Add creatine phosphokinase to a final concentration of 0.4 mg/mL. Lysates with 70–80 OD/mL at 260 nm should be obtained (*see* **Note 10**).
15. Dispense the cell lysate into micro-tubes, and freeze them in liquid nitrogen. Store frozen aliquots at −80 °C.

3.2 Translation Reaction and Recapitulation of RNA Interference (RNAi)

Basically, the translation protocol described below can be applicable for any capped mRNA transcripts (*see* **Note 11**). In order to examine posttranscriptional control, we typically use two reporter mRNAs, one encodes a firefly luciferase and the other encodes *Renilla* luciferase.

3.2.1 Translation and Luciferase Assay

1. Prepare mRNAs by in vitro transcription followed by purification. RNA transcripts should be dissolved in RNase-free water.
2. Set up translation reactions with (v/v): 60 % S2 cell lysate, 20 % translation master mix, 2 % FluoroTect™ $\text{Green}_{\text{Lys}}$ tRNA (if labeled proteins are desired), and 18 % mRNA solution. Before mixing, keep all of the reagents in ice. Combine the ingredients in the following order: water, translation master mix, FluoroTect™ $\text{Green}_{\text{Lys}}$ tRNA (if required), mRNA, and cell lysate. Mix each reaction mixture by gentle pipetting.
3. Incubate the tubes at 27 °C for 60 min.
4. Chill the tubes on ice, and withdraw a 5 μL aliquot for the reporter assay.
5. For luciferase assays, measure the luciferase activity using an appropriate assay system and equipment (e.g., the Promega luciferase assay system, dual luciferase assay system, and the Berthold Technologies MiniLumat LB 9506 bioluminometer). The withdrawn 5 μL sample usually must be diluted 10- to 100-fold with an appropriate buffer (e.g., 1× Passive Lysis Buffer from the Promega luciferase assay system) to attain a light unit level of luminescence within the linear range of detection of the instrument used.
6. For the analysis of proteins labeled with the FluoroTect™ $\text{Green}_{\text{Lys}}$ in vitro Translation Labeling System, fractionate the samples by SDS-polyacrylamide gel electrophoresis (SDS-PAGE) and detect the labeled proteins by using a fluorescent image analyzer (e.g., BIO-RAD Molecular Imager FX).

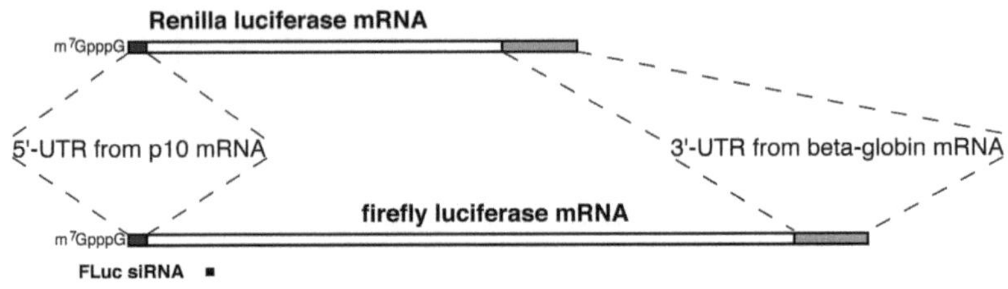

Fig. 3 Schematic representation of the luciferase mRNAs. Both the firefly and *Renilla* luciferase mRNAs contain the 5′-untranslated region (UTR) from baculovirus p10 mRNA and the 3′-UTR from *Xenopus* beta-globin mRNA. The position of the siRNA is shown as a black bar. Modified from ref. [4]

3.2.2 Recapitulation of RNA Interference (RNAi)

As an example of using the *Drosophila* S2 cell-free system to recapitulate RNA interference, a protocol for siRNA-mediated knockdown of firefly luciferase (FLuc) mRNA is presented. Two siRNAs, GL2 and GL3, were tested to examine whether FLuc synthesis is inhibited in a sequence-specific manner. The GL3 siRNA shares perfect sequence homology with the FLuc mRNA coding region, corresponding to positions 155–173 relative to the initial A of the start codon, whereas the GL2 siRNA contains three bases mismatched to the FLuc sequence. As a control of translation reaction, *Renilla* luciferase (RLuc) mRNA was used. The construction of the mRNAs is shown in Fig. 3:

1. Preincubate the S2 cell lysate with either the GL3 or GL2 siRNA, at final concentrations of 0–570 nM, for 30 min at 27 °C.
2. Program the translation reaction, including 4 ng/μL each of the FLuc and RLuc mRNAs, using the above cell lysates, in the presence of the FluoroTect™ Green$_{\mathrm{Lys}}$ tRNA.
3. Fractionate the samples by SDS-PAGE, and analyze the gel with a fluorescent image analyzer.

Typical results are shown in Fig. 4. While an increase in the GL2 siRNA concentration does not alter FLuc synthesis, it is remarkably decreased in response to an increase in the GL3 siRNA concentration. The RLuc synthesis is not inhibited by the GL2 and GL3 siRNAs. The assays of the FLuc and RLuc activities also confirmed the sequence-specific inhibition of the FLuc synthesis.

3.3 Coupled Transcription and Translation System

For recombinant protein expression, a coupled transcription and translation (TnT) system is more convenient than an mRNA-based in vitro translation system. An important parameter in establishing an efficient system is the construction of the mRNA, especially the sequence of the 5′-untranslated region (5′-UTR). Translation initiation on most eukaryotic mRNAs occurs by a 5′-terminal 7-methyl-G cap structure-dependent mechanism [7].

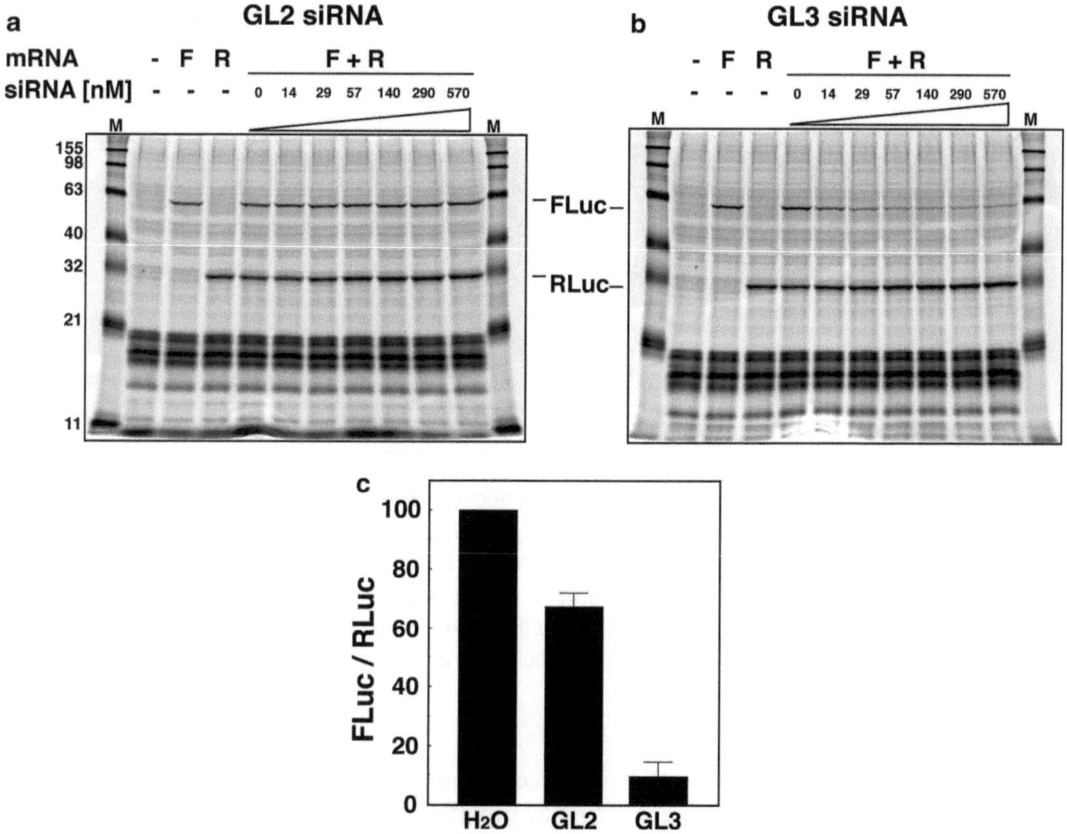

Fig. 4 SiRNA-mediated inhibition of protein synthesis in the S2 cell-free system. (**a**, **b**) *F* and *R* denote firefly luciferase (FLuc) and *Renilla* luciferase (RLuc), respectively. The positions of the synthesized proteins are indicated on the side. A hyphen (–) indicates "not added." The concentration of the siRNA indicated above each lane refers to the final concentration in the translation reaction mixture. (**c**) The S2 lysates were preincubated with 240 nM of the FLuc siRNA (final concentration 140 nM in the translation reaction mixture) or water (as a control) for 30 min. The FLuc and RLuc activities were measured, and their ratios (FLuc to RLuc) were normalized to a control. The data shown constitute an average of three independent experiments with standard deviations. Modified from ref. [4]

While cellular mRNAs are capped in nuclei, the mRNAs transcribed in vitro in a T7 RNA polymerase-based TnT system lack the cap. Some viral RNA sequences are known to improve the translational efficiency of the uncapped transcripts. We employed the sequence of the 5′-UTR derived from baculovirus p10 mRNA [8]. A TnT reaction using a plasmid encoding firefly luciferase, pT7p10-Luc-gb (*see* **Note 2**), is performed as described below:

1. For a typical 100 μL TnT reaction (*see* **Note 12**), mix 20 μL of TnT master mix with an appropriate volume of water, 20 ng/μL DNA, and 10 units/μL T7 RNA polymerase (TOYOBO).

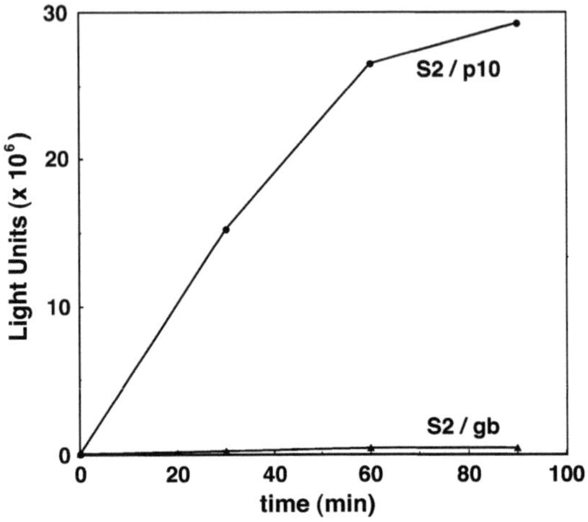

Fig. 5 The p10 5′-UTR enhances mRNA translation in the S2 TnT system. S2 lysates were programmed with 0.02 μg/μL of plasmids, pT7p10-Luc-gb (p10) and pSP36T7-Luc-gb (gb, ref. [9]) linearized with *SalI*. Translation reactions were performed at 27 °C, and aliquots were removed at the time points indicated. A 5 μL aliquot of the reaction mixture was diluted to 1 % with cell culture lysis reagent (Promega), and luciferase assays were performed. Modified from ref. [5]

2. Add 60 μL of S2 cell lysate to the abovementioned pre-mixture, and gently mix by pipetting.
3. Incubate the TnT reaction mixture at 27 °C.
4. For luciferase assays, withdraw a 5 μL aliquot at the appropriate time point.

The p10 5′-UTR has been shown to enhance the translation of uncapped mRNA in an insect cell lysate. The comparison of the 5′-UTR sequences derived from the p10 mRNA (p10) and *Xenopus* beta-globin mRNA [9] is shown in Fig. 5. We have synthesized various proteins using the pT7p10 vector. For detection and purification, FLAG-tag is useful (*see* **Note 13**).

4 Notes

1. D.Mel2 cells (Invitrogen) were used previously, which are adapted to *Drosophila* SFM medium (Invitrogen). However, Invitrogen no longer provides this cell line and *Drosophila* SFM.
2. We constructed the pT7p10-Luc-gb and pT7p10-RLuc-gb [4]. These plasmids contain the 5′-untranslated leader sequence from the baculovirus p10 mRNA and the 3′-untranslated sequence from *Xenopus* beta-globin mRNA.

3. When using a 500 mL Erlenmeyer flask, dispense 100 mL of culture medium.
4. We use an automatic cell viability analyzer (Vi-CELL XR, Beckman Coulter) in counting cell numbers. At this cell density the viability of cells should be better than 95 %.
5. Break up cell clumps by careful pipetting.
6. While the inner volume of the cell disruption bomb 4639 is 45 mL, a good working volume of the cell suspension is in the range of 5–10 mL.
7. Several minutes after setting the initial pressurization, the pressure of nitrogen gas will drop slightly. Readjust the pressure back to 100 psi.
8. Retrieve cell lysate from the valve indicated in Fig. 1. Cell disruption occurs as the homogenate is released to atmospheric pressure through the discharge valve. Discharge at a speed that allows you to count the drops.
9. Take a fraction from the clear layer. Centrifuge again if the lipid layer is contaminated.
10. If the OD value at 260 nm is around 50, check nitrogen gas leakage. Also, suspect cell clumps in the suspension (*see* **Note 5**). The nitrogen gas pressure can be increased up to 200 psi.
11. In the *Drosophila* S2 cell-free system, a 5′-terminal cap structure enhances translation 10–20 times. Polyadenylation to the capped mRNA synergistically stimulates translation. In our experiences, capped and polyadenylated mRNA is more than five times efficient than mRNA with cap only.
12. Magnesium concentration is critical. In a TnT reaction, we use higher magnesium concentration than that of mRNA-based translation reaction. This is because T7 RNA polymerase requires magnesium ions.
13. Several strong nonspecific bands are seen when using anti-FLAG M2 monoclonal antibody.

Acknowledgments

We would like to thank Yoko Kaitsu for excellent technical assistance. This work was supported by the RIKEN Structural Genomics/Proteomics Initiative (RSGI), National Project on Protein Structural and Functional Analysis, Targeted Proteins Research Program (TPRP), from the Ministry of Education, Culture, Sports, Science and Technology (MEXT), Japan.

References

1. Tuschl T, Zamore PD, Lehmann R et al (1999) Targeted mRNA degradation by double-stranded RNA in vitro. Genes Dev 13: 3191–3197
2. Elbashir SM, Lendeckel W, Tuschl T (2001) RNA interference is mediated by 21- and 22-nucleotide RNAs. Genes Dev 15:188–200
3. Elbashir SM, Harboth J, Lendeckel W et al (2001) Duplexes of 21-nucleotide RNAs mediate RNA interference in cultured mammalian cells. Nature 411:494–498
4. Wakiyama M, Kaitsu Y, Yokoyama S (2006) Cell-free translation system from Drosophila S2 cells that recapitulates RNAi. Biochem Biophys Res Commun 343:1067–1071
5. Wakiyama M, Kaitsu Y, Matsumoto T et al (2010) Coupled transcription and translation from polymerase chain reaction-amplified DNA in Drosophila Schneider 2 cell-free system. Anal Biochem 400:142–144
6. Schneider I (1972) Cell lines derived from late embryonic stages of *Drosophila melanogaster*. J Embryol Exp Morph 27:353–365
7. Gingras AC, Raught B, Sonenberg N (1999) eIF4 initiation factors: effectors of mRNA recruitment to ribosomes and regulators of translation. Annu Rev Biochem 68:913–963
8. Scheper GC, Vries RG, Broere M et al (1997) Translational properties of the untranslated regions of the p10 messenger RNA of Autographa californica multicapsid nucleopolyhedrovirus. J Gen Virol 78:687–696
9. Wakiyama M, Takimoto K, Ohara O et al (2007) Let-7 microRNA-mediated mRNA deadenylation and translational repression in a mammalian cell-free system. Genes Dev 21:1857–1862

Chapter 18

Cell-Free Membrane Protein Expression

Tomomi Kimura-Soyema, Mikako Shirouzu, and Shigeyuki Yokoyama

Abstract

Cell-free protein synthesis is advantageous for membrane protein production because of the potential lethality of membrane proteins synthesized in cell-based systems. Additionally the ability to adapt the membrane composition for each membrane protein increases the likelihood of functional expression. In this chapter we describe methods for membrane protein production by the cell-free protein synthesis system assisted by detergents and lipids.

Key words Membrane protein, Cell-free protein synthesis, Detergent, Liposome, Dialysis

1 Introduction

Membrane proteins play vital roles in living organisms such as energy generation, nutrient transport, and signal transduction through the cell membrane. Currently, 50 % of all known drugs are targeted to membrane proteins, including G-protein-coupled receptors [1]. Furthermore, the transport function of membrane proteins can be used to convey pharmaceutical compounds from outside to inside the cell. To accelerate rational drug development, structural and functional studies of membrane proteins are therefore quite valuable. However, high-level membrane protein expression is sometimes difficult by conventional in vivo expression systems using heterologous cells, because of the diversity in the lipid composition and/or membrane insertion system. This represents a major bottleneck that has delayed biochemical and structural studies of membrane proteins in comparison with those of soluble proteins.

Cell-free protein synthesis systems have various advantages in membrane protein expression, such as that a PCR product can be used as a template without subcloning into a vector and that the components of the system are easy to modify with the necessary additives [2]. Membrane proteins can be synthesized in the presence of appropriate detergents and in the natural lipid environments [3].

Kirill Alexandrov and Wayne A. Johnston (eds.), *Cell-Free Protein Synthesis: Methods and Protocols*, Methods in Molecular Biology, vol. 1118, DOI 10.1007/978-1-62703-782-2_18, © Springer Science+Business Media, LLC 2014

The inclusion of inverted membrane vesicles or microsomal fractions in the synthesis was also reported, but resulted in low yield [4].

On the other hand, artificial liposomes added to the protein synthesis reaction can provide a suitable lipid environment containing the proper lipid content, adjustable for each membrane protein. Moreover, relatively pure membrane proteins can be obtained using cell-free expression in the presence of detergents and lipids, because the membrane fraction should be predominated by the target protein [5].

In this chapter, we describe our recently developed method for cell-free membrane protein synthesis in the presence of detergents and lipids [5, 6]. In this method the membrane proteins are synthesized in parallel with the formation of a lipid bilayer, thus limiting their misfolding.

2 Materials

Prepare all solutions using sterilized ultrapure water and analytical grade reagents.

1. QIAGEN Plasmid Plus Maxi Kit (QIAGEN: 12965).
2. 1.6 M $Mg(OAc)_2$.
3. LMCPY mixture: 0.75 g/L tyrosine, 160 mM HEPES-KOH (pH 7.5), 10.70 % PEG8000, 534 mM K-glutamate, 5 mM DTT, 1.07 mM ATP, 2.40 mM each of ATP, GTP, CTP, UTP, 96 mg/L folic acid, 1.78 mM cAMP, 74 mM ammonium acetate, 214 mM creatine phosphate (*see* **Notes 1** and **2**).
4. 20 mM each of 19 amino acids, without tyrosine (*see* **Note 1**).
5. 3.75 mg/mL creatine kinase.
6. 17.5 mg/mL tRNA (*E. coli* MRE600-derived, Roche Applied Sciences, cat.109550).
7. *E. coli* S30 extract (OD_{260} ~200) in S30 buffer (60 mM KOAc, 10 mM Tris-OAc, pH 8.2, 16 mM $Mg(OAc)_2$, 1 mM DTT) (*see* **Note 1**).
8. S30 buffer (60 mM KOAc, 10 mM Tris-OAc, pH 8.2, 16 mM $Mg(OAc)_2$, 1 mM DTT) (*see* **Note 1**).
9. T7 RNA polymerase, 10 mg/mL.
10. NaN_3 stock solution, 5 % in water.
11. Digitonin (Wako Pure Chemical Industries, Ltd., Japan), 15 % stock in water.
12. Brij-78 (Sigma-Aldrich, P4019), 10 % stock in water.
13. Sodium cholate (Nacalai Tesque, Japan), 20 % stock in water.
14. Egg yolk phosphatidylcholine (Avanti Polar Lipids, Inc. 840051C, >99 % purity), 100 mg/mL stock suspension in water (*see* **Note 3**).

15. Dialysis tube: Spectra/Por7 (MWCO, 15k; Sealing Width, 23 mm, SPECTRUM LABORATORIES).
16. 15 mL conical centrifuge tubes (IWAKI).
17. Rotary Shaker.

3 Methods

Perform all procedures at room temperature, unless otherwise specified.

3.1 Preparation of Template DNA for E. coli Cell-Free Expression

1. Prepare the PCR reaction mixture, including the target protein-encoding plasmid and primers containing the linker sequences and the 5′- and 3′-terminal sequences of the target [2]. Perform the PCR reaction using a programmable PCR instrument.
2. Prepare the second PCR reaction mixture, including the five-fold diluted first PCR product, and the N- and C-terminal double-stranded fragments including promoter and terminator sequences, respectively [2]. Perform the PCR reaction using a programmable PCR instrument (*see* **Note 4**).
3. Purify the second PCR fragment by fractionating it on an agarose gel, excising the DNA band with the expected length, and using a desalting spin column.
4. Clone the second PCR fragment into pCR2.1-TOPO (Invitrogen) or another suitable cloning vector (*see* **Note 5**).
5. Amplify and purify the plasmid using a Plasmid Plus Maxi Kit, according to the manufacturer's instructions. Concentrate the plasmid solution to 1–3 μg/μL, if necessary.

3.2 Cell-Free Protein Synthesis Reaction in the Presence of Detergent Micelles

This is a typical protocol, using 0.9 mL of reaction mixture and 9 mL of feeding solution. The reaction scale can be adjusted for smaller or larger scale purposes such as expression screening or large-scale purification, respectively. The volume of the feeding solution should be at least ten times larger than that of the reaction mixture.

1. Mix the feeding solution reagents listed in Table 1 and the detergent (typically 1 % Brij-78 or 0.1–1 % digitonin, *see* **Note 6**). Place the mixture in a 15 mL conical tube.
2. Mix the reaction mixture reagents ($Mg(OAc)_2$, LMCPY, amino acid mixture, creatine kinase, tRNA and NaN_3) listed in Table 1. To this mixture, sequentially add T7 RNA polymerase, detergent, and the template plasmid, and gently mix by pipetting thoroughly (*see* **Note 7**).
3. Tie one end of the dialysis tube (Spectra/Por 7, MWCO = 15,000), place the reaction mixture in the tube, and

Table 1
Contents of the reaction mixture and the feeding solution

Reaction mixture, 900 μL	Feeding solution, 9 mL
Template plasmid 3.6 μg	–
10 mM $Mg(OAc)_2$	10 mM $Mg(OAc)_2$
LMCPY 336 μL	LMCPY 3.36 mL
1.5 mM 19 amino acids without tyrosine	1.5 mM 19 amino acids without tyrosine
0.25 mg/mL creatine kinase	–
0.175 mg/mL tRNA	–
S30 extract (OD_{260}, ~200) 270 μL	S30 buffer 2.7 mL
66.7 μg/mL T7 RNA polymerase	–
0.05 % NaN_3	0.05 % NaN_3
water up to 900 μL	water up to 9 mL
Detergent >2× cmc With or without lipid 6 mg	With or without detergent >2× cmc

Each stock solution was prepared as indicated in Subheading 2
cmc critical micelle concentration

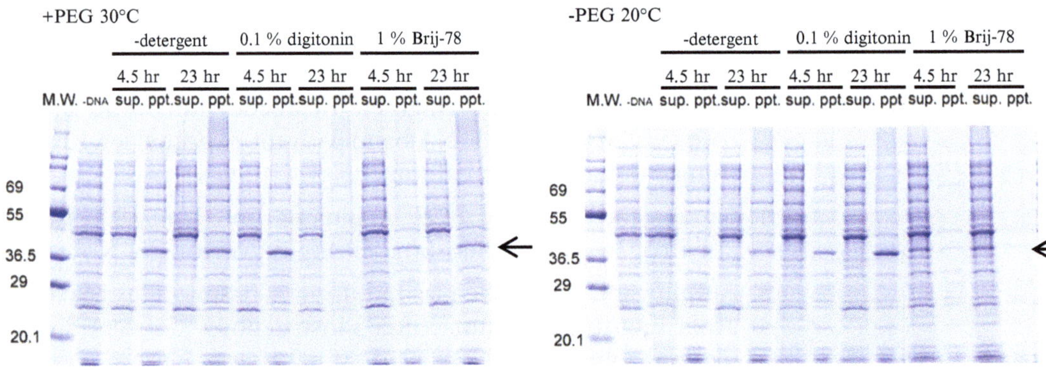

Fig. 1 Cell-free synthesis of a human chemokine receptor in the presence of detergents. Proteins were synthesized in the presence of the indicated concentrations of detergents and 4 μg/mL of plasmid DNA, in 30 μL reaction mixtures for the durations (hours) indicated above the lanes. Each lane of the 12.5 % SDS-polyacrylamide gel contains 0.5 μL of either the supernatant (sup.) or the precipitate (ppt.) of the reaction mixture. The precipitate was suspended in the same volume as the reaction mixture. – DNA represents the supernatant of reaction mixture without template DNA. *Arrows* indicate the molecular weight of the chemokine receptor. In this case, the soluble expression was higher in the absence of PEG at 20 °C than in the presence of PEG at 30 °C in 1 % Brij-78, whereas 0.1 % digitonin does not seem to be suitable for the soluble expression of this protein

tie the other end. Submerge the dialysis tube into the 15 mL tube containing the feeding solution (**step 1**). Shake (480 rpm) the 15 mL tube for 5–6 h at 20, 25, or 30 °C (*see* **Note 8** and Fig. 1).

4. Transfer the reaction mixture from the dialysis tube into microtubes, and save a small portion of it for electrophoresis. Centrifuge at 17,400 ×*g* for 30 min, and collect the supernatant. The remaining precipitate should be suspended in the

same volume of an appropriate buffer for the estimation by SDS-PAGE. The target membrane protein should be present in the supernatant.

3.3 Cell-Free Protein Synthesis Reaction in the Presence of Detergents and Lipids

1. Prepare 10–20 % (w/v) detergent solution and 100 mg/mL lipid suspension in water (*see* **Note 3**). Combine 45 μL of 20 % sodium cholate or 24 μL of 15 % digitonin with 60 μL of 100 mg/mL egg yolk phosphatidyl choline (egg PC) or other lipids, to obtain a final concentration of either 1 % sodium cholate or 0.4 % digitonin with 6.7 mg/mL (final) of egg PC in the reaction mixture, and then incubate for 1 h at room temp.
2. Mix the feeding solution reagents listed in Table 1. Do not add detergents or lipids. Put the mixture into a 15 mL conical tube.
3. Mix the reaction mixture reagents ($Mg(OAc)_2$, LMCPY, amino acid mixture, creatine kinase, tRNA and NaN_3) listed in Table 1. To this mixture, sequentially add T7 RNA polymerase, and either 105 μL (sodium cholate) or 84 μL (digitonin) of the mixed detergent/lipid micelles prepared in **step 1**, and the template plasmid, and gently mix by pipetting thoroughly (*see* **Note 7**).
4. Tie one end of the dialysis tube (the same as above), place the reaction mixture in the tube, and then tie the other end. Submerge the dialysis tube in the 15 mL tube containing the feeding solution (**step 2**). Shake (480 rpm) the 15 mL tube for 5–6 h at 30 °C.
5. Transfer the reaction mixture into microtubes, and save a small portion of it for SDS-PAGE analysis. Centrifuge at 87,000 × *g* for 30 min by ultracentrifugation, and then save the supernatant for electrophoresis. Suspend the precipitate by pipetting in 900 μL of 50 mM Tris–HCl (pH 6.8) and 10 mM EDTA, and then repeat the centrifugation. Repeat the membrane wash step twice with 50 mM Tris–HCl (pH 6.8) and 400 mM NaCl. Resuspend the final precipitate with a Dounce homogenizer in 900 μL of 50 mM Tris–HCl (pH 6.8) and 400 mM NaCl. The target membrane protein should be present in this membrane fraction (Fig. 2).

4 Notes

1. This method requires the *E. coli* S30 extract, which can either be produced in house or be obtained from a commercial kit. We use the extract prepared from *E. coli* BL21 bearing the pMINOR plasmid [7] as described previously [8]. If using a commercial *E. coli* cell-free synthesis kit (such as the Remarkable Yield Translation System Kit (Protein Express)), follow the instructions of the manufacture.

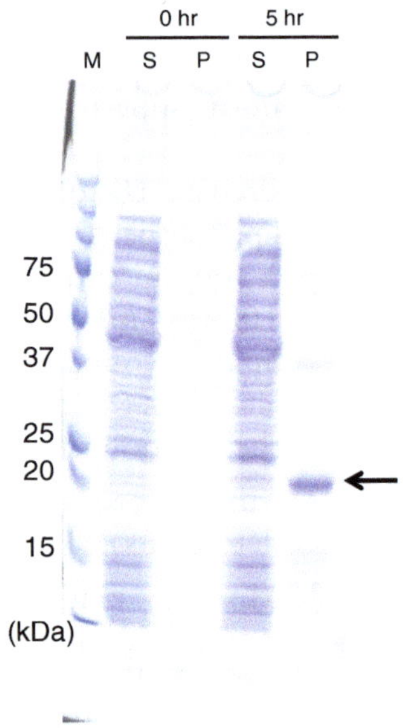

Fig. 2 Cell-free synthesis of *Acetabularia* rhodopsin in the presence of detergents and lipids. Proteins were synthesized in the presence of 0.4 % digitonin, 6.7 mg/mL egg PC, and 100 μM all-*trans* retinal, in a 9 mL reaction mixture for 5 h. The synthesized proteins were collected in the precipitate fraction as the main band. After solubilization with DDM and purification, the protein was crystallized in the lipidic mesophase and the 3D structure was solved [6]

2. PEG8000 can be excluded when the production levels were enough without PEG8000. It is recommended to check the production levels in the presence and absence of PEG8000.
3. The lipids can be chosen to suit the target protein. For example, cholesterol can be mixed with the egg PC. Natural lipids, such as a brain polar lipid extract, can also be used. Lipid(s) should first be solubilized in chloroform and then dried completely by using a rotary evaporator. Then add sterilized water to the dried lipid film, and suspend using an ultrasonic bath sonicator and a vortex mixer. For pipetting the lipid suspension, use low-binding tips (e.g., Maximum-Recovery Tip (BIO-BIK)). When using more than two kinds of lipids, mix them at the chloroform solubilization stage. If each lipid suspension is prepared in the aqueous solvent and then combined, the lipids will not completely mix with each other.
4. Encode tag sequences and/or protease recognition sequences into PCR fragments, if necessary. We usually use a modified HAT tag at the N-terminus which was observed to improve expression [2].
5. This step is not necessary if you perform the expression with a PCR fragment as the template for translation. Plasmids bearing T7 promoter sequences, such as pET series vectors, can be

used. However, we usually use the pCR2.1 vector which appears more suitable for efficient membrane protein expression.

6. Other suitable detergents include Brij-35, Brij-58, and dodecylmaltoside (DDM) [9], as they do not inhibit the transcription and translation reactions.
7. Since the synthesis reaction will be started by adding the DNA, perform the next step as quickly as possible after this step. The concentration of DNA can be adjusted for maximum expression. Never allow the dialysis tube to dry, as the efficiency of dialysis will decrease.
8. Due to the presence of detergent in the reaction mixture, the synthesized membrane proteins should be produced as solubilized proteins, However, the synthesized proteins may precipitate in some cases, for instance, when the temperature of the reaction is too high, although protein synthesis levels tend to increase at higher temperatures. Therefore, screening for optimal temperature among 20, 25, and 30 °C is recommended.

Acknowledgements

We thank Dr. Noboru Ohsawa for comments on the manuscript. This work was supported by the RIKEN Structural Genomics/ Proteomics Initiative (RSGI), the National Project on Protein Structural and Functional Analyses, and the Targeted Proteins Research Program (TPRP) of the Ministry of Education, Culture, Sports, Science and Technology of Japan.

References

1. Drews J (2000) Drug discovery: a historical perspective. Science 287:1960–1964
2. Yabuki T, Motoda Y, Hanada K et al (2007) A robust two-step PCR method of template DNA production for high-throughput cell-free protein synthesis. J Struct Funct Genomics 8:173–191
3. Schneider B, Junge F, Shirokov VA et al (2010) Membrane protein expression in cell-free systems. Methods Mol Biol 601:165–186
4. Lyford LK, Rosenberg RL (1999) Cell-free expression and functional reconstitution of homo-oligomeric alpha7 nicotinic acetylcholine receptors into planar lipid bilayers. J Biol Chem 274:25675–25681
5. Shimono K, Goto M, Kikukawa T et al (2009) Production of functional bacteriorhodopsin by an *Escherichia coli* cell-free protein synthesis system supplemented with steroid detergent and lipid. Protein Sci 18:2160–2171
6. Wada T, Shimono K, Kikukawa T et al (2011) Crystal structure of the eukaryotic light-driven proton-pumping rhodopsin, Acetabularia rhodopsin II, from marine alga. J Mol Biol 411: 986–998
7. Chumpolkulwong N, Sakamoto K, Hayashi A et al (2006) Translation of "rare" codons in a cell-free protein synthesis system from *Escherichia coli*. J Struct Funct Genomics 7: 31–36
8. Kigawa T, Yabuki T, Matsuda N et al (2004) Preparation of *Escherichia coli* cell extract for highly productive cell-free protein expression. J Struct Funct Genomics 5:63–68
9. Ishihara G, Goto M, Saeki M et al (2005) Expression of G protein coupled receptors in a cell-free translational system using detergents and thioredoxin-fusion vectors. Protein Expr Purif 41:27–37

Chapter 19

The PURE System for Protein Production

Yoshihiro Shimizu, Yutetsu Kuruma, Takashi Kanamori, and Takuya Ueda

Abstract

In the field of molecular biology or biochemistry, preparation and use of purified proteins involved in a certain biological system is crucial for understanding their mechanisms and functions in cells or organisms. The recent progress in a cell-free translation system allows us to prepare proteins in a test tube directly from cDNAs that encode the amino acid sequences. The use of the reconstituted cell-free translation system termed PURE (Protein synthesis Using Recombinant Elements) for these purposes is effective in several applications. Here we describe methods of recombinant protein expression using the PURE system for molecular biological or biochemical studies.

Key words Cell-free protein synthesis system, In vitro translation system, PURE system, Ribosome, tRNA

1 Introduction

The preparation of recombinant proteins is one of the key processes in molecular biological or biochemical research. Structural determination of proteins by using X-ray crystallography, cryo-electron microscopy, or nuclear magnetic resonance (NMR) is effective for visualization of molecular mechanisms of the biological systems. The reconstitution of the biological systems from their components is beneficial for understanding their mechanisms and functions, both qualitatively and quantitatively.

A cell-free protein synthesis system is a useful technique for such protein production [1]. The system not only allows us to synthesize cytotoxic or unstable proteins that cannot be expressed in living cells but also offers several advantages such as labeling with isotopes at specific positions, site-specific incorporation of unnatural amino acids [2], and simplified purification processes. The usefulness is further amplified by the fact that gene synthesis technologies have undergone remarkable development enabling access to the custom-designed templates [3].

Kirill Alexandrov and Wayne A. Johnston (eds.), *Cell-Free Protein Synthesis: Methods and Protocols*, Methods in Molecular Biology, vol. 1118, DOI 10.1007/978-1-62703-782-2_19, © Springer Science+Business Media, LLC 2014

Combination of these technologies enables generation of proteins directly from the sequence information even when lacking access to physical genome.

A reconstituted cell-free protein synthesis system entitled PURE (Protein synthesis Using Recombinant Elements) is a unique cell-free protein synthesis format, which consists of purified factors and enzymes of *E. coli* that constitute the minimal protein synthetic machinery [4]. Compared with the *E. coli* S30 system (the general format that utilizes the cell extract called as the S30 fraction), the PURE system possesses higher controllability because all of the components are defined. Reducing the presence of nucleases or proteases is advantageous for RNA/protein complex formation, the production of the proteins or peptides prone to degradation, and development of screening assays [5]. By reorganizing the components of the system such as the release factors and tRNAs, reconfiguration of the genetic code in the system for the incorporation of unnatural amino acid is achievable [4, 6].

In this report, we describe how to use the PURE system for the protein production. We show the methods for the preparation of a variety of proteins using the PURE system. The affinity purification of produced products using Strep-tag II, the synthesis of proteins with disulfide bonds, and the protein complex formation in the membrane are described.

2 Materials

2.1 Affinity Purification of the Strep-Tag II Fusion Protein Synthesized in the PURE System

1. Plasmid vector pUC19 (Takara Bio).
2. Restriction enzyme *Nde*I.
3. Restriction enzyme *Bam*HI.
4. T7 forward primer: 5′-GGGCC<u>TAATACGACTCACTATA</u> G-3′. The underlined represents a T7 promoter sequence.
5. pUC19-specific reverse primer: 5′-ATGACATTAACCTATAAA AATAGG-3′.
6. *Taq* DNA polymerase.
7. DNA purification kit such as Wizard® SV Gel and PCR Clean-Up System (Promega).
8. The PURE system is commercially available from several companies (BioComber, New England BioLabs, and GeneFrontier) (*see* **Note 1**).
9. Strep-Tactin® Sepharose® (IBA).
10. Ultrafree®-MC Durapore®-PVDF 0.45 μm (Millipore).
11. Tube rotator RT-50 (Taitec).
12. Wash buffer: 100 mM Tris, pH 8.0, and 150 mM NaCl (*see* **Note 2**).

13. Elution buffer: 100 mM Tris, pH 8.0, 150 mM NaCl, and 2.5 mM D-desthiobiotin (IBA).
14. SYPRO Orange protein gel stain (Bio-Rad).
15. 7.5 % acetic acid.
16. Typhoon FLA 7000 (GE Healthcare).

2.2 Synthesis of Proteins with Disulfide Bonds in the PURE System

1. T7 forward primer: 5′-GGGCCTAATACGACTCACTATAG-3′. The underlined represents a T7 promoter sequence.
2. Gene-specific reverse primer.
3. KOD-plus-DNA polymerase (Toyobo (Japan)).
4. Wizard® SV Gel and PCR Clean-Up System (Promega).
5. PURE*frex*™ *SS* (GeneFrontier (Japan)).

2.3 Affinity Purification of the Membrane Protein Complex

1. F_1F_o-deficient *E. coli* strain, the strain DK8 [*bglR*, *thi-1*, *rel-1*, HfrP01, Δ(*uncB-uncC*), *ilv*::Tn*10*] [7] is used here.
2. 2× YT medium.
3. 100 mg/mL ampicillin.
4. Complete protease inhibitor cocktail, EDTA-free (Roche).
5. French Press: high-pressure homogenizer (EmulsiFlex-C5, Avestin).
6. Teflon homogenizer for 200 mL volume.
7. Ultrasonic water bath (Elma).
8. Ultrasonicator tip type.
9. HPLC.
10. Anion-exchange chromatography: ResourceQ (GE Healthcare).
11. Ni-Sepharose™ 6 Fast Flow (GE Healthcare).
12. Empty Column (Bio-Rad).
13. BCA protein assay kit (Pierce).
14. Amicon Ultra 50 K (Millipore).
15. Himac CP80WX (Hitachi Koki).
16. S55A rotor (Hitachi Koki).
17. PURE*frex*™ (GeneFrontier (Japan)).
18. MagneHis Protein Purification system (Promega).
19. anti-F_o-*a* subunit antibody.
20. PA3 buffer: 10 mM Hepes-KOH, pH 7.5, 5 mM $MgCl_2$, and 10 % glycerol.
21. M-buffer: 20 mM KPi, pH 7.5, and 100 mM KCl.
22. Wash buffer: M-buffer supplemented with 20 mM imidazole and 0.15 % DM (decyl-maltoside).
23. Elution buffer: M-buffer supplemented with 200 mM imidazole and 0.15 % DM.

24. A-buffer: 10 mM Hepes-KOH, pH 7.5, 0.2 mM EDTA, and 0.15 % DM.
25. B-buffer: A-buffer containing 1 M Na_2SO_4.
26. Lysis buffer: 50 mM Tris–KCl, pH 8.0, 300 mM NaCl, and 5 mM $MgCl_2$.

3 Methods

3.1 Affinity Purification of the Strep-Tag II Fusion Protein Synthesized in the PURE System

1. The vector construction for the synthesis of Strep-tag II fusion protein in the PURE system is shown in Fig. 1a. Clone the gene encoding a desired protein into this vector by using the restriction enzymes *Nde*I and *Bam*HI (*see* **Note 3**).
2. Using a forward primer containing a T7 promoter sequence and a reverse primer complementary to the sequence downstream of the T7 terminator, amplify the gene encoding the target protein by PCR (*see* **Notes 4** and **5**).
3. By using the Wizard® SV Gel and PCR Clean-Up System, purify the amplified gene according to the manufacture's protocol and dilute it to 0.1 pmol/μL (*see* **Note 6**).

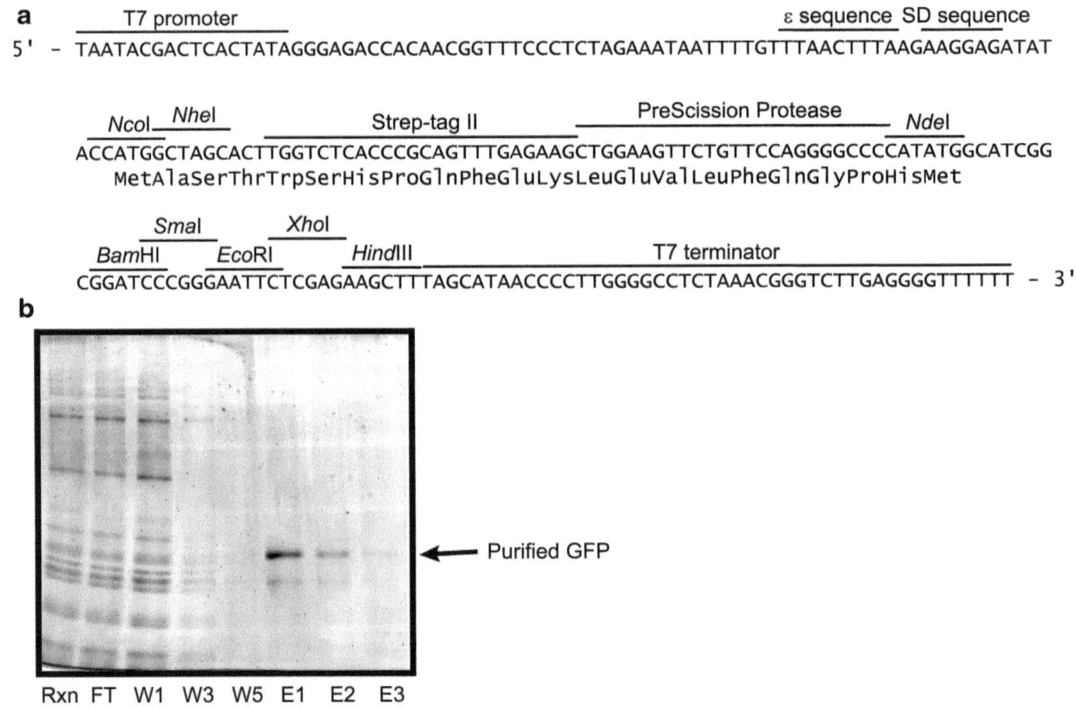

Fig. 1 Affinity purification of the Strep-tag II fusion protein. (**a**) Vector construction for the synthesis of Strep-tag II fusion protein in the PURE system. (**b**) Affinity purification of the Strep-tag II fusion GFP. Reaction mixtures (Rxn), a flow-through fraction (FT), wash fractions (W1, W2, W3), and elute fractions (E1, E2, E3) are shown. An *arrow* indicates the purified GFP

4. Synthesize the protein by mixing 20 μL amplified gene (2 pmol) with the PURE system to make 500 μL reaction mixture (*see* **Note 7**). Multiple aliquots can be set up. Incubate it at 37 °C for 2 h. Store 5 μL of the reaction mixtures for SDS-PAGE analysis.
5. Add 200 μL Strep-Tactin® Sepharose® (100 μL resin) to the Ultrafree®-MC Durapore®-PVDF 0.45 μm filtration tube and centrifuge at 1,500 × *g* for 1 min to remove the resin storage buffer.
6. Equilibrate the column twice by adding of 200 μL wash buffer to the resin and centrifuging it at 1,500 × *g* for 1 min.
7. Apply all of the reaction mixtures to the column and gently mix the content of the column by using a tube rotator for 20 min at 4 °C to allow the synthesized Strep-tag II fusion proteins to bind to the resin. Centrifuge the tube at 1,500 × *g* for 1 min. Collect the eluate as a flow-through fraction (FT).
8. Wash the column five times by adding 100 μL of wash buffer to the resin and centrifuging it at 1,500 × *g* for 1 min. Collect each eluate as wash fractions (W1–W5).
9. Add three times 100 μL elution buffer to the resin and centrifuge at 1,500 × *g* for 1 min to elute the synthesized Strep-tag II fusion proteins. Collect each eluate as elute fractions (E1–E3).
10. Subject the reaction mixtures FT, W1, W3, W5, E1, E2, and E3 to the SDS-PAGE. Stain the gel with SYPRO Orange protein gel stain, wash with 7.5 % acetic acid, and visualize with Typhoon FLA 7000 (Fig. 1b) (*see* **Note 8**).

3.2 Synthesis of Proteins with Disulfide Bonds in the PURE System

1. Amplify the gene encoding a desired protein, which has disulfide bonds by two-step PCR (*see* **Notes 3–5**).
2. By using the Wizard® SV Gel and PCR Clean-Up System, purify the amplified gene according to the manufacture's protocol and dilute it to 0.1 pmol/μL (*see* **Note 6**).
3. Synthesize the protein. Mix 1 μL of the amplified gene solution (0.1 pmol) with PURE*frex*™ *SS*, which is designed for the synthesis of the protein with disulfide bonds (*see* **Note 9**) to make 25 μL reaction mixtures. Incubate it at 37 °C for 2–4 h.
4. Analyze or use the synthesized products. For example, mix 2.5 μL of the reaction mixtures containing the synthesized protein with an appropriate chromogenic substrate and monitor the change of absorbance (Fig. 2).

3.3 Affinity Purification of the Membrane Protein Complex

1. Construct the plasmid which expresses *Bacillus* PS3 F_1F_o without F_o-*a* subunit ($F_1F_o\Delta a$) under the control of *E. coli* promoter-based systems. The plasmid pTR-ISBS2Δ*a* [8] is used in this example.

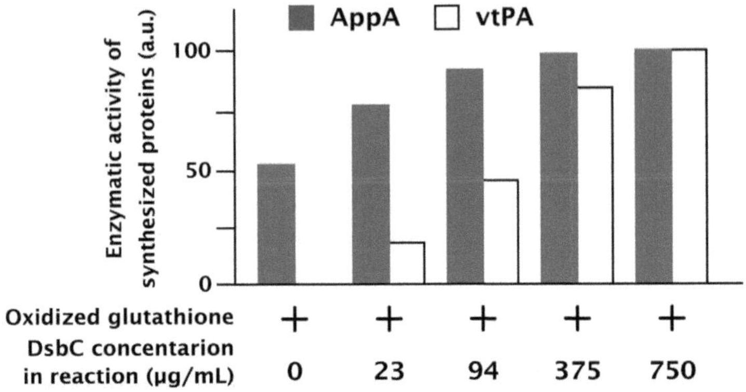

Fig. 2 The synthesis of truncated version of tissue plasminogen activator (vtPA) and AppA (*E. coli* acid phosphatase) as model proteins with multiple disulfide bonds. The gene encoding vtPA and AppA were translated in the PURE system supplemented with the indicated concentration of DsbC at 37 °C for 4 h. After the synthesis of vtPA or AppA, their enzymatic activities were measured with their substrate. The activity was compared as relative value, in which the maximum activity was set to 100

2. Transform an *E. coli* DK8 cells with the obtained plasmid (*see* **Note 10**).
3. Cultivate the transformed cells in 12 L of 2× YT medium containing 100 μg/mL ampicillin at 37 °C for 20 h. Harvest cells by centrifugation (*see* **Note 11**).
4. Suspend the 30 g cells in 210 mL of PA3 buffer and disrupt them by French Press (12–15 MPa, twice). After removing the cell debris, precipitate the membrane vesicles by ultracentrifugation at 160,000 ×*g* for 1 h at 4 °C.
5. Suspend the precipitated membrane in 210 mL of PA3 buffer using Teflon homogenizer.
6. Precipitate again the membrane vesicles by ultracentrifugation at 160,000 ×*g* for 45 min at 4 °C.
7. Resuspend the precipitated membrane vesicles in 30 mL of PA3 buffer, and use as inverted membrane vesicle (IMV) sample (*see* **Note 12**).
8. Add 1/500 v/v of beta-mercaptoethanol (32 μL to 16 mL) to IMVs.
9. Precipitate the IMVs by ultracentrifugation at 153,000 ×*g* for 15 min at 4 °C.
10. Solubilize the precipitated IMVs using Teflon homogenizer in 53 mL PA3 buffer containing 2 % Triton X-100, 0.5 % cholate, and 1/100 volume of inhibitor cocktail, and incubate for 30 min at 30 °C with gentle shaking.
11. Ultracentrifuge at 153,000 ×*g* for 20 min at 30 °C.

12. Mix the supernatant (53 mL) with M-buffer (280 mL) containing 20 mM imidazole/HCl (pH 7.5) and 1/100 volume of inhibitor cocktail.
13. Subsequently, mix the resulting sample with 16 mL of Ni-Sepharose FF previously equilibrated with wash buffer, and incubate for 30 min on ice with gentle mixing.
14. Pour the solution into an empty column, and wash with the 10-column volume of wash buffer.
15. Elute with 200 mL elution buffer and fractionate the elution (10 mL/tube × 20 tubes).
16. Check the concentration by BCA protein assay kit to identify the fractions of F_1F_o.
17. Collect the fractions containing F_1F_o to proceed to anion-exchange chromatography by HPLC.
18. Dilute the eluted fraction threefold with A-buffer and apply to ResourceQ (6 mL) at 6 mL/min.
19. Elute by a liner gradient of Na_2SO_4 with B-buffer, i.e., 0–400 mM Na_2SO_4 for 40 min at flow rate 4 mL/min.
20. Collect two large peaks appeared around 150 mM Na_2SO_4, respectively (*see* **Note 13**).
21. Concentrate the eluted fraction by Amicon centrifugal concentrator (50 K) to a final volume of ≤500 μL.
22. Aliquot the solution and frozen by liquid N_2 before storing at −80 °C.
23. Suspend PC lipid powder in PA3 buffer at 44 mg/mL, and stir for 30 min at room temperature (*see* **Note 14**).
24. Sonicate by tip sonicator for 40 s avoiding sample heating.
25. Mix in the purified F_1F_o at protein concentration of 1 mg/mL.
26. Repeat two times freeze and thaw by liquid N_2.
27. Process a bath sonication briefly, and use as proteoliposomes (PLs) containing F_1F_o.
28. Prepare 50 μL reaction mixture of PURE system containing PLs at 0.12 mg/mL membrane proteins (*see* **Note 15**).
29. Synthesize protein by inserting the gene encoding F_o-*a* subunit downstream of T7 promoter (*see* **Notes 3–5**) for 30 min (*see* **Note 16**) at 30 °C.
30. Dilute the mixture with 100 μL lysis buffer containing 1 % DDM (*n*-dodecyl-β-D-maltoside) and place on ice for 30 min.
31. Mix with 30 μL magnet beads pre-equilibrated with lysis buffer.
32. Incubate for 1 h at room temperature with gentle mixing.
33. Bind the magnet beads to an equipped magnet according to the manufacture's protocol, and remove supernatant.

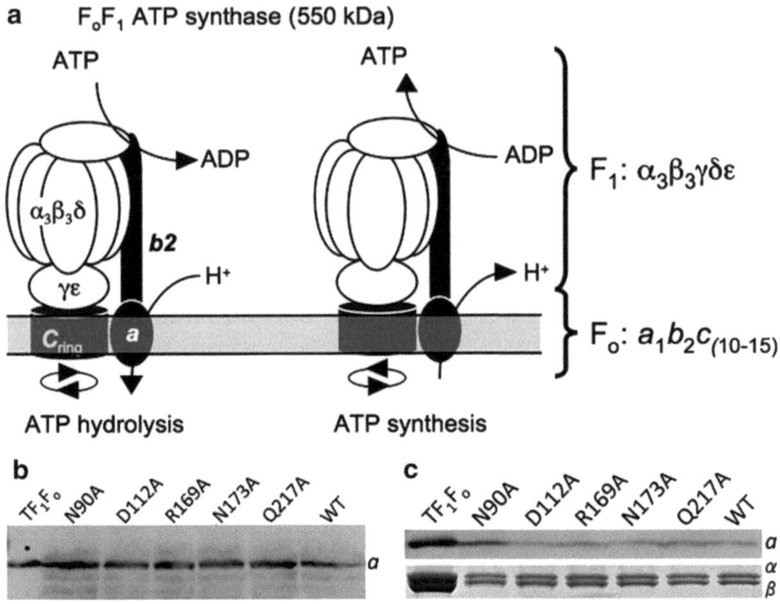

Fig. 3 The complex formation of F_1F_0-ATP synthase by the synthesis of F_0-*a* subunit in the presence of F_0-*a* subunit-less F_1F_0 complex ($F_1F_0\Delta a$) in the membrane. (**a**) Schematics of F_1F_0-ATP synthase (F_1F_0). F_1F_0 complex is composed of the soluble complex F_1 ($\alpha_3\beta_3\gamma\delta\varepsilon$) and the membrane complex F_0 ($a_1b_2c_{10-15}$). The complex catalyzes ATP synthesis by the H^+ gradient or H^+ translocation by the ATP hydrolysis. (**b**) Immunoblotting analysis of the synthesized F_0-*a* subunit in the PURE system, where a series of mutant F_0-*a* subunits were analyzed. The F_0-*a* subunits were synthesized in the presence of PLs-$F_1F_0\Delta a$ and processed by SDS-PAGE to analyze with anti-*a* antibody. (**c**) Immunoblot analyses of the purified F_0-*a*. The synthesized F_0-*a* subunits are co-purified through ten histidine-tag at the β subunit (*upper panel*) as a complete F_1F_0 complex. The lower panel shows that the equal amounts of complexes were purified for the mutants and the wild-type (WT) version of proteins by Ni-charged magnet beads. TF1Fo indicates a control of F_1F_0 purified from cell membrane. The positions of F_0-*a*, F_1-α, and β subunit are indicated beside the gels (**a–c** were reproduced with slight modifications from refs. [11, 8] with permission of Springer and Portland Press, respectively)

34. Wash twice with 150 μL lysis buffer containing 20 mM imidazole and 0.05 % DDM.
35. Elute twice with 150 μL lysis buffer containing 500 mM imidazole and 0.05 % DDM.
36. Add 1 mL acetone to 300 μL (150 μL × 2) eluted fraction and place on ice for 30 min.
37. Centrifuge at 20,000 × *g* for 30 min at 4 °C.
38. Dissolve the precipitant in SDS sample buffer and analyzed by SDS-PAGE.
39. Proceed to immunoblotting analysis using anti-*a* subunit antibody in order to detect incorporation of the synthesized F_o-*a* subunit into the F_1F_o complex (Fig. 3b, c) (*see* **Note 17**).

4 Notes

1. You can prepare the system yourself by purifying all of the his-tagged PURE system components and the ribosome fraction [9].
2. Unless otherwise specified, all solutions should be prepared in water that has a resistivity of 18.2 MΩ cm.
3. The DNA should at least encode T7 promoter, Shine-Dalgarno sequence, and the open reading frame (ORF) of the protein with start and stop codons. Strep-tag II sequence is necessary for the affinity purification. An epsilon sequence originated from bacteriophage T7 [10] enhances protein synthesis.
4. Two-step PCR, in which the ORF of the target protein is amplified by the first-step PCR and regulatory sequences on 5′ untranslated region (UTR) are attached by the second PCR, is also available for the preparation of the DNA template for the protein synthesis [9].
5. The attachment of the T7 terminator sequence makes the transcribed mRNA more stable because the terminator sequence forms a stem and loop structure.
6. Other kits for the purification of the PCR product can also be used.
7. For example, if you use WakoPURE kit, mix 250 μL solution A, 100 μL solution B, 130 μL water, and 20 μL amplified gene in a reaction tube. Please follow the manufacture's protocol.
8. Any other equipment that can detect the fluorescence can also be used.
9. The PURE system for the protein synthesis with disulfide bonds is constructed from the PURE system without reducing agents but with 3 mM oxidized glutathione and *Escherichia coli* DsbC. The optimum concentration of DsbC depends on the characteristics of the target protein (Fig. 2).
10. The complex formation and stoichiometry of F_1F_o-ATP synthase is described in Fig. 3a. The sub-complex formation by the co-synthesis of F_o-*c* subunit and UncI proteins in the lipid bilayer and F_1 sub-complex formation by the co-synthesis of α, β, γ, and ε subunit were described previously [11].
11. Typically ≈30 g of cells are harvested from 12 L culture.
12. IMVs (inverted membrane vesicles) are used for the subsequent steps for F1Fo purification.
13. Both contain F_1F_o complex, but the latter has higher activity.
14. The PC lipid should be pretreated by washing with acetone and ether.
15. Use a water volume of PURE system mixture to add the PL sample.

16. The reaction time of 30 min is sufficient to synthesize F_o-*a*.
17. Since ten histidine-tag is fused at the N-terminus of β subunit of F_1F_o complex, purification can be performed by the affinity purification using Ni^{2+}-charged magnetic beads. Using the purified F_1F_o sample, the incorporation of the in vitro-synthesized F_o-*a* into the F1Fo final complex can be detected by anti-*a* antibody.

Acknowledgements

We are thankful to the RIKEN Center for Developmental Biology (CDB) for the use of bio-imaging analyzer. This work was supported by the RIKEN Quantitative Biology Center (QBiC) (Y.S.).

References

1. Shimizu Y, Kuruma Y, Ying BW et al (2006) Cell-free translation systems for protein engineering. FEBS J 273:4133–4140
2. Hartman MC, Josephson K, Lin CW et al (2007) An expanded set of amino acid analogs for the ribosomal translation of unnatural peptides. PLoS One 2:e972
3. Xiong AS, Peng RH, Zhuang J et al (2008) Chemical gene synthesis: strategies, softwares, error corrections, and applications. FEMS Microbiol Rev 32:522–540
4. Shimizu Y, Inoue A, Tomari Y et al (2001) Cell-free translation reconstituted with purified components. Nat Biotechnol 19: 751–755
5. Ohashi H, Shimizu Y, Ying BW et al (2007) Efficient protein selection based on ribosome display system with purified components. Biochem Biophys Res Commun 352: 270–276
6. Murakami H, Ohta A, Ashigai H et al (2006) A highly flexible tRNA acylation method for non-natural polypeptide synthesis. Nat Methods 3:357–359
7. Klionsky DJ, Brusilow WS, Simoni RD (1984) In vivo evidence for the role of the epsilon subunit as an inhibitor of the proton-translocating ATPase of *Escherichia coli*. J Bacteriol 160: 1055–1060
8. Kuruma Y, Suzuki T, Ono S et al (2012) Functional analysis of membranous F_o-*a* subunit of F_1F_o-ATP synthase by in vitro protein synthesis. Biochem J 442:631–638
9. Shimizu Y, Ueda T (2010) PURE technology. Methods Mol Biol 607:11–21
10. Olins PO, Rangwala SH (1989) A novel sequence element derived from bacteriophage T7 mRNA acts as an enhancer of translation of the *lacZ* gene in *Escherichia coli*. J Biol Chem 264:16973–16976
11. Kuruma Y, Suzuki T, Ueda T (2010) Production of multi-subunit complexes on liposome through an *E. coli* cell-free expression system. Methods Mol Biol 607:161–171

Chapter 20

A Cell-Free Protein Synthesis System from Insect Cells

Toru Ezure, Takashi Suzuki, and Eiji Ando

Abstract

The newly developed Transdirect in vitro translation system for mRNA templates utilizes an extract from cultured *Spodoptera frugiperda* 21 (Sf21) insect cells. An expression vector, pTD1, which includes a 5′-untranslated region (UTR) sequence from a baculovirus polyhedrin gene as a translational enhancer, designed to obtain maximum performance from the insect cell-free protein synthesis system. The combination of insect cell extract and the expression vector results in protein productivity of about 50 μg/mL of the translation reaction mixture. This is the highest protein productivity yet recorded among commercialized cell-free protein synthesis systems based on animal extracts.

Key words Translation, *Spodoptera frugiperda* 21, Cell-free protein synthesis system, pTD1 vector, Insect cell extract

1 Introduction

Most of the recombinant proteins produced in a baculovirus expression system have been shown to be functionally similar to authentic proteins, because the insect cells can carry out many types of posttranslational modifications. Therefore, we developed a cell-free protein synthesis system from *Spodoptera frugiperda* 21 (Sf21) insect cells, which are widely used as the host for baculovirus expression systems, and commercialized it under the brand name of Transdirect *insect cell*.

We have demonstrated that this insect cell-free protein synthesis system is one of the most effective protein synthesis systems among those based on animal extracts [1]. Furthermore, it has the potential to perform eukaryote-specific protein modifications such as protein *N*-myristoylation [2], prenylation [3], disulfide formation [4], ubiquitination [5], acquisition of metal ions [6], and formation of protein complexes [7]. In addition, core glycosylation and cleavages of signal peptides were observed after the addition of microsomal membranes to the reaction mixture (unpublished data). Thus, we expect that the insect cell-free protein synthesis system

Kirill Alexandrov and Wayne A. Johnston (eds.), *Cell-Free Protein Synthesis: Methods and Protocols*, Methods in Molecular Biology, vol. 1118, DOI 10.1007/978-1-62703-782-2_20, © Springer Science+Business Media, LLC 2014

will find application for post-genomic studies. In this chapter, we describe the standard protocols to synthesize proteins of interest using the insect cell-free protein synthesis system.

2 Materials

2.1 Construction of an Expression Clone

1. Primers for amplification of the target cDNA (*see* Subheading 3.1): Store at –20 °C.
2. KOD-plus DNA polymerase (TOYOBO, Kyoto, Japan): Store at –20 °C (*see* **Note 1**).
3. Agarose for electrophoresis: Store at room temperature.
4. 50× TAE buffer: Mix 242 g of Tris base, 57.1 mL of glacial acetic acid, and 100 mL of 0.5 M EDTA (pH 8.0), and adjust to 1,000 mL with water. Store at room temperature.
5. 1 kb DNA Ladder (BIONEER, Korea): Store at –20 °C.
6. Ethidium bromide: Store at 4 °C.
7. Phenol/chloroform/isoamyl alcohol (25:24:1) saturated with 10 mM Tris, pH 8.0, and 1 mM EDTA (Sigma, St. Louis, MO): Store at 4 °C.
8. Chloroform: Store at room temperature.
9. 3 M sodium acetate, pH 5.2: Store at room temperature.
10. Ethanol: Store at room temperature.
11. 70 % (v/v) ethanol: Store at room temperature.
12. T4 polynucleotide kinase (TOYOBO): Store at –20 °C.
13. Restriction endonucleases (*Eco*RV, *Eco*RI, *Sac*I, *Kpn*I, *Bam*HI, and *Xba*I): Store at –20 °C.
14. MinElute PCR Purification Kit (QIAGEN, Maryland, USA): Store at room temperature.
15. pTD1 vector, a component of the Transdirect *insect cell* kit (Shimadzu, Kyoto, Japan): Store at –20 °C or below. The map of the pTD1 vector is shown in Fig. 1.
16. Quick Ligation™ Kit (NEW ENGLAND BioLabs, Ipswich, MA): Store at –20 °C.
17. Chemically competent cells *Escherichia coli* DH5α (TAKARA Bio, Shiga, Japan): Store at –80 °C.
18. Ampicillin sodium salt (Sigma) is dissolved in distilled water at 100 mg/mL and stored at –20 °C.
19. Luria-Bertani (LB, 1.0 % polypeptone, 0.5 % yeast extract, 1.0 % NaCl, pH 7.0) medium containing 100 μg/mL ampicillin (LB-amp) and LB-amp agar medium (LB containing 1.0 % agar): Store at 4 °C.

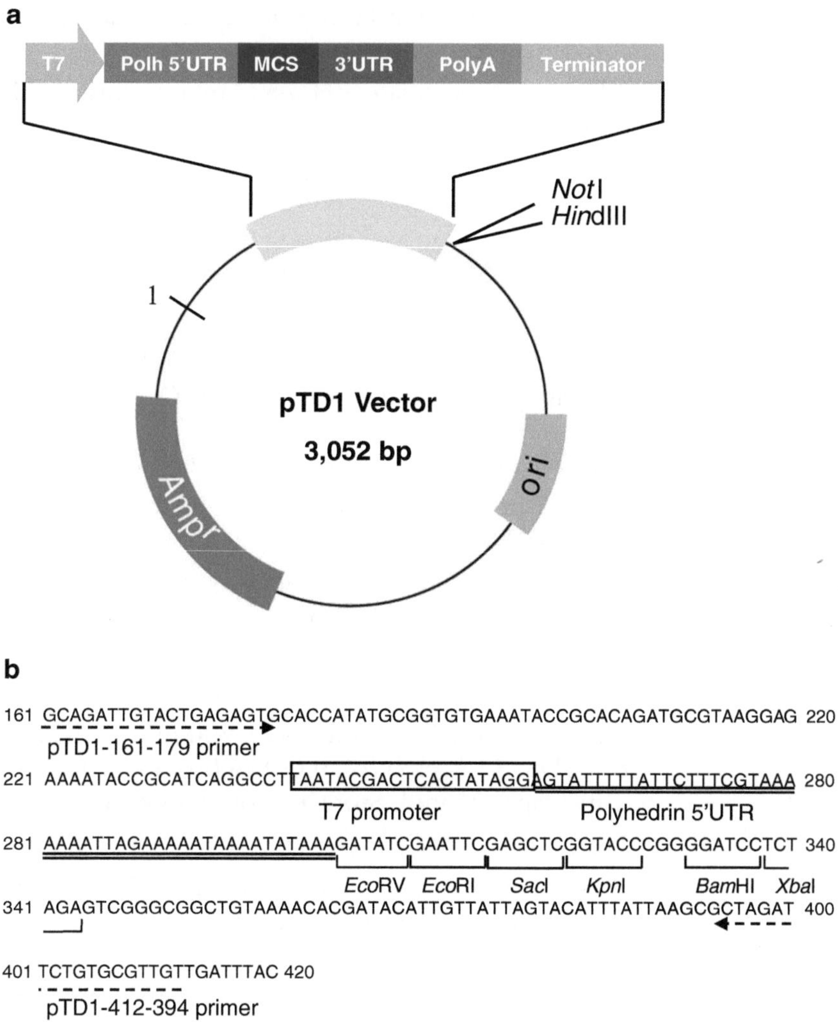

Fig. 1 The expression vector pTD1. (**a**) pTD1 vector map. (**b**) DNA sequence of the pTD1 vector around the multiple cloning sites

20. GenElute plasmid miniprep kit (Sigma): Store at room temperature. Resuspension solution should be stored at 4 °C after the addition of RNase A.
21. BigDye Terminator v3.1 Cycle Sequencing Kit (Applied Biosystems, Warrington, UK): Store at −20 °C.
22. Primers for DNA sequencing (*see* Subheading 3.1): Store at −20 °C.

2.2 Preparation of mRNA

1. Restriction endonucleases (*Hin*dIII and *Not*I) are stored at −20 °C.
2. T7 RiboMAX™ Express Large Scale RNA Production System (Promega, Madison, WI) (*see* **Note 2**): Store at −20 °C.

3. NICK Columns (GE Healthcare, Buckinghamshire, UK): Store at room temperature.
4. 3 M potassium acetate, pH 5.5 (Ambion, Austin, TX): Store at room temperature.
5. TE buffer (10 mM Tris–HCl, pH 8.0, 1 mM EDTA): Store at room temperature.
6. MOPS buffer (20×): 400 mM MOPS, 100 mM NaOAc, and 20 mM EDTA (adjust to pH 7.0 by NaOH). Store at 4 °C in dark conditions.
7. 37 % formaldehyde: Store at room temperature.
8. Deionized formamide: Store at −20 °C.

2.3 In Vitro Translation and Detection of Synthesized Proteins

1. Transdirect *insect cell* (Shimadzu): Store at −80 °C.
2. Purified mRNA (*see* Subheading 3.2 Preparation of mRNA): Store at −80 °C.
3. FluoroTect™ $\text{Green}_{\text{Lys}}$ in vitro Translation Labeling System (Promega): Store at −80 °C.
4. SDS-PAGE running buffer (10×): 250 mM Tris, 1.92 M glycine, and 1 % SDS. Store at room temperature.
5. SDS-PAGE loading buffer (4×): 200 mM Tris–HCl (pH 6.8), 8 % SDS, 32 % glycerol, 0.008 % bromophenol blue, and 8 % 2-mercaptoethanol. Store at room temperature.
6. Precast gel: c-PAGEL (ATTO, Tokyo, Japan). Store at 4 °C.
7. Prestained molecular weight markers: Full-Range RAINBOW (GE Healthcare). Store at −20 °C.

3 Methods

The following methods are designed to work with the Transdirect *insect cell* kit, an in vitro translation system for mRNA templates. We developed and optimized a method to prepare the insect cell extract, the concentrations of the reaction components, and an expression vector pTD1 [1, 8]. The pTD1 vector contains all factors involved in mRNA and protein synthesis, including the T7 promoter sequence required for mRNA synthesis, the polyhedrin 5′-UTR which enhances the translation reaction, and multiple cloning sites (MCS) (*see* Fig. 1). The complete DNA sequence of the pTD1 vector is registered in the following DNA databank: DDBJ/GenBank®/EMBL with the accession number AB194742.

To obtain maximal protein productivity, it is necessary to construct an expression clone in which a protein coding region (open reading frame, mature region, domain, etc.) is inserted into the MCS of the pTD1 vector. Typically, the expression of the target protein at about 35–50 μg/mL of the translation reaction mixture can be obtained by using mRNA transcribed from the expression clone and the Transdirect *insect cell* kit.

Although intended for use with the Transdirect *insect cell* kit, the expression clone described in this protocol can be effectively combined with other eukaryotic cell-free protein synthesis systems, such as rabbit reticulocyte lysate or wheat germ system (*see* **Note 3**).

3.1 Construction of the Expression Clone for the Insect Cell-Free Protein Synthesis System

1. Procedures for construction of the expression clone follow classical molecular cloning methods. The overall cloning strategy is shown in Fig. 2.
2. Design and synthesize two primers, an N-terminal primer and a C-terminal primer, for amplification of protein coding region of the target cDNAs. The N-terminal primer should have the

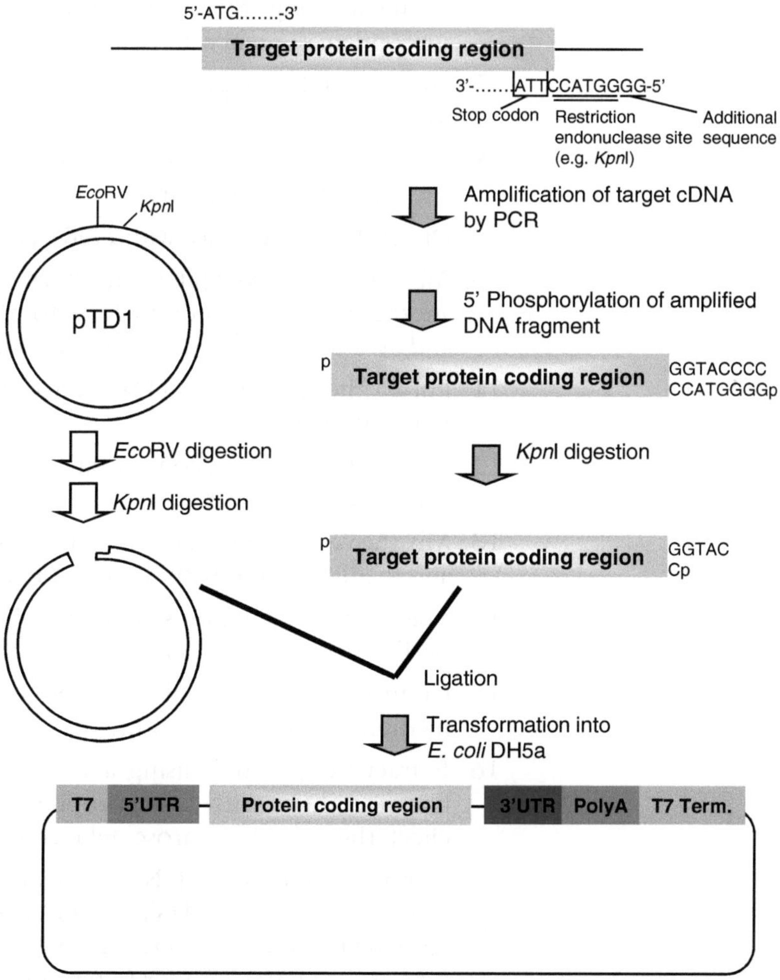

Fig. 2 Strategies for construction of an expression clone. This figure shows procedures to construct an expression clone when a *KpnI* site is introduced into a C-terminal primer

initiation codon at its 5′-terminus. In the case of the C-terminal primer, the restriction endonuclease recognition sequence should be introduced upstream of the stop codon, and an additional sequence (at least two bases) should be added to the 5′-end (*see* **Note 4**).

3. Perform PCR using these primers, KOD-plus, and cDNA of the target gene as the template and check the size of amplified DNA by gel electrophoresis.
4. Purify the amplified DNA fragment by phenol/chloroform extraction and ethanol precipitation.
5. Treat the purified DNA fragment by T4 polynucleotide kinase at 37 °C for 1 h and then purify the DNA fragment by ethanol precipitation.
6. Suspend the precipitate in distilled water, and digest the DNA fragment using the restriction enzyme that recognizes the appropriate sequence in the C-terminal primer at 37 °C for 2 h.
7. Purify the DNA fragment using the MinElute PCR Purification Kit.
8. Quantitate the DNA fragment by absorbance at 260 nm using a spectrophotometer, and use it as an insert.
9. Digest the pTD1 vector with *Eco*RV at 37 °C for 2 h (*see* **Note 5**).
10. After ethanol precipitation, digest the pTD1 vector with another restriction endonuclease to produce the same cohesive end as that introduced in the C-terminal primer.
11. Purify the digested pTD1 vector using a MinElute PCR Purification Kit.
12. Quantitate the DNA concentration using a spectrophotometer, and use it as the vector.
13. Mix the vector and the insert at a ratio of about 1:10 (mol:mol), and incubate them with T4 DNA ligase at 25 °C for 5 min.
14. Transform the ligation sample into *E. coli* DH5α and incubate on LB-amp agar plates at 37 °C overnight (*see* **Note 6**).
15. Cultivate single colonies of the transformants in LB-amp medium at 37 °C overnight.
16. Extract the plasmids using a GenElute plasmid miniprep kit, linearize them with an appropriate restriction enzyme, and check their size by agarose gel electrophoresis.
17. Confirm the plasmid DNA sequence using a pTD1-161-179 primer (5′-GCAGATTGTACTGAGAGTG-3′) for N-terminal sequencing and a pTD1-412-394 primer (5′-ACAACGCA CAGAATCTAGC-3′) for C-terminal sequencing. The annealing temperature of these primers is 50 °C (*see* **Notes 7** and **8**).

3.2 Preparation of mRNA

1. Linearize the expression clone using an appropriate restriction endonuclease downstream from the T7 terminator sequence (*see* **Notes 9** and **10**).
2. Purify the digested expression clone by phenol/chloroform extraction and ethanol precipitation (*see* **Note 11**).
3. Dissolve the pellet in sterilized distilled water and quantitate the DNA concentration by spectrophotometer, and then use as the template for mRNA synthesis (*see* **Note 12**).
4. Perform the in vitro transcription reaction using the T7 RiboMAX™ Express Large Scale RNA Production System (*see* **Note 13**) at 37 °C for 30 min (*see* **Note 14**). Use 5 μg of DNA template for 100 μL of transcription reaction. Typically, about 500 μg of purified mRNA is obtained from 100 μL of transcription reaction. This yield corresponds to about 1.5 mL of translation reaction.
5. After the incubation, the synthesized mRNA should be purified immediately using NICK™ Columns, which are gel filtration columns. This purification step is necessary to remove salts and unincorporated NTPs. We recommend performing this treatment in order to achieve stable and highly reproducible translation reactions.
6. Procedures to set up the NICK™ Columns are as follows. First, remove the column cap and pour off the excess liquid. Rinse the column with 3 mL of sterilized distilled water. Remove the bottom cap and place it in a column stand. Equilibrate the gel with 3 mL of sterilized distilled water and flush completely. These procedures should be carried out during the in vitro transcription reaction.
7. Apply 100 μL of the transcriptional reaction mixture on top of the gel, and flush completely. If the reaction scale of the in vitro transcription is less than 100 μL, fill up the reaction mixture to 100 μL with sterilized distilled water before applying to the column.
8. Add 400 μL of sterilized distilled water, and then flush completely. Discard the eluate.
9. Before elution of the mRNA fraction, place a new 1.5 mL tube under the column.
10. Add 400 μL of sterilized distilled water, and collect the eluate.
11. Add 40 μL of 3 M potassium acetate (*see* **Note 15**) and 950 μL of ethanol to the eluate. Mix thoroughly and centrifuge for 20 min at 15,000 rpm at 4 °C.
12. Discard the supernatant and then rinse the pellet with 70 % ethanol. Do not dry the pellet completely, so it will dissolve

mRNA in water easily. Dissolve the pellet in 100 μL of sterilized distilled water. If the reaction scale of the in vitro transcription is less than 100 μL, dissolve the pellet with sterilized distilled water in an equal volume of the in vitro transcription reaction.

13. After the purification, measure the absorbance at 260 nm of the purified mRNA solution using a spectrophotometer. Dilute 2 μL of the purified mRNA solution into 500 μL of TE buffer, and then measure the absorbance. Use TE as the blank. The mRNA concentration is determined by the following equation: mRNA concentration (mg/mL) = A_{260} value × 0.04 × 250.
14. About 3–6 mg/mL mRNA is usually obtained by the above method.
15. Confirm the purity and size of synthesized mRNA by gel electrophoresis as described in the following protocol steps.
16. Prepare a 1.0 % agarose gel in 1× TAE (*see* **Note 16**).
17. Mix 10 μL of 20× MOPS buffer, 30 μL of 37 % formaldehyde, and 80 μL of deionized formamide, and use this as an RNA sample buffer.
18. Mix 8 μg of the mRNA sample and 11 μL of the RNA sample buffer, and adjust to 20 μL with sterilized distilled water.
19. Treat the mRNA sample at 65 °C for 15 min, and immediately place it on ice.
20. Perform electrophoresis, and visualize the mRNA sample by ethidium bromide staining. If the RNA band is smeared or not visible, possible causes may be degradation of mRNA by RNase contamination.

3.3 In Vitro Translation

1. Kit components of the Transdirect *insect cell* kit are the insect cell extract (yellow cap) (*see* **Notes 17–19**), reaction buffer (blue cap), 4 mM methionine (red cap), pTD1 vector (green cap), and the control DNA (white cap).
2. Procedures to set up a translation reaction mixture should be carried out on ice.
3. Thaw the reaction buffer, 4 mM methionine, and insect cell extract. The reaction buffer and 4 mM methionine can be thawed at room temperature.
4. Assemble the reaction components (*see* Table 1) (*see* **Note 20**). Gently mix by pipetting up and down. If necessary, centrifuge briefly to return the sample to the bottom of the tube.
5. Incubate the translation reaction mixture at 25 °C for 5 h.
6. The protein productivity of the Transdirect *insect cell* kit is about 50 μg/mL of the translation reaction mixture (*see* **Note 21**).

Table 1
Reaction components

mRNA	16 μg
4 mM methionine	1 μL
Reaction buffer	15 μL
Insect cell extract	25 μL
Sterilized distilled water adjust to	50 μL

3.4 Detection of a Synthesized Protein

1. Generally, it is difficult to detect synthesized proteins by Coomassie Brilliant Blue staining. To confirm the expression of the target protein, we usually perform fluorescent labeling of the in vitro translation products using the FluoroTect™ $\text{Green}_{\text{Lys}}$ in vitro Translation Labeling System (FluoroTect) (*see* **Note 22**).
2. Add 1 μL of the FluoroTect solution to 50 μL of the translation reaction mixture described in Table 1, and incubate at 25 °C for 5 h.
3. After the translation reaction, add 2 μL of SDS-PAGE loading buffer (4×) to 6 μL of the reaction mixture. Incubate at 70 °C for 3 min.
4. Resolve the sample by SDS-PAGE.
5. Detect the fluorescent-labeled protein using a laser-based fluorescent scanner. An experimental example is shown in Fig. 3 (*see* **Note 23**).

4 Notes

1. Other high-fidelity DNA polymerases that do not have terminal transferase activity also can be used.
2. It has been confirmed that protein synthesis can be performed effectively using mRNA prepared using the following kits: AmpliScribe™ T7-*Flash*™ Transcription Kit (Epicentre), Ampli Scribe™ T7 Transcription Kit (Epicentre), CUGA® 7 in vitro Transcription Kit (Nippon Gene), MEGAscript® T7 High Yield Transcription Kit (Ambion), RiboMAX™ Large Scale RNA Production System-T7 (Promega), RNAMaxx™ High Yield Transcription Kit (Stratagene), and the ScriptMAX™ Thermo T7 Transcription Kit (TOYOBO).
3. The pTD1 vector should not be used as the expression vector for an *E. coli* cell-free protein synthesis system because this vector does not contain Shine–Dalgarno sequence.

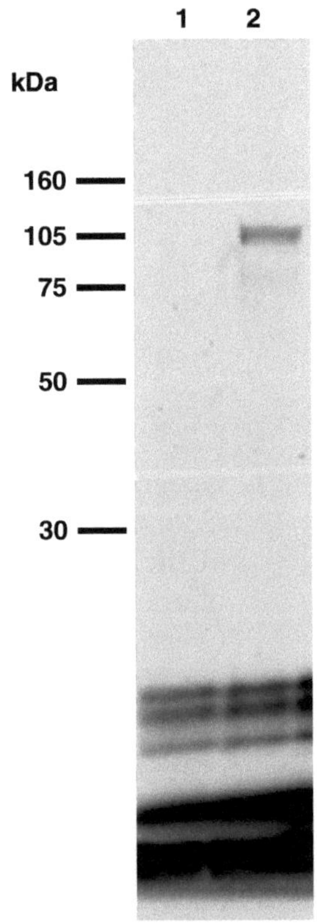

Fig. 3 Detection of a synthesized protein by fluorescent labeling. Cell-free protein synthesis was carried out with or without the use of mRNA transcribed from a linearized expression clone containing the β-galactosidase gene, and the synthesized protein was labeled by FluoroTect. The translational reaction mixtures were resolved on 12.5 % SDS-PAGE. Detection of labeled protein was performed using a laser-based fluorescent scanner (FX Pro, Bio-Rad, Hercules, CA). *Lanes 1* and *2* represent negative control (absence of mRNA) and β-galactosidase, respectively

4. We recommend introducing the *Kpn*I recognition sequence into the C-terminal primer if the target cDNA does not have a *Kpn*I site, because this strategy has been shown to have the highest cloning efficiency.
5. Generally, translation efficiency gradually decreases depending on the length between the initiating codon of the target cDNA and the polyhedrin 5′-untranslated region, which contains a translational enhancer sequence. To obtain the highest translation efficiency, the initiating codon of the target cDNA should be inserted into the *Eco*RV site of the pTD1 vector.
6. This system does not utilize blue–white selection.

7. Deletion mutation of the initiating codon (especially "A") has sometimes been observed when using this cloning strategy.
8. To clarify whether mutations occur during PCR, the overall nucleotide sequence of the insert DNA should be confirmed.
9. For linearization of expression clones, we recommend using *Hind*III or *Not*I. Restriction enzymes *Cfr* 10I, *Eco* 52I, *Eco* T14I, *Nde* I, *Pvu* II, *Sca* I, and *Stu* I may be used.
10. PCR-generated DNA templates can be used in transcription reactions. In such cases, the following primers are recommended: A pTD1-161-179 primer (5′-GCAGATTGTACT GAGAGTG-3′) and a pTD1-845-827 primer (5′-GGAAACA GCTATGACCATG-3′). Their annealing temperature is 50 °C.
11. This step is very important to avoid RNase contamination.
12. At least 125 μg/mL of the linearized DNA template is required for in vitro transcription reactions.
13. Before use, RiboMAX™ Express T7 2× Buffer must be dissolved completely by warming the buffer at 37 °C and mixing well.
14. The suggested incubation times should be adhered to. In the case of a long template (more than about 2 kb), an excessive reaction time may cause precipitation. In this case, mRNA will not be collected. To avoid this problem, (a) shorten the reaction time to 20 min, (b) decrease the quantity of DNA template to 70–80 %, and (c) use a PCR-generated DNA template.
15. Do not use sodium acetate for precipitation of synthesized mRNA. Sodium ion inhibits the translation reaction.
16. We usually use a non-denaturing agarose gel.
17. We confirmed that the insect cell extract is stable against freeze-thawing up to eight times. After use, the extract should be immediately stored at −80 °C.
18. Insect cell extract is sensitive to CO_2. After opening the package, avoid prolonged exposure to CO_2 (e.g., dry ice).
19. We confirmed that the insect cell extract has the ability to perform eukaryote-specific protein modifications, such as *N*-myristoylation [2] and prenylation [3]. To obtain such modified proteins effectively, specific substrates for each protein modification should be added to the translation reaction mixture.
20. The addition of RNase inhibitor (50 units) to the translation reaction mixture (50 μL) may improve translational efficiency (Recommended products: Promega Code No. N2611).
21. Generally, it is difficult to synthesize membrane proteins having multiple transmembrane domains.

22. Radioisotope labeling or Western blotting may also be used to detect synthesized proteins.
23. For detection of proteins having molecular masses less than 20 kDa, it is necessary to treat the translation reaction mixture with RNase A to degrade unincorporated FluoroTect tRNA migrating at about 20 kDa.

Acknowledgements

We wish to thank Dr. Osamu Ohara and Dr. Takahiro Nagase, Kazusa DNA Research Institute, for valuable discussions. We would also like to thank Dr. Sumiharu Nagaoka, Kyoto Institute of Technology; Dr. Toshihiko Utsumi, University of Yamaguchi; and Dr. Susumu Tsunasawa, Shimadzu Corporation, for helpful discussions.

References

1. Ezure T, Suzuki T, Higashide S et al (2006) Cell-free protein synthesis system prepared from insect cells by freeze-thawing. Biotechnol Prog 22:1570–1577
2. Suzuki T, Ito M, Ezure T et al (2006) N-Terminal protein modifications in an insect cell-free protein synthesis system and their identification by mass spectrometry. Proteomics 6:4486–4495
3. Suzuki T, Ito M, Ezure T et al (2007) Protein prenylation in an insect cell-free protein synthesis system and identification of products by mass spectrometry. Proteomics 7:1942–1950
4. Ezure T, Suzuki T, Shikata M et al (2007) Expression of proteins containing disulfide bonds in an insect cell-free system and confirmation of their arrangements by MALDI-TOF MS. Proteomics 7:4424–4434
5. Suzuki T, Ezure T, Ando E et al (2010) Preparation of ubiquitin-conjugated proteins using an insect cell-free protein synthesis system. J Biotechnol 145:73–78
6. Ezure T, Suzuki T, Ando E et al (2009) Expression of human Cu, Zn-superoxide dismutase in an insect cell-free system and its structural analysis by MALDI-TOF MS. J Biotechnol 144:287–292
7. Masuda T, Goto F, Yoshihara T et al (2007) Construction of homo- and heteropolymers of plant ferritin subunits using an in vitro protein expression system. Protein Expr Purif 56: 237–246
8. Suzuki T, Ito M, Ezure T et al (2006) Performance of expression vector, pTD1, in insect cell-free translation system. J Biosci Bioeng 102:69–71

Chapter 21

A Cell-Free Expression Platform for Production of Protein Microarrays

Xristo Zárate and David W. Galbraith

Abstract

Self-assembling protein microarrays have recently emerged as a particularly useful and flexible platform supporting high-throughput analyses of proteins and their interactions. They are produced by printing an array containing a tag-specific antibody. The array is then covered with a solution containing a cell-free expression (linked transcription–translation) system and DNA templates encoding tagged fusion proteins. The proteins synthesized in situ are immobilized by the capture antibody at each array element. An efficient cell-free protein expression system is therefore a critical component in the production of these arrays. Here we describe the methodology for the construction of autofluorescent protein microarrays, by transcription and translation of chimeric proteins containing a C-terminal green fluorescent protein (GFP) tag using different cell-free expression systems, based on either an *Escherichia coli* S30 extract, a wheat germ extract, or a hybrid system which combines both. The method is followed by a modified version that describes the use of fluorescence amplification as a means for detection of protein interactions. This protocol provides more sensitivity for protein detection on the arrays and allows the choice of different protein tags and/or fluorescent dyes.

Key words Protein microarrays, Cell-free protein expression, Green fluorescent protein, S30 extract, Wheat germ extract, Hybrid cell-free protein expression, Tyramide signal amplification

1 Introduction

Protein microarrays are attractive platforms for the development of high-throughput protein assays [1, 2]. Self-assembling protein microarrays are produced by applying a DNA template and a cell-free expression system, in a transcription–translation-coupled format, onto a tag-specific antibody microarray. The synthesized tagged proteins then are immobilized by the antibody [3, 4]. Having a microarray containing proteins tagged at the C-terminus with the green fluorescent protein (GFP) offers advantages: it allows protein immobilization by an easily available anti-GFP antibody and direct quantification of folded proteins by measuring the fluorescence intensity on each array element without the need for detection using exogenous fluorescent dyes [5].

Kirill Alexandrov and Wayne A. Johnston (eds.), *Cell-Free Protein Synthesis: Methods and Protocols*, Methods in Molecular Biology, vol. 1118, DOI 10.1007/978-1-62703-782-2_21, © Springer Science+Business Media, LLC 2014

Here, we present a full protocol for the production of autofluorescent protein microarrays. The arrays are printed on each substrate in an 8 × 3 format (giving a total of 24 subarrays per slide), which accommodates physically separated four slides inside a 24 × 4 chamber mimicking a 96-well plate. To produce proteins in situ, a cell-free expression solution containing the template DNA is added into each well, and after incubation, the GFP-tagged proteins are automatically immobilized at the locations of the array elements by the capture antibodies. For autofluorescent arrays, or any array for that matter, the choice of cell-free expression system depends on the origin of the proteins of interest and downstream applications envisioned after protein synthesis; here, three different cell-free expression systems are described. Using the *E. coli* S30 extract generally provides excellent protein yields, but multidomain eukaryotic proteins can display low yields as reflected by the fluorescence values of the C-terminal fluorescent protein. With the wheat germ extract, fluorescence from a greater proportion of the products of the input templates is observed, but overall levels are low since the wheat germ extract tends to produce low amounts of protein. The third system described combines the S30 and wheat germ extracts. It shows the higher fluorescence values, combined with higher proportions of signals from the products of the input templates. The hybrid system appears optimal for the production of microarrays for downstream applications requiring high protein yield such as arrays for mass spectrometric analyses [6].

A final alternative protocol is included, in which different arrays containing a much lower amount of capture antibody are printed, using the same 8 × 3 format. After cell-free expression using the wheat germ extract, and extra blocking steps, a second anti-GFP antibody coupled with horseradish peroxidase (HRP) is incubated on the array and tyramide signal amplification (TSA) is applied to label array elements with fluorescent dyes. This modified protocol provides more sensitivity, enables the possibility of using different protein tags, and allows utilization of the common Cy3 and Cy5 dyes for detection.

2 Materials

Prepare all solutions with ultrapure water and analytical grade reagents.

2.1 Components for Microarray Printing

1. CodeLink activated slides (SurModics, Eden Prairie, MN). Keep slides inside a desiccator at room temperature once the package has been opened. Print the slides before the expiration date indicated on the box.
2. 10× PBS: phosphate-buffered saline, 10× concentrate.

3. Stock phosphate buffer for printing: 500 mM sodium phosphate, pH 8.5. Dissolve 7.1 g of Na_2HPO_4 and 0.08 g of NaH_2PO_4 in 100 mL of water. Filter through a 0.2 μm membrane. Store at room temperature.
4. Capture antibody: mouse monoclonal anti-GFP antibody (Rockland Immunochemicals Inc., Gilbertsville, PA). It comes in 1× PBS buffer (*see* **Note 1**).
5. Bovine serum albumin (BSA) solution: 30 mg/mL BSA in 1× PBS. Mix 0.2 mL of 10× PBS in 1.8 mL of water. Add 60 mg of BSA and mix until dissolved. Store in 0.5 mL aliquots at −20 °C.
6. Saturated NaCl solution. Dissolve 40 g of NaCl per 100 mL of water, and heat if necessary.
7. Slide humidity chamber: any container with a tight-fitting lid is appropriate.
8. Slide blocking buffer: 100 mM Tris–HCl, 50 mM ethanolamine, pH 9.0. Add 800 mL of water to a glass beaker. Weigh 12.1 g of Tris base and transfer to the beaker, add 3 mL of ethanolamine, mix, and adjust the pH to 9.0 with HCl. Make up to 1 L with water and store at room temperature.
9. Coplin jars.
10. Slide spinner.

2.2 Components for Cell-Free Protein Expression

1. Blocking solution: StabilGuard Choice (SurModics, Eden Prairie, MN). Store at 4 °C.
2. ArraySlide 24-4 chamber (The Gel Company, San Francisco, CA).
3. Recombinant DNA: 1 μg of plasmid DNA dissolved in water for each cell-free reaction. The circular DNA should contain both a T7 RNA polymerase promoter and T7 transcription terminator elements. The vector pIVEX2.3d suggested in the *E. coli* cell-free expression kit works very well. Clone the gene for the protein of interest with an in-frame C-terminal GFP tag (*see* **Note 2**).
4. Nonidet P-40, 10 % solution. Nonidet P-40 (NP-40) is a nonionic, non-denaturing detergent. Store at 4 °C.
5. *Escherichia coli* S30 cell-free expression system: RTS 100 *E. coli* HY kit (5 PRIME, Gaithersburg, MD). The kit contains *E. coli* lysate (S30 extract), reconstitution buffer, amino acid mixture (minus methionine), reaction mix, and methionine. The freeze-dried components need to be reconstituted with the reconstitution buffer, following the manufacturer's instructions. Store lysate at −70 °C and the rest of the components at −20 °C (*see* **Note 3**).
6. T7 RNA polymerase. Store at −20 °C.

7. Wheat germ extract cell-free expression system: TNT T7 coupled wheat germ extract system (Promega, Madison, WI). It contains wheat germ extract, reaction buffer, T7-WG RNA polymerase, amino acid mixture (minus methionine), and amino acid mixture (minus leucine). Store all components at −70 °C (*see* **Note 4**).
8. RNasin ribonuclease inhibitor (Promega, Madison, WI). Store at −20 °C.
9. Thermal cycler.
10. 96-well plate cover foil.
11. HEPES wash buffer: 50 mM HEPES buffer, pH 7.8. Dissolve 11.9 g of HEPES in 800 mL of water; adjust the pH to 7.8 with NaOH. Make up to 1 L with water. Store at 4 °C.

2.3 Components for Array Signal Amplification

1. 10 % Tween-20: dissolve 1 mL of Tween-20 in 9 mL of water. Store at room temperature.
2. 10× TBS buffer. For 0.5 L dissolve 30.3 g Tris base and 43.8 g NaCl in 400 mL of water, adjust the pH to 7.8 with HCl, and bring volume up to 500 mL. 1× TBS will be 50 mM Tris–HCl, 150 mM NaCl, and pH 7.8. Store at room temperature.
3. Blocking buffer A: 5 % milk and 0.1 % Tween-20 in 1× PBS. For 100 mL dissolve 5 g of nonfat dry milk in 80 mL of water and add 10 mL of 10× PBS and 1 mL of 10 % Tween-20. Bring volume to 0.1 L. Use fresh.
4. Blocking buffer B: 1 % milk and 0.1 % Tween-20 in 1× TBS. For 100 mL dissolve 1 g of nonfat dry milk in 89 mL of water and add 10 mL of 10× TBS and 1 mL of 10 % Tween-20. Use fresh.
5. Anti-GFP–HRP-labeled antibody. Store at −20 °C.
6. TBS-Tween (TBST) buffer. For 100 mL, mix 89 mL of water with 10 mL of 10× TBS and 1 mL of 10 % Tween-20. Store at room temperature.
7. TSA plus cyanine 5 system or cyanine 3 if preferred (PerkinElmer, Waltham, MA). The kit comes with TSA reagent and dilution buffer. Store both at 4 °C.

2.4 Microarray Analysis

Microarray scanner.

3 Methods

3.1 Microarray Printing

1. Prepare anti-GFP antibody solution for printing. For 100 μL solution, mix 15 μL of stock phosphate buffer for printing, anti-GFP antibody, and water if necessary (based on the concentration that the anti-GFP antibody comes in) to have

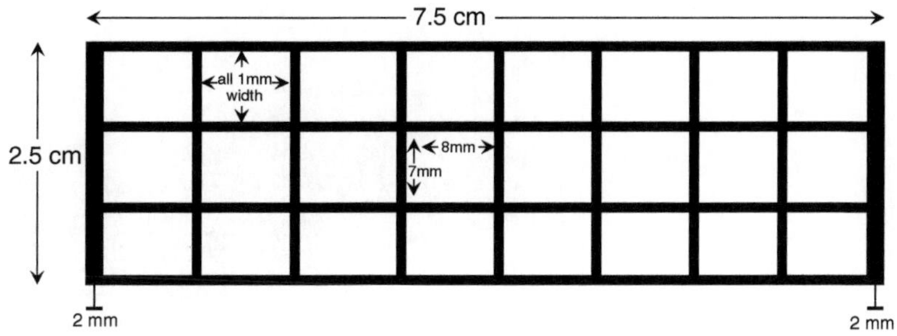

Fig. 1 Footprint for microarray printing. Up to 24 wells per slide can be printed. Each subarray should not cover more than 8 × 7 mm as shown in order to fit inside a well once the 24-4 chamber is assembled. The gasket will occupy the 1 and 2 mm spaces in order to separate each subarray physically

0.1 mL of 1.5 mg/mL antibody in 75 mM phosphate buffer, pH 8.5.

2. Program microarray printer and print slides with the layout shown in Fig. 1 (*see* **Note 5**).
3. After printing, place the slides into a slide storage box. Put the uncovered box on a rack inside the humidity chamber that contains the saturated NaCl solution at the bottom. Seal chamber and incubate overnight (16–20 h).
4. Place printed slides in a Coplin jar and add slide blocking buffer. Incubate (on a shaker) at room temperature for 45 min to 1 h (*see* **Note 6**).
5. Discard slide blocking buffer and rinse slides thoroughly with water.
6. Centrifuge slides to dryness using the slide spinner; a few seconds are enough.
7. Store slides inside a desiccator at 4 °C until needed.

3.2 Cell-Free Expression

1. Block slides by incubating in blocking solution, 20 mL per slide at room temperature for 30 min.
2. Rinse slides with water and spin-dry.
3. Insert slides into ArraySlide 24-4 chamber (shown in Fig. 2) and secure them with the screws; assemble the whole chamber (*see* **Note 7**).
4. For each *Escherichia coli* S30 cell-free expression reaction, combine in a 1.5 mL tube 12 μL lysate, 10 μL substrate mix, 12 μL amino acid mixture, 1 μL methionine, 5 μL reconstitution buffer, 0.5 μL (5 units) of T7 RNA polymerase, 0.5 μL of 10 % NP-40, and 1 μg of plasmid DNA, and water up to a final volume of 60 μL (*see* **Note 8**). Mix reaction thoroughly and add to the wells. Cover chamber with a 96-well plate cover foil and place it inside the thermocycler (*see* **Note 9**).

Fig. 2 Photograph of the ArraySlide 24-4 chamber. The 96-well gasket clamps tightly onto the slide holder, creating 96 sealed wells over four microarray slides

Program the machine to run 3 h at 24 °C and 30 min at 8 °C. *See* Fig. 3 for a microarray image using the S30 extract.

5. For wheat germ-based protein expression, combine in a 1.5 mL tube 25 μL wheat germ extract, 2 μL reaction buffer, 1 μL T7-WG RNA polymerase, 0.5 μL of amino acid mixture (minus methionine), 0.5 μL of amino acid mixture (minus leucine), 1 μL (40 units) of recombinant RNasin ribonuclease inhibitor, 0.5 μL 10 % NP-40, and plasmid DNA (1 μg), add water up to a total volume of 60 μL. Cover chamber with a 96-well plate cover foil and place it inside the thermocycler. Program the machine to run 90 min at 30 °C, followed by 1 h at 15 °C.
6. For hybrid cell-free expression reaction, include the following reagents from the RTS 100 *E. coli* HY kit—6 μL of *E. coli* S30 extract, 5 μL of substrate mix, 6 μL amino acid mixture (except methionine), 0.5 μL of methionine, and 2.5 μL of reconstitution buffer—and from the wheat germ TNT Promega kit: 12.5 μL wheat germ extract, 1 μL reaction buffer, 0.25 μL amino acid mixture (minus methionine), 0.25 μL amino acid mixture (minus leucine), and 0.5 μL T7-WG RNA polymerase; also add 0.25 μL T7 RNA polymerase (Fermentas), 0.5 μL RNasin, 0.5 μL 10 % NP-40, and 14.25 μL of water. 10 μL of circular DNA (1 μg total) was added for a final reaction volume of 60 μL. Cover chamber with a 96-well plate cover foil and place it inside the thermocycler. Program the machine to run at 24 °C for 3 h, 8 °C for 30 min, and 15 min at 4 °C.

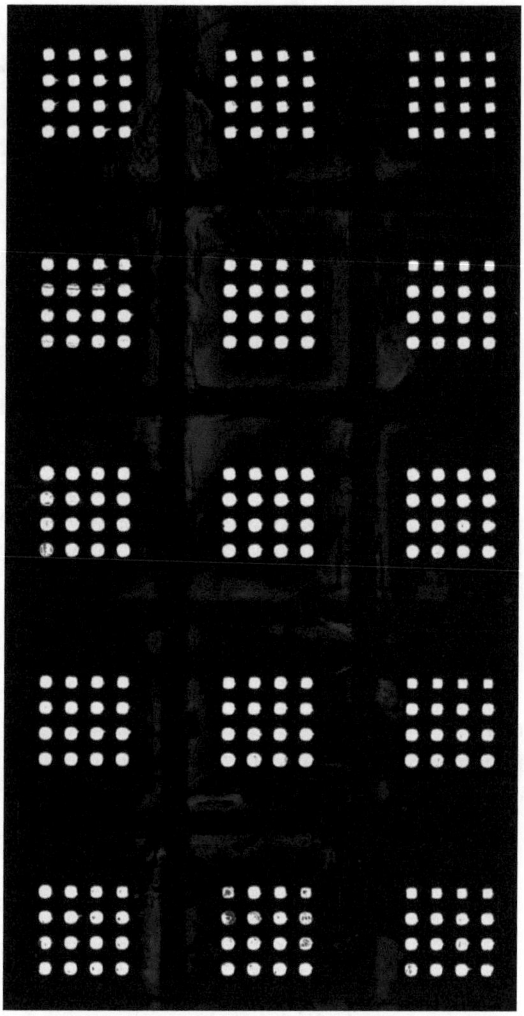

Fig. 3 Production of autofluorescent protein microarrays. 15 GFP-tagged proteins were expressed using *E. coli* S30 extract. The microarray was imaged using the MDS GenePix autoloader 4200 AL

See Fig. 4 for a microarray image that compares the expression levels using the S30 extract and hybrid systems.

7. After any of the cell-free expression reactions mentioned above, and the cold treatment, remove reaction mixtures directly from the frame wells (*see* **Note 10**).
8. Rinse each subarray well with 100 μL ice-cold HEPES wash buffer on a shaker for 5 min.
9. Remove wash buffer with a pipette.
10. Repeat the wash two more times.
11. Remove slides from chamber and rinse them with cold water inside a Coplin jar three times.
12. Spin-dry slides.

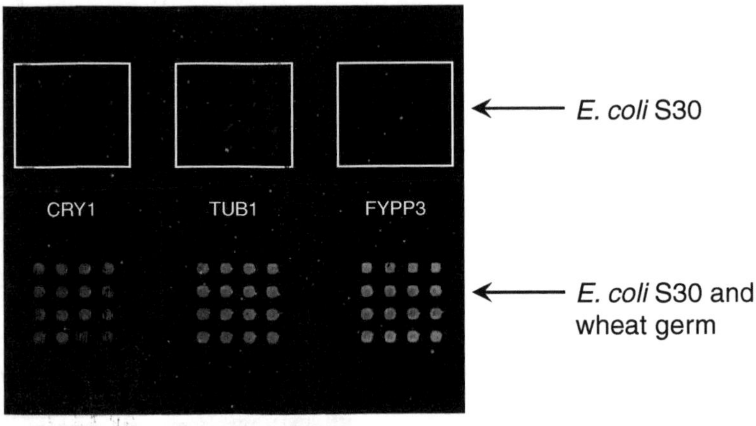

Fig. 4 Comparison of the cell-free expression levels using an *E. coli* S30 extract and the hybrid system that combines wheat germ and S30 extracts for the *Arabidopsis* proteins cryptochrome 1 (CRY1), beta-tubulin (TUB1), and phytochrome-associated serine–threonine protein phosphatase 3 (FYPP3)

3.3 Protein Microarrays with Signal Amplification

1. Prepare BSA–anti-GFP antibody solution for printing. For 100 μL solution, mix 15 μL of stock phosphate buffer for printing, 5 μL of BSA solution, anti-GFP antibody, and water to obtain 0.1 mL of 1.5 mg/mL BSA and 100 μg/mL antibody in 75 mM phosphate buffer, pH 8.5.
2. Follow the same protocol for array printing, blocking, and storage (Subheading 3.1, **steps 2** to 7).
3. Block slides in blocking buffer A (20 mL per slide) at room temperature for 1 h.
4. Rinse slides with water and spin-dry.
5. Insert slides into the ArraySlide 24-4 chamber and perform wheat germ-based expression as described in **step 5** in Subheading 3.2.
6. Remove the cell-free reaction solutions and rinse each well with 100 μL of blocking buffer B for 5 min. Repeat rinse four more times.
7. Add 100 μL per well of blocking buffer B and incubate at room temperature for 1 h.
8. Dilute anti-GFP–HRP-labeled antibody 1,000× in blocking buffer B (60 μL will be needed for each well).
9. Remove blocking buffer B from each well and add the 60 μL of anti-GFP–HRP antibody. Incubate at room temperature for 1 h.
10. Remove the anti-GFP–HRP antibody solution.
11. Wash three times, for 5 min each with 100 μL 1× TBST buffer.

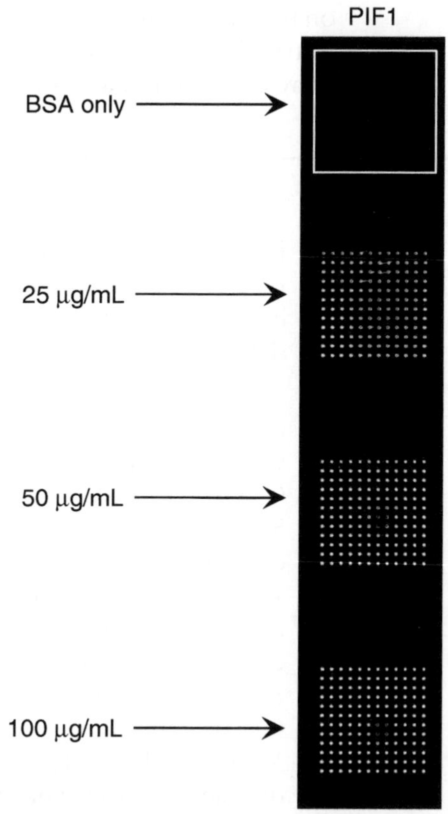

Fig. 5 Self-assembling protein microarrays with proteins detected with tyramide signal amplification. BSA only or BSA–anti-GFP antibody mixture (anti-GFP antibody concentration indicated on the left, all contain 1.5 mg/mL BSA) was printed. After wheat germ expression, anti-GFP–HRP-labeled antibody was used to catalyze Cy3 or Cy5 deposition on array elements using tyramide signal amplification. The *Arabidopsis* protein phytochrome-interacting factor 1 (PIF1) was expressed

12. Prepare TSA Cy5 (or Cy3) solution: for 1.5 mL mix 1,472.5 μL of dilution buffer, 7.5 μL of 10 % Tween-20, and 20 μL of TSA reagent (*see* **Note 11**).
13. Remove TBST buffer and add 60 μL of TSA solution in each well. Incubate for 2 min (*see* **Note 12**).
14. Remove slides from slide chamber and place them inside a Coplin jar with cold water (*see* **Note 13**).
15. Rinse with water three times.
16. Spin-dry the slides. Figure 5 shows a microarray image after tyramide signal amplification.

3.4 Microarray Scanning and Analysis

Detect the fluorescence signals using a microarray scanner. For GFP-tagged proteins set the excitation at 488 nm and the emission detected at 511 nm at the preferred resolution (10–5 μm, depending on the size of each array element). For Cy3 and Cy5 follow scanner instructions. Extract the fluorescence median values from each array element and average for each subarray/protein.

4 Notes

1. If available, a different anti-GFP antibody can be used. It is very important to avoid the use of Tris buffer, since it will interfere with the binding of the antibody on the CodeLink slides.
2. Amplify mutant S65T of green fluorescent protein, GFP, with primers 5′-CTGACTTCCGGAATGGTGAGCAAGGGCGAGG-3′ (BspEI, forward) and 5′-ACTGAAGATCTTTATTCGTGCCATTCGATTTTC-3′ (BglII, reverse). Prepare the 50 μL PCR reaction by mixing 10 ng of template DNA that contains GFP sequence, 50 pmol of each primer, 2 μL of 10 mM dNTP mix, and 2.5 units of a high-fidelity DNA polymerase in its 1× reaction buffer. The recommended thermocycler conditions are 95 °C for 2 min; 30 cycles of 95 °C (30 s), 56 °C (30 s), and 72 °C (1 min); and a final extension at 72 °C for 10 min. Linearize pIVEX2.3d vector (5 PRIME) with XmaI and BamHI, and ligate GFP into it after BspEI/BglII double digestion. This approach retains most of the pIVEX2.3d multiple cloning site for further manipulations to produce in-frame C-terminal GFP-tagged proteins. Proteins can be cloned into pIVEX2.3d-GFP with 5′-NcoI, 5′-NotI, and 3′-XhoI restriction sites (use other sites if necessary).
3. It is highly recommended, once a new kit has been opened and all components reconstituted, to divide the *E. coli* lysate into 12 or 24 μL aliquots and store them at −70 °C to maintain activity by avoiding freeze–thaw cycles.
4. The TNT T7 wheat germ extract system from Promega contains 1 mL of wheat germ extract in 5 × 200 μL aliquots; if less than 200 μL will be used, divide the leftover extract in 25 μL aliquots and store back at −70 °C to avoid freeze–thaw cycles.
5. Print slides with less than 50 % humidity; dehumidifiers can be used in the printing room. Print slides in less than 4 h (recommended by the manufacturer).
6. The manual for the CodeLink activated slides indicates an incubation temperature of 50 °C for DNA arrays. Since a protein is printed, room temperature is recommended here and a longer incubation time (45 min to 1 h).
7. Handle the slides by the edges and do not to touch the surface where the arrays are printed. Place up to four slides for a total of 96 cell-free reactions, one protein per subarray, and one

subarray per well. If using less than four slides, just replace printed slides for regular microscope slides. It is recommended to use the middle positions of the chamber for cell-free reactions. Make sure to adjust the gasket correctly to the chamber's upper structure and tighten the top four screws to avoid leaking.

8. A master mix can be prepared by combining all ingredients except the plasmid DNA. Once the master mix has been prepared, divide it in single reactions and add 1 μg of plasmid DNA and water up to the 60 μL.
9. Place the whole chamber on top of the thermocycler and close the lid; it will fit. Program the thermocycler as indicated.
10. Place the tip of the pipette at a corner of each well to avoid scratching the arrays.
11. The dilution here (75×) is greater than that recommended in the TSA kit (*see* ref. 7). The Tween-20 (final concentration 0.05 %) is necessary for even spreading of the solution inside each well.
12. The incubation time is shorter than that recommended in the TSA kit as well (*see* ref. 7).
13. Just separate the whole chamber (it does not matter if the TSA solution spreads all over) and place the slides in water.

References

1. Kaushansky A, Allen JE, Gordus A et al (2010) Quantifying protein-protein interactions in high throughput using protein domain microarrays. Nat Protoc 5:773–790
2. Gong W, He K, Covington M et al (2008) The development of protein microarrays and their applications in DNA–protein and protein–protein interaction analyses of Arabidopsis transcription factors. Mol Plant 1:27–41
3. Ramachandran N, Hainsworth E, Bhullar B et al (2004) Self-assembling protein microarrays. Science 305:86–90
4. Ramachandran N, Raphael JV, Hainsworth E et al (2008) Next-generation high-density self-assembling functional protein arrays. Nat Methods 5:535–538
5. Zarate X, Henderson DC, Phillips KC et al (2010) Development of high-yield autofluorescent protein microarrays using hybrid cell-free expression with combined Escherichia coli S30 and wheat germ extracts. Proteome Sci 8:32
6. Kimzey MJ, Zarate X, Galbraith DW et al (2011) Optimizing microarray-based in situ transcription and translation of proteins for matrix-assisted laser desorption ionization mass spectrometry. Anal Biochem 414:282–286
7. Bourzac KM, Rounseville MP, Zarate X et al (2011) A high-density quantitative nuclease protection microarray platform for high-throughput analysis of gene expression. J Biotechnol 154:68–75

ERRATA

Production of Eukaryotic Cell-Free Lysate from *Leishmania tarentolae*

Wayne A. Johnston and Kirill Alexandrov

Kirill Alexandrov and Wayne A. Johnston (eds.), *Cell-Free Protein Synthesis: Methods and Protocols*, Methods in Molecular Biology, vol. 1118, DOI 10.1007/978-1-62703-782-2

DOI 10.1007/978-1-62703-782-2_22

The publisher regrets that in Chapter 1 of the print and online versions of this title, the following errors occurred. The corrections are given below.

Page 3, Item 4, final line: "to TB medium at 0.5 % v/v just prior to use" should read "to TB medium at 0.2 % v/v just prior to use"

Page 4, Section 3.1, Line 7: "0.25-0.4" should read "0.25-4.0".

The online version of the original chapter can be found at http://dx.doi.org/10.1007/978-1-62703-782-2_1

Cell-Free Membrane Protein Expression

Tomomi Kimura-Soyema, Mikako Shirouzu, and Shigeyuki Yokoyama

Kirill Alexandrov and Wayne A. Johnston (eds.), *Cell-Free Protein Synthesis: Methods and Protocols*,
Methods in Molecular Biology, vol. 1118, DOI 10.1007/978-1-62703-782-2

DOI 10.1007/978-1-62703-782-2_22

The publisher regrets that in Chapter 18 of this title, the primary author's family name is incorrectly listed as Kimura-Soyema; this should read Kimura-Someya. This error occurs in the chapter opening page, running heads, Table of Contents, and the Contributors list of the print and online versions, as well as in the ebook records on SpringerLink.

The online version of the original chapter can be found at
http://dx.doi.org/10.1007/978-1-62703-782-2_18

INDEX

Kirill Alexandrov and Wayne A. Johnston (eds.), *Cell-Free Protein Synthesis: Methods and Protocols*, Methods in Molecular Biology, vol. 1118, DOI 10.1007/978-1-62703-782-2_2, © Springer Science+Business Media, LLC 2014

MIX
Papier aus verantwortungsvollen Quellen
Paper from responsible sources
FSC® C105338

If you have any concerns about our products,
you can contact us on
ProductSafety@springernature.com

In case Publisher is established outside the EU,
the EU authorized representative is:
Springer Nature Customer Service Center GmbH
Europaplatz 3, 69115 Heidelberg, Germany

Printed by Libri Plureos GmbH
in Hamburg, Germany